Cognition and Survey Research

Cognition and Survey Research

Edited by

MONROE G. SIRKEN
National Center for Health Statistics

DOUGLAS J. HERRMANN
Indiana State University

SUSAN SCHECHTER
National Center for Health Statistics

NORBERT SCHWARZ
University of Michigan

JUDITH M. TANUR
State University of New York at Stony Brook

ROGER TOURANGEAU
The Gallup Organization

A Wiley-Interscience Publication
JOHN WILEY & SONS, INC.
New York • Chichester • Weinheim • Brisbane • Singapore • Toronto

This book is printed on acid-free paper.♾

Copyright © 1999 by John Wiley & Sons, Inc. All rights reserved.

Published simultaneously in Canada.

No part of this publication may be reproduced, stored in a retrieval system or transmitted in any form or by any means, electronic, mechanical, photocopying, recording, scanning or otherwise, except as permitted under Sections 107 or 108 of the 1976 United States Copyright Act, without either the prior written permission of the Publisher, or authorization through payment of the appropriate per-copy fee to the Copyright Clearance Center, 222 Rosewood Drive, Danvers, MA 01923, (978) 750-8400, fax (978) 750-4744. Requests to the Publisher for permission should be addressed to the Permissions Department, John Wiley & Sons, Inc., 605 Third Avenue, New York, NY 10158-0012, (212) 850-6011, fax (212) 850-6008, E-Mail: PERMREQ@WILEY.COM.

Library of Congress Cataloging-in-Publication Data:
Cognition and survey research / Monroe G. Sirken ... [et al.].
    p.   cm. — (Wiley series in probability and statistics. Survey methodology section)
    Includes bibliographical references and index.
    ISBN 0-471-24138-5 (alk. paper)
    1. Cognitive psychology—Research.   2. Cognitive science—Research.   3. Scientific surveys.   I. Sirken, Monroe G.
    II. Series.
    BF201.C625   1998
    153′.07′2—dc21   98-38408

Printed in the United States of America.

10 9 8 7 6 5 4 3 2 1

Contents

Contributors

Murray Aborn, National Science Foundation (ret.)

Frederick G. Conrad, Bureau of Labor Statistics

Mick P. Couper, University of Michigan and Joint Program in Survey Methodology

Theresa DeMaio, Bureau of the Census

Charles J. Fillmore, University of California at Berkeley

Michael Friendly, York University

Eleanor R. Gerber, Bureau of the Census

Arthur C. Graesser, The University of Memphis

Robert M. Groves, University of Michigan and Joint Program in Survey Methodology

Brian Harris-Kojetin, The Arbitron Company

Douglas J. Herrmann, Indiana State University

Tina Kennedy, The University of Memphis

Stephan Lewandowsky, University of Western Australia

Elizabeth A. Martin, Bureau of the Census

Jeffrey C. Moore, Bureau of the Census

Colm O'Muircheartaigh, University of Chicago

Victor Ottati, Purdue University

Lance J. Rips, Northwestern University

Susan Schechter, National Center for Health Statistics

Michael F. Schober, New School for Social Research

Norbert Schwarz, University of Michigan

Michael S. Shum, Northwestern University

Monroe G. Sirken, National Center for Health Statistics

Eliot R. Smith, Purdue University

Linda L. Stinson, Bureau of Labor Statistics

Judith M. Tanur,   State University of New York at Stony Brook
Roger Tourangeau,   The Gallup Organization
Clyde Tucker,   Bureau of Labor Statistics
Edward J. Welniak Jr.,   Bureau of the Census
Peter Wiemer-Hastings,   The University of Memphis
Gordon B. Willis,   Research Triangle Institute

Preface

This book tells the story of efforts by survey researchers to use cognitive psychology and other sciences to improve the quality of the data collected in surveys. It offers a review of the early work in cognition and survey research, an update on the initiatives currently under way, and a glimpse into the future of interdisciplinary work on survey methods. The book seeks to capture, describe, and reinvigorate the movement that has come to be known as the Cognitive Aspects of Survey Methodology or CASM. The movement is still in its infancy and promises to grow considerably in the years to come.

In 1993, after the first CASM Seminar passed its 10th anniversary, staff at the National Center for Health Statistics realized that the accomplishments and status of the movement needed to be evaluated and assessed in order to secure its place as an established interdiscipline. As these needs were discussed with and among prominent leaders in the survey research community as well as with the experienced practitioners most familiar with CASM-related state of the art, it became clear that a three-pronged approach would be most likely to succeed in stimulating the research agenda.

First, communication among those conducting CASM research had to be facilitated and enhanced, and new lines of communication had to be established with those conducting related research in different disciplines. Clearly, many new avenues of CASM-oriented research were being pursued and many successes had been achieved, but there was no mechanism in place to share the knowledge gained and to systematically incorporate research findings into new work. Thus, the first initiative was to convene a Second Advanced Research Seminar in the Cognitive Aspects of Survey Methodology (CASM II). Papers would be commissioned for plenary sessions and working groups would be established to promote small-group discussions resulting in reports that would describe the critical problems in need of further research attention.

Second, dissemination of CASM research findings and related literature needed to be expanded to a wider audience both within and beyond the survey research community. Many books and articles had already been published that serve as excellent sources for better understanding of CASM theory and application (see especially Hippler, Schwarz, and Sudman, 1987; Jobe and

Loftus, 1991; Schwarz and Sudman, 1992, 1994, 1996; Sudman, Bradburn, and Schwarz, 1996; Tanur, 1992). The CASM II Seminar would contribute to this literature since its papers and reports were targeted to reflect on and evaluate past work, summarize the state of the art, and, perhaps most importantly, demonstrate the exciting potential of a focused interdisciplinary research agenda. The hope was that a careful and deliberately-planned approach to widely disseminating the knowledge and ideas shared at CASM II would stimulate creativity, thoughtfulness, and innovation for the future. To that end, highlights of the CASM II Seminar were presented at an invited session of the 1997 Joint Statistical Meetings and a 1997 Symposium sponsored by the Washington Statistical Society and the DC Chapter of the American Association for Public Opinion Research. Publication of this book as well as a proceedings of the CASM II Seminar were also planned.

Third, resources and infrastructure to support future research initiatives had to be planned for and garnered. Those who were involved in the beginning stages of CASM research knew that the financial support of the National Science Foundation and other federal statistical agencies enabled the movement to demonstrate and accelerate its potential. Developing funding sources for the future research agenda would be critical in pushing forward the movement and in encouraging innovation in research. Thus, an initiative was launched to establish a consortium of federal statistical agencies to collaborate with the National Science Foundation in an effort to supply funding and leverage existing funds.

Deciding how to achieve these three goals—improving communication, expanding dissemination of research findings, and ensuring future resources— was itself a challenging and intensive task. By the end of 1995, the Seminar organizing committee had been formed, chaired by Monroe Sirken and cochaired by Susan Schechter and consisting of Douglas Herrmann, Elizabeth Martin, Norbert Schwarz, Judith Tanur, Roger Tourangeau, and Clyde Tucker. In addition, an advisory group including Thomas Jabine, Robert Groves, and Norman Bradburn was established. Editorship of this book, as well as the proceedings, were decided upon and responsibilities were delineated for planning the themes and direction of the Seminar as a whole, the specific Seminar sessions, and the focus of the eight working groups.[1]

This book follows the same organization as the plenary sessions of the Seminar. It consists of four sections, each edited by the same person who had organized the corresponding session at the Seminar:

Section A, edited by Judith Tanur, provides an overview of the history and status of the CASM movement and a critical appraisal of its achievements and some remaining gaps.

---

[1]The proceedings (Sirken, Jabine, Willis, Martin, and Tucker, in press) focuses primarily on the eight working group reports and also includes summaries and abstracts of papers given in each plenary session.

Section B, edited by Norbert Schwarz, presents chapters that review and discuss the effects of the CASM movement on cognitive theory and survey measurement.

Section C, edited by Roger Tourangeau, contains chapters that examine the potential roles of disciplines other than classical cognitive psychology in survey methods research.

Section D, edited by Douglas Herrmann, concludes with chapters that suggest new areas of CASM research beyond the phase of data collection.

Each session organizer identified and recruited nationally- and internationally-known researchers as potential Seminar participants and commissioned monograph authors. In addition, a selected list of Seminar invitees was developed with the intention of keeping the Seminar limited to approximately 50 attendees. The format of the Seminar was designed to facilitate intensive interactions among a relatively small number of distinguished participants. Plenary sessions were held each morning and working group meetings were scheduled during the remainder of the day. The Seminar site was selected with intentions that it would provide a contemplative environment as free of intrusions as possible.

This project was fortunate in having both financial and professional support from a number of key agencies and individuals. A great deal of gratitude goes to the National Center for Health Statistics and the National Science Foundation for funding the CASM II Seminar. We were also very lucky to receive considerable professional and technical support from the Bureau of the Census, the Bureau of Labor Statistics, the Joint Program in Survey Methodology, and last and foremost, from the National Center for Health Statistics.

Regarding production of this book, we are especially grateful to Gordon Willis and Paul Beatty who assisted us in reviewing many of the chapters and to Karen Whitaker who provided considerable editorial support in finalizing this book. Thanks are also due to Steve Quigley from John Wiley & Sons, Inc. who worked closely with us on all phases of this monograph project and also to Alison Bory from Wiley, who ably dealt with the various administrative requirements. Last, our employing organizations also deserve our thanks and appreciation for supporting our activities in conducting the Seminar and assembling this monograph: The National Center for Health Statistics, Indiana State University, the University of Michigan, the State University of New York at Stony Brook, and The Gallup Organization.

MONROE SIRKEN
DOUGLAS HERRMANN
SUSAN SCHECHTER
NORBERT SCHWARZ
JUDITH TANUR
ROGER TOURANGEAU

REFERENCES

Hippler, H. J., Schwarz, N., and Sudman, S. (Eds.). (1987). *Social Information Processing and Survey Methodology.* New York: Springer-Verlag.

Jobe, J., and Loftus, E. (Eds.). (1991). Cognition and survey measurement. Special issue of *Applied Cognitive Psychology, 5(1).*

Schwarz, N., and Sudman, S. (Eds.). (1992). *Context Effects in Social and Psychological Research.* New York: Springer-Verlag.

Schwarz, N., and Sudman, S. (Eds.). (1994). *Autobiographical Memory and the Validity of Retrospective Reports.* New York: Springer-Verlag.

Schwarz, N., and Sudman, S. (1996). *Answering Questions: Methodology for Determining Cognitive and Communicative Processes in Survey Research.* San Francisco: Jossey-Bass.

Sirken, M., Jabine, T., Willis, G., Martin, E., and Tucker, C. (in press). A new agenda for interdisciplinary survey research methods: *Proceedings of the CASM II Seminar.* National Center for Health Statistics.

Sudman, S., Bradburn, N. M., and Schwarz, N. (1996). *Thinking About Answers: The Application of Cognitive Processes to Survey Methodology.* San Francisco: Jossey-Bass.

Tanur, J. M. (Ed.). (1992). *Questions About Questions: Inquiries into the Cognitive Bases of Surveys.* New York: Russell Sage Foundation.

CHAPTER 1

Interdisciplinary Survey Methods Research

Monroe G. Sirken and Susan Schechter
National Center for Health Statistics

1.1 INTRODUCTION

The spectacular growth in survey research during the twentieth century would have been impossible without interdisciplinary survey methods research to facilitate transfers of new knowledge and technology from the social sciences, statistics, and computer sciences. Similarly, new thrusts of interdisciplinary survey methods research will be critical in meeting the expanded needs of surveys during the 21st century.

Research on the cognitive aspects of survey methods (CASM) is a relatively new area of interdisciplinary survey methods research that emerged at the Advanced Research Seminar on Cognitive Aspects of Survey Methodology (CASM I Seminar) in June 1983. It is noteworthy that the CASM I Seminar initiated a deliberately fostered effort to bridge the communication gaps between survey research and the cognitive and social sciences, and to initiate CASM research that would benefit survey applications as well as basic cognitive research. The fostered effort was recently reviewed and renewed at the Second Advanced Research Seminar on Cognitive Aspects of Survey Methodology (CASM II Seminar) in June 1997.

This book contains the commissioned papers that were presented at four plenary sessions of the CASM II Seminar, with introductory remarks by the plenary session chairs. Judith Tanur's Section A and Norbert Schwarz' Section B provide critical appraisals of CASM research performed prior to the CASM II

Cognition and Survey Research, Edited by Monroe G. Sirken, Douglas J. Herrmann, Susan Schechter, Norbert Schwarz, Judith M. Tanur, and Roger Tourangeau.
ISBN 0-471-24138-5 © 1999 John Wiley & Sons, Inc.

Seminar, and discuss its effects on cognitive theory and survey measurement. Section C, edited by Roger Tourangeau, looks at the potential benefits of bringing additional disciplines into CASM research. Section D, edited by Douglas Herrmann, concludes with chapters that address future research opportunities that go beyond the data collection phase of surveys. In the last section, Martin and Tucker (Chapter 22) offer a summary and analysis of new CASM research proposals that were developed during intensive small-group discussions at the CASM II Seminar. The proceedings of the Seminar (Sirken, Jabine, Willis, Martin, and Tucker, forthcoming) elaborate further on these project proposals and other features of the CASM II Seminar.

In this introductory chapter, we present a broad overview of CASM research, retrospectively and prospectively. The retrospective assessment deliberately focuses on questionnaire design research conducted in collaboration with cognitive psychologists, because that was the principal area of CASM research prior to the CASM II Seminar. Prospectively, we sketch a roadmap for CASM research during the coming decade that expands the inventory of possible survey methods research areas beyond questionnaire design, and seeks to involve disciplines other than cognitive psychology.

1.2   IMPACT OF CASM RESEARCH

In bridging the culture gap between survey researchers and cognitive psychologists, the CASM I Seminar fostered a shift in survey response research from the behaviorist to the cognitive paradigm (see Herrmann, Chapter 17, for further discussion). Generally, the paradigm shift implied that the two-stage stimulus/response sequence postulated by behaviorist theory was intersected by a cognitive phase in which respondents perform a series of mental tasks in responding to survey questions. In effect this shift expanded the survey researcher's view of the survey response process from a two-stage stimulus/response process to a three-stage stimulus/cognition/response process, as illustrated by the flowchart in Figure 1.1. The flowchart implies that flawed questions are cognitively burdensome for survey respondents to process and can cause response errors. The paradigm switch inspired two kinds of CASM research. Both investigate cognitive process failure in survey response but for different purposes:

Applied CASM research improves questionnaire design by using cognitive methods to detect and repair cognitively burdensome survey questions;

Basic CASM research increases fundamental knowledge about cognition by investigating the causes of cognitive difficulties in survey response.

Applied and basic CASM research are symbiotically related (Sirken and Herrmann, 1996). Identification of survey response phenomena worthy of being

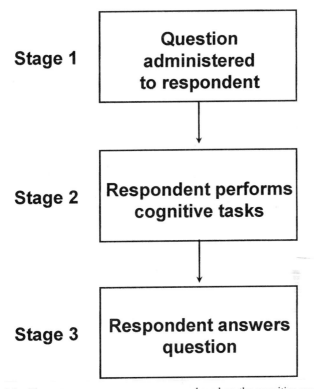

Figure 1.1   Three-stage survey response process based on the cognitive paradigm.

investigated in basic CASM research is a potential by-product of applied CASM research. Fundamental knowledge about cognition contributed by basic CASM research may be used in applied CASM research and can result in advances in survey technology.

1.2.1 Applied CASM Research

There were two major achievements in the use of cognitive technology to improve questionnaire design during the past decade: (1) demonstration of the utility of laboratory-based intensive interview methods in detecting and repairing the cognitive glitches caused by flawed questions, and (2) institutionalization of the cognitive research laboratory in official statistical agencies and survey research centers as a place for designing and testing new and revised survey questionnaires.

In the mid-1980's, Lessler, Tourangeau, and Salter (1989) conducted an experiment that compared questionnaires designed and tested by laboratory-based intensive interviewing methods with questionnaires developed by conventional field testing methods. Laboratory methods included techniques such as think-aloud interviews and intensive probing (DeMaio, 1983; Forsyth and Lessler, 1991). The interviews were administered to purposively-selected subjects in a makeshift laboratory setting. The experiment decisively demonstrated that intensive interviewing methods administered in this manner had considerable potential for the detection and repair of cognitively burdensome survey questions, and that these problems were frequently missed in conventional field testing. Equally important, the experiment demonstrated benefits in field testing that were lacking in laboratory testing. Consequently, the study concluded that cognitive laboratory methods complimented rather than superseded conventional field methods and that both methods were essential components in the process of designing and testing survey questionnaires.

Encouraged by these findings, the staff of the National Center for Health Statistics and other statistical agencies and organizations began to develop the infrastructure for permanent cognitive laboratories, giving attention to space, budget and staff requirements, hardware and software needs, administrative procedures for recruiting and paying subjects, and protocols for interviewing laboratory subjects in iterative rounds. The objectives were to (1) design laboratories suitable for routinely and systematically applying cognitive interviewing methods as an adjunct to conventional field testing methods; (2) preserve the integrity of the qualitative interviewing process; and (3) meet organizational needs to test questionnaires in a timely and efficient manner (see Sirken, 1991, for a description of this infrastructure).

Today, the cognitive research laboratory should still be considered in its infancy. It is rapidly outgrowing its infrastructure, and refinement of this infrastructure should be a priority in the work of applied CASM research for some time to come. For example, relatively little attention has been given to the infrastructural features needed for testing questionnaires of establishment surveys, and for testing computer-assisted questionnaires that are administered in the personal interviewing mode (CAPI), the telephone-assisted mode (CATI), or the self-interviewing mode (CASI). Also, the laboratories' quality control and quality assurance methods need strengthening, and these procedures need to develop means for the systematic review of the cognitive phenomena observed during laboratory testing (see Willis, DeMaio, and Harris-Kojetin, Chapter 9, for discussion of these issues).

The cognitive research laboratory is the outstanding success story of the CASM research effort. Since the first permanent laboratory was established in 1985 (Royston, Bercini, Sirken, and Mingay, 1986), others have been established in several official statistics agencies in the United States and elsewhere. Furthermore, academic survey research centers as well as survey research firms in the private sector have also established cognitive laboratories or have started to use cognitive interviewing methods on a less formal basis. This trend is continuing throughout much of the survey community. The laboratory method of designing and testing questionnaires has forever changed both the professional *attitude* that questionnaire design is simply an art rather than a science, and the professional *practice* of relying solely on conventional field pretesting to finalize a survey instrument.

1.2.2 Basic CASM Research

Basic CASM research is problem-oriented and hypothesis-driven experimental research that investigates cognitive processes in survey response in a search for clues as to how the processes generally work. It has potential for affecting questionnaire design either directly by explaining why the cognitive process is not working as intended and suggesting ways of redesigning survey questions to make it work, or indirectly by stimulating basic research that creates fundamental knowledge about cognition with ultimate benefits to survey applications.

Basic CASM research has focused primarily on two kinds of survey phenomena, error-prone heuristics of survey respondents and error-prone survey questions. Findings about error-prone heuristics are more likely to benefit survey applications indirectly by creating fundamental knowledge about how cognition works, whereas findings about error-prone questions seem more likely to benefit survey applications directly by explaining why the cognitive process does not work and suggesting ways of redesigning questions to make it work. The following are examples of each type of phenomenon.

Classic experiments on error-prone survey heuristics were conducted by Bradburn, Rips, and Shevell (1987) and by Schwarz and Strack (1991). Bradburn et al. (1987) investigated the survey heuristic referred to as *forward telescoping*. This heuristic refers to the memory phenomenon in which events occurring prior to a bounded calendar period are reported as having occurred within the bounded period. Investigating forward telescoping provided clues about how autobiographical memory works. Schwarz and Strack (1991) investigated *context effects*, a survey heuristic referring to the response effects of the ordering of survey questions or response categories. These investigations provided clues about the process of forming judgments.

Experiments on error-prone survey questions were undertaken by Smith (1991) and Means, Nigam, Zarrow, Loftus, and Donaldson (1989). Smith investigated failures in long-term dietary recall of foods eaten, and their frequencies and judgments of portion size. He concluded that if what is needed is a set of items typical of a person's diet, it would be better to have respondents report their generic dietary knowledge rather than specific dietary memories. Means et al. (1989) investigated autobiographical memory for health-related events. Their experiments indicated that respondents have difficulty in recalling specific instances of recurring events. However, they concluded that recall enhancement interventions, such as constructing personal time lines, substantially improved reporting for recurring events.

Basic cognitive research has successfully demonstrated its utility by providing interesting technical findings about how cognition works in general, and why it sometimes does not work when people respond to survey questions. This success more than warrants extending the research to other survey error-prone heuristics, including anchoring effects[1] and seam effects,[2] and to a host of survey error-prone topics including complex social concepts such as disability, unemployment, poverty, and complex economic indicators such as the gross domestic product and the consumer price index.

However, the ultimate success of basic CASM research will be measured not in terms of its technical findings, but in terms of the impact these findings have on survey applications. So far these technical findings have had very little impact on survey practices either directly by improving questionnaire designs of current surveys or indirectly by stimulating basic research that seems likely to improve the technology for designing the next generation of survey questionnaires. However, Shum and Rips (Chapter 7) offer some encouragement with respect to the latter. They suggest that survey research has already been a factor in accelerating basic cognitive research on autobiographical memory. Perhaps the basic cognitive research that they refer to will ultimately lead to improvements in survey technology. Nevertheless, it is quite clear that fostered efforts will be needed in the future to convert the technical findings of basic CASM research into useful survey applications.

1.3   FUTURE DIRECTIONS OF CASM RESEARCH

The CASM II Seminar produced an inventory of priority survey needs during the next decade and beyond for new kinds of interdisciplinary survey methods research (Sirken et al, forthcoming). The inventory increases many fold potential areas of research beyond questionnaire design, and extends interdisciplinary networking beyond cognitive psychology and the social sciences.

1.3.1   Expanding the Inventory of CASM Research Projects

The expanded inventory includes several new questionnaire design research projects. For example, a recurring theme heard at the CASM II Seminar and found in recent work by Beatty (1995), Schaeffer and Maynard (1996), Schober and Conrad (1997), and Suchman and Jordan (1990) is the critical need to investigate alternatives to standardized interviewing methods for conducting field interviews. The conversational aspects of questionnaire design, that is, the degree to which survey discourse should be conversational rather than standardized and inflexible, has provoked considerable discussion and interest in some areas of the survey research community (see O'Muircheartaigh, Chapter 4, for commentary on this issue). A conversationally-designed questionnaire raises especially challenging problems for national surveys that are conducted by large numbers of interviewers

---

[1]Anchoring refers to the phenomenon in which responses are biased to the initial value or starting point (Tversky and Kahneman, 1990). In surveys, the initial starting point is given by the survey question.

[2]The seam effect refers to the phenomenon observed in panel surveys in which retrospectively reported autobiographical events are over reported for time periods overlapping the juncture points of successive interviews (Jabine, King, and Petroni, 1990).

who vary considerably in their interviewing capabilities. Adopting the conversational mode in such surveys would require more intensive interviewer training and in most cases, a more professional interviewing staff, and therefore would probably add considerably to overall survey costs.

Underutilization of the potential of the computerized interview to ease the cognitive workloads of interviewers and respondents was another recurring theme at the CASM II Seminar. Utilizing computer potential could, for example, make the conversational questionnaire more feasible, especially in large national surveys. Some usability testing of computerized questionnaires is currently underway in cognitive research laboratories or elsewhere. So far, though, the research has focused more on adaptations of the interviewer to the software than adaptations of the software to the interviewer (see Couper, Chapter 18).

The expanded inventory also includes many new CASM research projects that are responsive to needs at stages of the survey measurement process other than data collection. The survey measurement process is viewed as encompassing six stages: data specification, data collection, data reduction, estimation, analysis, and data dissemination (Schechter, Sirken, Tanur, Martin, and Tucker, in press). There was considerable interest at the CASM II Seminar, for example, in continuing the research recently begun by Pickle and Herrmann (1994) on the cognitive aspects of designing statistical maps and extending it to graphs, tables, and other modes of data presentation. Pickle and Herrmann's work was a direct outgrowth of the CASM movement and serves as an excellent example of interdisciplinary research that applied the CASM model to a new research endeavor (see also papers in this volume by Friendly, Chapter 20, and Lewandowsky, Chapter 21).

Another recurring theme at the CASM II Seminar was the possibility of improving data analysis by incorporating cognitive and social theories as parameters in statistical error models. Groves (Chapter 15) provides several examples of how cognitive and social theories and their experimental findings can improve estimation and control of survey measurement bias and variance. Press (1996, 1998) proposes a Bayesian hierarchical model to improve point and interval estimation for retrospectively reported survey events. The model involves asking respondents an additional question that yields an interval in which the respondent believes his/her true value lies. Press and Tanur (1998) report preliminary results of experiments to evaluate the efficacy of the model.

1.3.2 Extending Interdisciplinary Networking

Most unmet needs for new kinds of CASM research are at the intersections of multiple scientific disciplines. For example, the projects noted earlier on conversational and computerized questionnaires are at intersections of the social sciences and computer sciences, and the projects on statistical maps and cognitive models of measurement errors are at intersections of the social sciences and the statistical sciences.

The Venn diagram in Figure 1.2 partitions interdisciplinary survey measure-

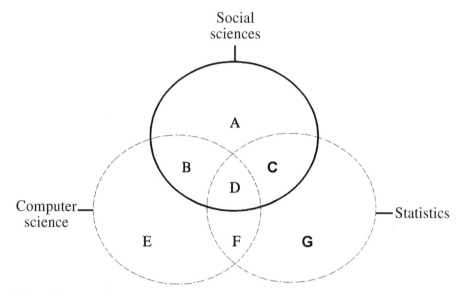

Figure 1.2 Intersection of disciplines in survey methods research. A, social sciences only; B, computer and social sciences; C, statistics and social science; D, computer and social sciences, and statistics; E, computer science only; F, computer science and statistics; G, statistics only.

ment research into seven interdisciplinary domains. The domains are formed by three circles representing the social sciences, statistical sciences, and computer science—the three sets of disciplines that have contributed most to interdisciplinary survey methods research during the past half century.

Prior to the CASM II Seminar, CASM research was fostered almost exclusively with social scientists, particularly cognitive psychologists. This interdisciplinary domain is represented by area A in the Venn diagram. The CASM II Seminar fostered extensions of interdisciplinary networking to domains that are at the intersections of the social sciences with statistics and/or computer science. The added domains are represented by areas B, C, and D in the Venn diagram. If in the future, serious consideration is given to merging CASM research into an integrated program of interdisciplinary survey methods research, interdisciplinary networking would be extended to domains in the Venn diagram that do not intersect with the social sciences. The added domains would be those represented by areas E, F, and G.

1.4 CONCLUDING REMARKS

Unless interdisciplinary survey methods research is fostered in the future, it is unlikely to evolve naturally in a way that would give it the capability of addressing priority survey needs for new thrusts of interdisciplinary research. Several excellent reports such as the Report of a Panel of the Institute of Mathematical

Statistics (1988) and the Report on the Transfer of Methodology Between Academic and Government Statisticians (1978) discuss the challenges and benefits of fostered interdisciplinary research. The CASM II Seminar clearly demonstrated the need for the cognitive and social sciences to form partnerships with the statistical and computer sciences. Although we feel strongly that this is the right direction for the future, in the long run it will be insufficient. What we think is needed to address the priority needs for the next generation of surveys is a deliberately orchestrated effort to establish an integrated interdisciplinary survey methods research program. Such a program would be a natural progression in the direction that CASM research is currently moving in linking the social sciences with statistics and the computer sciences.

ACKNOWLEDGMENTS

The opinions expressed herein are the authors and do not necessarily represent the official views or positions of the National Center for Health Statistics.

REFERENCES

American Statistical Association (1978, March). *Conference on Transfer of Methodology between Academic and Government Statisticians*. Washington, DC: American Statistical Association.

Beatty, P. (1995). Understanding the standardized/non-standardized interviewing controversy. *Journal of Official Statistics, 11,* 147–160.

Bradburn, N. M., Rips, L. J., and Shevell, S. K. (1987). Answering autobiographical questions: The impact of memory and inference on surveys. *Science, 236,* 151–167.

DeMaio, T. J. (Ed.) (1983). *Approaches to developing questionnaires*. Statistical Policy Working Paper No. 10. Washington, DC: Office of Management and Budget.

Forsyth, B. H., and Lessler, J. T. (1991). Cognitive laboratory methods: A taxonomy. In P. P. Biemer, R. M. Groves, L. E. Lyberg, N. A. Mathiowetz, and S. Sudman (Eds.), *Measurement Errors in Surveys,* pp. 393–418. New York: Wiley.

Institute of Mathematical Statistics (1988, September). *Cross-disciplinary research in the statistical sciences.*

Jabine, T. B., King, K. E., and Petroni, R. J. (1990). *Survey of income and program participation quality profile,* 2nd ed. Washington, DC: U.S. Department of Commerce.

Lessler, J., Tourangeau, R., and Salter, W. (1989). Questionnaire design in the cognitive research laboratory. *Vital and Health Statistics,* Series 6, No. 1 (DHHS Publication No. PHS 89-1076). Washington, DC: U.S. Government Printing Office.

Means, B., Nigam, A., Zarrow, M., Loftus, E., and Donaldson, M. S. (1989). Autobiographical memory for health-related events. *Vital and Health Statistics,* Series 6, No. 2 (DHHS Publication No. PHS-89-1077). Washington, DC: Government Printing Office.

Pickle, L. W., and Herrmann, D. J. (1994). The process of reading statistical maps: The effect of color. *Statistical Computing and Graphics Newsletter, 5,* 1, 12–15.

Press, S. J. (1996). *Bayesian recall: A cognitive Bayesian modeling approach to surveying a recalled quantity* (Tech. Rep. No. 236). Riverside, CA: University of California, Department of Statistics.

Press, S. J. (1998). *Modeling Bayesian recall in sample surveys.* Manuscript submitted for publication, University of California, Department of Statistics.

Press, S. J., and Tanur, J. M. (1998). *Experimenting with Bayesian Recall* (Tech. Rep. No. 254). Riverside, CA: University of California, Department of Statistics.

Royston, P., Bercini, D., Sirken, M. G., and Mingay D. (1986). Questionnaire Design Research Laboratory. *Proceedings of the Section on Survey Research Methods of the American Statistical Association,* 703–707.

Schaeffer, N.C., and Maynard D. W. (1996). From paradigm to prototype and back again: Interactive aspects of cognitive processing in standardized survey interviews. In N. Schwarz and S. Sudman (Eds.), *Answering Questions: Methodology for Determining Cognitive and Communicative Processes in Survey Research,* pp. 65–90. San Francisco: Jossey-Bass.

Schechter, S., Sirken, M, Tanur, J. Martin, E., and Tucker, C. (1997). CASM II: Current and future directions in interdisciplinary research. *Proceedings of the Section on Survey Research Methods of the American Statistical Association,* 1–10.

Schober, M. F., and Conrad, F. G. (1997). Does conversational interviewing reduce survey measurement error? *Public Opinion Quarterly, 60,* 576–602.

Schwarz, N., and Strack, F. (1991). Context effects in attitude surveys: Applying cognitive theory to social research. In W. Stroebe and M. Hewstone (Eds.), *European Review of Social Psychology,* Vol 2, pp. 31–50. Chichester, England: Wiley

Sirken, M. G. (1991). The role of a cognitive laboratory in a statistical agency. In *Office of Management and Budget Seminar on Quality of Federal Data, Statistical Policy Working Paper 20,* pp. 268–277, Washington, DC: Statistical Policy Office.

Sirken, M. G., and Herrmann, D. J. (1996). Relationships between cognitive psychology and survey research. *Proceedings of the Section on Survey Research Methods, American Statistical Association,* 245–249.

Sirken, M., Jabine, T., Willis, G., Martin, E., and Tucker, C. (in press). A new agenda for interdisciplinary survey research methods: *Proceedings of the CASM II Seminar.* National Center for Health Statistics.

Smith, A. F. (1991). Cognitive processes in long-term dietary recall. *Vital and Health Statistics,* Series 6, No. 4 (DHHS Publication No. PHS 92-1079). Washington, DC: U.S. Government Printing Office.

Suchman, L., and Jordan, B. (1990). Interactional troubles in face-to-face survey interviews, *Journal of the American Statistical Association, 85(409),* 232–241.

Tversky, A., and Kahneman, D. (1990). Judgment under uncertainty: Heuristics and biases. In D. Kahneman, P. Slovic, and A. Tversky (Eds.), *Judgment Under Uncertainty: Heuristics and Biases,* pp. 3–20, Cambridge, England: Cambridge University Press.

SECTION A

CHAPTER 2

Looking Backwards and Forwards at the CASM Movement

Judith M. Tanur
State University of New York at Stony Brook

Both the contributors to this section allude to the history of the movement to study cognitive aspects of survey methodology (CASM), but neither describes it in any detail. Hence, a brief historical sketch may be in order here. In the United States the roots of the movement grew from rising nonresponse rates to surveys during the 1970's and from increasing concerns during the same years about the validity of the survey data on which much government policy and academic research was being based. A Panel of the National Academy of Sciences reflected these concerns in a report on the National Crime Survey (Penick and Owens, 1976). A report of an interdisciplinary conference on Perspectives on Attitude Assessment Surveys and Their Alternatives, appeared in 1976. It was jointly sponsored by the Office of Naval Research and the Naval Personnel Research and Development Center (Sinaiko and Broedling, 1976). In response to these concerns, both in the United States and in England, conferences were called with the express purpose of examining cognitive issues as they relate to surveys. In England, the recall method was the topic (see Moss and Goldstein, 1979), while in the United States, under the guidance of Albert Biderman and the auspices of the Bureau of Social Science Research, survey researchers, cognitive scientists, and statisticians were brought together to explore cognitive issues in the National Crime Survey (Biderman, 1980). These efforts began the conversations across disciplinary boundaries, but few actual collaborations or published research seem directly traceable to them.

The "official" beginning of the CASM movement is usually dated as June

Cognition and Survey Research, Edited by Monroe G. Sirken, Douglas J. Herrmann, Susan Schechter, Norbert Schwarz, Judith M. Tanur, and Roger Tourangeau.
ISBN 0-471-24138-5 © 1999 John Wiley & Sons, Inc.

1983 in the United States and July 1984 in Germany. Under the auspices of the Committee on National Statistics of the U.S. National Academy of Sciences/National Research Council, an Advanced Research Seminar on Cognitive Aspects of Survey Methodology was convened, including statisticians, survey researchers, researchers from the cognitive sciences, and U.S. government agency staff. The name of the seminar was chosen deliberately so that the resulting acronym, CASM, would signify the deep interdisciplinary chasms that would have to be bridged if the proposed collaborative efforts were to be successful. The report of the Seminar (see Jabine, Straf, Tanur, and Tourangeau, 1984) embodied concrete research proposals and envisaged a two-way street via which not only would the theories and methods of the cognitive sciences be applied to help solve the long-standing problems of survey research, but also large-scale surveys would serve as extensions of cognitive laboratories, permitting workers in the cognitive sciences to generalize their results to broader populations and more realistic situations. In Germany, a conference on Social Information Processing and Survey Methodology was held at ZUMA in Mannheim—its proceedings also constituted a watershed (see Hippler, Schwarz, and Sudman, 1987).

In the aftermath of these originating conferences, many research projects involving cross-disciplinary collaboration were proposed, funded, carried out, and published. Results of many of these are referenced later in this section and throughout the volume. But perhaps the most notable impact of the CASM movement so far is on U.S. government statistical agencies. The initial Advanced Research Seminar had, by design, focused on the National Health Interview Survey (NHIS) and invited senior personnel from the agency responsible for that survey, the National Center for Health Statistics (NCHS). Perhaps as a result of that focus and surely because of the vision of one of the participants in the Seminar, Monroe Sirken, a National Laboratory for Collaborative Research in Cognition and Survey Measurement was established at NCHS shortly thereafter. That program had two components. The first was the NSF-funded Collaborative Research Program which supported university scientists in conducting problem-oriented basic research on cognitive issues germane to improving surveys. The second was the NCHS-supported Questionnaire Design Research Laboratory which conducts applied cognitive research in developing, designing, and testing survey questionnaires and has become a standard pretesting venue for NCHS surveys. In 1988, the U.S. Congress appropriated funds for the Bureau of Labor Statistics (BLS) to establish a similar laboratory, and at about the same time the Bureau of the Census continued its long tradition of research on nonsampling errors by also establishing a cognitive laboratory. Shortly thereafter, a Protocol of Cooperation was signed by the heads of these three statistical agencies, ensuring that the cooperation already established at the staff level would continue and increase.

In Europe, cognitive work does not enjoy such formal status, but is carried out, among other places, by researchers at the national statistical agencies of The Netherlands and Sweden, and at the London School of Economics and Political Science, and ZUMA.

Another noteworthy institutional arrangement in the United States that contributed to the dissemination of CASM ideas was the establishment, again with NSF funding, of a Committee on Cognition and Survey Research of the Social Science Research Council (SSRC). During its lifetime the Committee invited experts in the cognitive sciences and applied fields to join survey researchers in exploring the cognitive underpinnings of such diverse topics as the semantics of the survey interview and the reporting of pain. These eight workshops were the occasions for participants to look for research opportunities concerning various aspects of the survey process and they encouraged broad interdisciplinary collaboration. More details about the Committee's work is available in its culminating volume (Tanur, 1992). Thus, in the United States, the CASM movement was from the beginning a fostered experiment in interdisciplinary collaboration. The initial Advanced Research Seminar was carefully orchestrated to facilitate communication among researchers from different disciplinary cultures. Extensive preparation of the participants contributed to this fostering. Prior to the Seminar, participants exchanged biographical information, received specially written background papers expositing the basics of the cognitive sciences (Tourangeau, 1984) and speculating on the contributions of cognitive research to survey questionnaire design (Bradburn and Danis, 1984), and served as respondents for the NHIS. At the beginning of the Seminar the participants collectively watched specially prepared videotapes of NHIS interviews. All of these activities gave participants a good start on developing a common language in which to discuss the new interdiscipline. The atmosphere of the Seminar itself was especially conducive to cross-disciplinary communication. Participants were purposely sequestered for a full week, away from offices and other professional distractions; in the absence of others to talk to, they were forced to talk to one another. These conversations were encouraged by a mix of formal and informal activities, described in the report of the CASM Seminar (Jabine et al., 1984).

The fostering of the CASM movement continued through NSF's funding, the SSRC Committee described above, and the government laboratories. It seems to me that the fostering had both extremely salutary and somewhat limiting results. On the positive side, although there is no way of knowing what interdisciplinary collaboration would have developed without the special impetus provided by the fostering, I believe that impetus is largely responsible for the birth, growth, and very existence of the movement. But the fostering also shaped the movement in ways that perhaps increased its depth but limited its breadth. The background papers and the major focus of the Seminar were on the contributions of the cognitive sciences, especially cognitive psychology, to survey questionnaire design. The model used by Tourangeau (1984), characterizing the respondent's task as involving comprehension, recall, judgment, and response selection was seized upon as canonical, and has guided much subsequent research. Its implications were explored for the construction of questions, for considerations of the optimum length of recall periods, for developing useful aids for recall and for judgment, among other topics. And these projects have yielded much practically useful information and even some theoretical insights. But the flow of research from this model, sometimes funded by outside agencies and often supported by the government laboratories, in some sense caused what might have been a broader stream to remain dammed. Few cognitive disciplines other than cognitive psychology were deeply involved, though there were some contributions from anthropology (e.g., Suchman and Jordan, 1992) and from linguistics (e.g., Clark and Schober, 1992). And few aspects of the survey process outside the questionnaire and interview itself were explored—there was little consideration of cognitive aspects of the conceptualization of survey concepts or of the data processing and presentation stages, and little attention paid to the cognitive aspects of the introduction of computer assisted interviewing.

Nevertheless, during the 1990's, the CASM movement grew and prospered. Work originating in the government laboratories as well as in nongovernmental survey organizations and academic institutions, increasingly addressed both cognitive issues and took for granted that cognitive pretesting and a cognitively-oriented stance were de rigeur for any serious survey enterprise. But the fostering function that had been carried out by the National Science Foundation's Program on Measurement Methods and Data Improvement (MMDI) Panel and the SSRC Committee was no longer obviously driving the movement. Nor had there been any recent evaluation of the movement's progress or prospects. Nor was there any strong impetus to incorporate other disciplines into the CASM dialogue or to examine cognitively the other phases of the survey process. Hence, the time seemed ripe for an evaluation, reinvigoration, and perhaps reorientation of the movement Thanks especially, once again, to the vision of Monroe Sirken, the CASM II Seminar and this book are the outcomes of that thinking.

Thus, the purpose of this book is to broaden the disciplinary base of the CASM movement to include such fields as anthropology, linguistics, and computer science and to broaden the application in survey research of ideas coming from other disciplines beyond questionnaire construction and interviewing to all phases of the survey process, from early conceptualization to final data presentation. This first section of this book is devoted to looking backwards to what has and has not been accomplished so far by the movement to study cognitive aspects of survey methodology, and forward to what can be accomplished in the future, starting perhaps with ideas generated in the interdisciplinary ferment reported in the following chapters themselves.

The first section of this book includes two contributions. The contributors were chosen for their familiarity with and long associations with the CASM movement and for their broad knowledge and statesmanlike understanding of survey research, statistics, and the social sciences. The first contributor, Murray Aborn could well be called the godfather of the CASM movement. He was the Program Officer for the MMDI Program at the National Science Foundation throughout the 1980's. It was during that time that the MMDI Program awarded funding to support the initial CASM Seminar under the auspices of the Committee on National Statistics of the National Academy of Sciences/National Research Council. Under Aborn's direction, the MMDI, program then helped support a good deal of the early research carried out under the CASM banner. He prepared an early evaluation of the movement (Aborn, 1989). Happily, his current evaluation "CASM Revisited" suggests that it is not only a look back at the movement, but a new look; it finds much more to be pleased with and optimistic about than did the earlier appraisal.

Our second contributor, Colm O'Muircheartaigh was, at the time of writing, Director of the Methodology Institute of the London School of Economics and Political Science. As well as being a productive researcher in his own right, inside and outside the CASM movement, O'Muircheartaigh is a perceptive commentator on the work of others, a skilled discussant, and a careful critic, able to meet the most difficult challenge to a discussant, to point out what has not been done, what is missing. Hence, he was a logical choice to make a contribution to CASM II for which the working title was "Gaps in CASM Achievements."

Three interrelated issues have often arisen as thoughtful participants and critics of the CASM movement ponder its content and contributions. First, is the movement really presenting something new, or is it merely a continuation of the long-term interest in understanding and solving problems of nonsampling errors, an instance of old wine in new bottles? Second, if indeed something new is being done, what is the evidence that what is new is really better? And third, why is the preponderance of CASM-related research operations driven, relying mainly on tools adapted from cognitive psychology to pretest questionnaires in the laboratory, and neither guided by nor feeding back into cognitive theory?

In one way or another, both our contributors to this section address these issues, and the themes echoed throughout the book. Consensus is hard to find, however. O'Muircheartaigh notes, for example, that the inclusion/exclusion model of assimilation and contrast effects that Norbert Schwarz and his colleagues (1992) have been working on and Jon Krosnick's model (1991) of the cognitive miser and satisficing are both theoretical formulations new to the survey world; they could also be considered serious contributions to cognitive psychology. Yet O'Muircheartaigh argues that the government laboratories function mostly in an applied mode, pretesting questionnaires in a production-line fashion. Indeed, he argues that the institutional constraints of the government agencies in which the most active of these laboratories are housed dictate that the work carried out there should be routinized rather than innovative. Workers in the government laboratories, while agreeing that a good deal of the work load consists of questionnaire pretesting, argue that such research is cumulative and indeed is being codified in order to lead to better understanding of response effects (see Sirken, Jabine, Willis, Martin, and Tucker, in press). They also point to partnerships between members of the laboratory staff and academics that have resulted in a good deal of basic research, some of it reported in this book. And they cite projects underway to evaluate the efficacy of laboratory procedures. In the same vein, Aborn sees no impact at all of the CASM movement on the cognitive sciences and thinks that perhaps that lack of impact will serve to protect survey research from public concerns about "mind-probing experiments" embedded in surveys and from coming to be considered merely a branch of cognitive psychology. On the other hand, some have argued that the CASM movement has indeed fed back into cognitive psychology and helped to guide some of its studies of autobiographical memory.

No doubt these controversies will continue; it might well be argued that such contentiousness is the sign of a healthy and lively interdiscipline. The material in this section provides some of the background for these controversies; the remainder of the book points to some of the new directions in which they will carry the movement to study cognitive aspects of surveys.

REFERENCES

Aborn, M. (1989). *Is CASM bridging the chasm? Evaluation of an experiment in cross-disciplinary survey research.* Paper presented at the American Statistical Association 1989 Winter Conference, San Diego, CA.

Biderman, A. (1980). *Report on a workshop on applying cognitive psychology to recall problems of the National Crime Survey.* Washington, DC: Bureau of Social Science Research.

Bradburn, N., and Danis, C. (1984). Potential contributions of cognitive research to survey questionnaire design. In T. Jabine, M. Straf, J. Tanur, and R. Tourangeau (Eds.), *Cognitive Aspects of Survey Methodology: Building a Bridge Between Disciplines,* pp. 101–129. Washington, DC: National Academy Press.

Clark, H. H., and Schober, M. F. (1992). Asking questions and influencing answers. In J. M. Tanur (Ed.), *Questions About Questions: Inquiries into the Cognitive Bases of Surveys,* pp. 15–48. New York: Russell Sage Foundation.

Hippler, H. J., Schwarz, N., and Sudman, S. (Eds.) (1987). *Social Information Processing and Survey Methodology.* New York: Springer-Verlag.

Jabine, T., Straf, M., Tanur, J., and Tourangeau, R. (Eds.). (1984). *Cognitive Aspects of Survey Methodology: Building a Bridge Between Disciplines.* Washington, DC: National Academy Press.

Krosnick, J. A. (1991). Response strategies for coping with the cognitive demands of attitude measures in surveys. *Applied Cognitive Psychology, 5,* 213–236.

Moss, L., and Goldstein, H. (Eds.) (1979). *The Recall Method in Social Surveys.* London: NFER Publishing.

Penick, B. K., and Owens, M. E. B. (1976). *Surveying Crime.* Washington, DC: National Academy of Sciences.

Schwarz, N., and Bless, H. (1992). Constructing reality and its alternatives: Assimilation and contrast effects in social judgment. In L. L. Martin and A. Tesser (Eds.), *The Construction of Social Judgement,* pp. 217–245. Hillsdale, NJ: Erlbaum.

Sinaiko, H. W., and Broedling, L. A. (Eds.) (1976). *Questions and Answers in Attitude Surveys: Experiments on Question Form, Wording, and Context.* New York: Academic Press.

Sirken, M., Jabine, T., Willis, G., Martin, E., and Tucker, C. (in press). A new agenda for interdisciplinary survey research methods: *Proceedings of the CASM II Seminar.* National Center for Health Statistics.

Suchman, L., and Jordan, B. (1992). Validity and the collaborative construction of meaning in face-to-face surveys. In J. M. Tanur (Ed.), *Questions About Questions: Inquiries into the Cognitive Bases of Surveys,* pp. 241–267. New York: Russell Sage Foundation.

Tanur, J. M. (Ed.). (1992). *Questions About Questions: Inquiries into the Cognitive Bases of Surveys.* New York: Russell Sage Foundation.

Tourangeau, R. (1984). Cognitive sciences and survey methods. In T. Jabine, M. Straf, J. Tanur, and R. Tourangeau (Eds.), *Cognitive Aspects of Survey Methodology: Building a Bridge Between Disciplines,* pp. 73–100. Washington, DC: National Academy Press.

CHAPTER 3

CASM Revisited

Murray Aborn
National Science Foundation (retired)

3.1 CASM IN ESSE AND IN POSSE

In the introduction to his comprehensive account of the cognitive revolution, the author remarks that: *"One might say that cognitive science has a very long past but a relatively short history"* (Gardner, 1985, p. 9). As put forth in the pages ahead and in the chapter that follows, the same might be said of CASM, a product of the cognitive revolution whose roots can be traced back to the early decades of this century but whose recognition as a discrete entity goes back less than fifteen years.

Given its historical youth, it is hardly surprising to find that in the short span of its existence, CASM's principal concern has been with the most readily attainable of its aims, namely, the adoption of contemporary concepts and techniques employed in the study of cognition to assist in mastering survey problems as old as survey research itself. Actually, this course of events is not at odds with the developmental agenda prescribed in the report of the first CASM Seminar (CASM I), which extensively describes the ways in which research performed collaboratively by cognitivists and survey practitioners could lead to significant improvements in survey methodology (Jabine, Straf, Tanur, and Tourangeau, 1984, especially pp. 10–21 and 73–100).

Many of the innovations in survey methodology introduced over the past fourteen years are attributable to the influence of the CASM I Seminar, and most of them can rightfully be regarded as improvements. However, the agenda emanating from CASM I calls for much more than the application of investigative tools from cognitive science to help deal with problems encountered in obtain-

Cognition and Survey Research, Edited by Monroe G. Sirken, Douglas J. Herrmann, Susan Schechter, Norbert Schwarz, Judith M. Tanur, and Roger Tourangeau.
ISBN 0-471-24138-5   © 1999 John Wiley & Sons, Inc.

ing information from survey respondents. The CASM I report places strong emphasis on the importance of involving cognitive specialists in CASM-engendered research, spelling out—as a matter of motivation if nothing else—the potential benefits such participation could bring to the cognitive sciences (ibid., pp. 6–10). Moreover, the central theme of the report speaks of building an *"interdisciplinary bridge"* between survey research and cognitive science, and goes so far as to contemplate the possible emergence of *"a whole new field"* of endeavor (ibid., pp. ix-2).

These far-reaching aims notwithstanding, the CASM I report openly acknowledges that serious obstacles stand in the way of collaboration between members of two such different disciplinary cultures—obstacles which in all probability explain why actual collaboration has thus far been limited. To aid in surmounting these obstacles, Sirken and Herrmann (1996, p. 245) suggest that the earlier conception of the survey research-cognitive science relationship be expanded to a broader model in which intermediate problem-oriented research activities serve as *"acculturating agents"* in adapting technological and knowledge transfers from one disciplinary culture to the other. This model is based upon twelve years of Sirken and Herrmann's experiences with cognitive research laboratories in the federal government. Cognitive scientists (primarily cognitive psychologists) work interactively with survey researchers in applying cognitive methods to the design and testing of survey questionnaires and, conversely, in developing ideas for conducting basic cognitive research germane to issues and problems which arise in the designing of survey questionnaires.

3.1.1   Changing Times

In showing the model as an example for future interdisciplinary research carried out under the CASM banner, Sirken and Herrmann (1996, p. 245) depart from the CASM I notion of *"encouraging"* cross-disciplinary collaboration (Jabine et al., 1984, pp. ix and 1) by adopting the more proactive stance of *"fostering"* it. The thought of fostering science or in any way directing the course of its development was anathema to the World War II scientists responsible for the creation of the National Science Foundation (NSF) and other "pure science" agencies. NSF's founding father (Vannevar Bush) was imbued with a philosophy perhaps best expressed by Marie Curie when she remarked: "Humanity . . . needs dreamers, for whom the unselfish following of a purpose is so imperative that it becomes impossible for them to devote much attention to the raw material benefit" (Nichols, 1977, p. 6). However, times are changing, and Sirken and Herrmann, although supported by an NSF grant, are in tune with the times.

Today, it is not uncommon to find the subject of changing times with respect to federally-funded pure science openly discussed in the pages of scientific journals and magazines. The demands of a balanced budget, the end of the cold war, and the changing climate of public attitudes toward tax-supported research and development (R&D) is making it increasingly difficult to justify large expenditures for basic scientific research in the absence of prespecified applications to practical ends.

According to one science historian, *"R&D has become D&R, or perhaps even 'development and research for development'"* (Petroski, 1997, p. 212). He goes on to provide a fairly detailed account of the rise and decline of the R&D doctrine, and to examine the gradual remodeling of its underlying precepts in conformance with great growth in societal urgencies and engineering capabilities. Some of Petroski's concluding remarks (p. 213) are addressed to those dependent upon government support for their research:

"The evolution to support for science and technology led by social and national needs means that funding seekers can no longer successfully propose simply to do what can be done. It is no longer sufficient to propose to do the next logical thing in basic research to advance or test a theory, to refine a theoretical model, or to pursue a new line of thinking with the promise of results sometime down the line being applicable to some practical end. Development and research must begin with a clear articulation of the developmental problem and justify any research in terms of it. The increasing desire of funding agencies to see multidisciplinary, multi-institutional, multi-investigator proposals, with ad- hoc teams assembled to address an objective rather than to exploit a capability or maintain an existing team, is a further manifestation of the strengthening and maturing D&R philosophy."

If the realities of federal funding are increasingly likely to favor goal-directed over unfettered basic research, cognitivists may soon find the CASM movement, with its clear-cut applicational objectives and its obvious relevance to national need, a much more attractive environment for advancing basic cognitive science than has up to this time been the case. This may bode well for the future of the CASM movement; however, it is more difficult to speculate on its effects with respect to the whole of societal and behavioral science given the checkered history of prior attempts to prespecify the practical outcomes of research in this domain. A few chosen examples of the ups and downs of interdisciplinary ventures in social science appear in the Historical Perspectives section of this chapter.

3.1.2 Ahead of Its Time

Changing times and shifts in patterns of federal funding notwithstanding, the 1983 CASM I was—if not exactly prescient—at least ahead of its time in two respects. First, in promulgating the mutual benefits to be derived from the interplay of field-observational and laboratory-analytical modes of studying human cognition; second, in recognizing that important questions about human cognition are virtually impossible to answer without undertaking population-wide assessments of the components of cognition, particularly those that relate to the nation's intellectual resources (Jabine et al., 1984, pp. 6–10 and 31). While the CASM I report makes no predictions concerning future developments in these two regards, it does, in retrospect, foreshadow them, as the following examples tend to demonstrate.

Ten years after the publication of the CASM I report, the eminent cognitivist Neal E. Miller, delivering the first in a series of lectures named in his honor, calls upon young clinical psychologists to participate in the development of neuroscience by furnishing workers in the laboratory with *"astute observations"* of the behavior of patients receiving therapy in the clinic, thus stimulating, in turn, controlled research which can refine or correct clinical observations and provide knowledge fundamental to the introduction of new clinical applications (Miller, 1995, p. 901). Miller gives numerous examples of the scientific value of combining behavioral and physiological techniques and in so doing, supplies a virtual hornbook for the establishment of field-laboratory interrelationships as a means of advancing knowledge in such cognitive realms as memory.

The recent publication of the book *The Bell Curve* by Herrnstein and Murray (1994) reignited a scientific and public policy controversy that had lain dormant for a quarter of a century, namely, the question of whether ongoing social and technological changes were dividing America into two camps: the intellectual haves and have-nots, with economic and political power destined to accrue to the former and subservience destined to be the fate of the latter. Underlying the debate are the twin uncertainties of the heritability and mutability of intelligence, and the makeup of its component cognitive capabilities. A lucid exposition of the complex conceptual, statistical, and cultural factors constituting these uncertainties was provided by Hunt (1995) not long after the Herrnstein and Murray book made its appearance. Recently, the multidisciplinary journal *Intelligence* published a special issue containing an even more thoroughgoing (but by no means antithetical) explication of the scientific and social policy problems involved in the controversy (Gottfredson, Guest Ed., 1997). Both the Hunt article and the journal issue draw conclusions based upon connections between the measurement of cognitive abilities and survey research data—the latter with respect to relationships between thinking skills and opinion polls (Gottfredson, 1997, pp. 249–269), the former with respect to the relationship between cognitive skills and economic outcomes (Hunt, 1995, p. 363).

For example, Hunt's conclusion regarding the importance of IQ in getting into a job or profession is in part based upon Herrnstein and Murray's argument concerning the strength of relationship between IQ and economic status, and Herrnstein and Murray's argument is, in turn, based upon their analysis of data from the National Longitudinal Survey of Labor Market Experience of Youth and a respondent's score on the Armed Forces Qualification Test. This nexus of the measurement of cognitive abilities and survey research echoes a theme sounded a dozen years earlier in the CASM I report, wherein Endel Tulving and S. James Press proposed the ways in which surveys could be used to gather systematic information on the distribution of cognitive skills across a range of demographic variables on a national basis (Jabine et al., 1984, pp. 44–60). Although the Tulving—Press proposal deals specifically with the development of national norms for the different kinds of memory, memorial capabilities are, after all, fundamental to any notion of intellectual competence whether viewed from a psychometric or cognitive psychology standpoint.

3.2 HISTORICAL PERSPECTIVES

CASM came to be designated a "movement" six years after the advent of CASM I. It came about as the result of a dispute which took place during a meeting of the Survey Research Methods Section of the American Statistical Association, in which the question of whether CASM should be designated a "revolution" or a "movement" was settled in favor of the latter. The settlement brought to light the importance of viewing any disciplinary development from the perspective of its historical predecessors, and in so doing moved the roots of CASM back several decades from the year usually celebrated as the date of its birth, namely, 1983, the year in which CASM I took place (Aborn, 1989b).

Although the designation as a movement endows CASM with a longer historical past than the date of its birth would suggest, it leaves unanswered the question of how far back that past extends. The answer to that question depends entirely upon whether one views CASM as (a) another effort in survey research's chronology of attempts to reduce nonsampling error, (b) another in social science's attempts to foster interdisciplinary research endeavors, or (c) a movement in the sense of the banding together of people to further a certain cause.

3.2.1 Reducing Nonsampling Error

If one thinks of CASM as belonging to the train of efforts to study, and ultimately to modulate, the sources of nonsampling error, then CASM's ancestry goes back at least as far as 1944, with the publication of Hadley Cantril's classic *Gauging Public Opinion*, which drew attention to the effects of question wording and question order on survey responses (Cantril, 1944). Not too long after, the survey-based mispredictions of the 1948 Presidential election and the ensuing public derision of surveys led the Social Science Research Council (SSRC) to sponsor two investigations of what threw the polls so far off. Badly biased sampling was the main culprit, but the resulting SSRC publications also dealt with problems stemming from interviewing methods (Hyman et al., 1954; Mosteller, Hyman, McCarthy, Marks, and Truman, 1949).

In the 30 years subsequent to the SSRC studies, both academic and governmental survey research centers produced much research on the nonsampling aspects of survey-taking, as is well-documented in the historical reviews contained in: the *Journal of Applied Cognitive Psychology* (Jobe and Mingay, 1991, pp. 175–191); the book *Questions About Questions* (Tanur, ed., 1992, pp. ix–xii and 3–12); and the book *Thinking About Answers* (Sudman, Bradburn, and Schwarz, 1996, pp. 4–14). The last of these works also furnishes comprehensive documentation of research on nonsampling error published in recent years.

There is one publication not cited in the historical literature which puts

CASM's lineage clearly in the nonsampling error reduction category, namely, the *Twelfth Annual Report of the National Science Board* (NSF, 1981). The Board is this nation's highest scientific advisory body and also serves as the governing body of NSF. Its annual reports are submitted to the President and through him to the Congress. They carry considerable weight both within and outside NSF.

In the *Twelfth Annual Report*, the Board exercises its mandate to *"appraise the impact of research on industrial development and upon the general welfare"* (ibid., p. v). The report does this by describing six different representative fields of science in which basic research is linked to practical application: computers and semiconductors, seismic exploration for oil and gas, pesticides and pest control, synthetic fibers, X rays and medical diagnosis, survey research and opinion polls.

In describing the statistical and scientific bases of surveys and polls, the Board report mentions the need for greater research and refinement in measuring the behavioral and social dimensions of survey-taking. Budgetary allocations within NSF are dominated by the funding needs of the regular fields of science, requiring strong justification for the funding of programs or projects that go beyond ordinary disciplinary boundaries. Therefore, it is no exaggeration to say that the Board report was instrumental in obtaining the budgetary increments that made it possible to support CASM I and subsequent research projects, as well as the cognitive research laboratory at the National Center for Health Statistics (NCHS).

3.2.2   Interdisciplinary Research Endeavors

The CASM I report begins by outlining the difficulties likely to be faced by any effort to achieve collaboration across disciplinary boundaries: practitioners in different disciplines live in different "cultures," see different things as important or trivial, use different research techniques, have different understandings of what is and what is not knowledge, and employ specialized terminologies in which words have different meanings in different disciplines (Jabine et al., 1984, p. ix). Therefore, if the CASM movement is seen to be an interdisciplinary research endeavor, it may be worth looking back at a few prior endeavors of this sort to see what, if anything, they can tell us.

Machine Translation   In the mid-1950's, with funds contributed by a number of other federal agencies, NSF embarked upon a multidisciplinary program of research aimed at automating the process of translating printed materials from one language to another. Nine large research centers were set up at major universities and equipped with state-of-the-art hardware. Staffing these facilities were linguists, computer scientists, language specialists, and a variety of technical experts working collaboratively toward accomplishing the specific objectives assigned to each site.

The program went on for about ten years, at which point a panel assembled by the National Academy of Sciences conducted an evaluation of the program's progress. The panel found then-current attempts at machine translation absurd and issued a decidedly negative report which, among other things, criticized the government for having invested in such a will-o'-the-wisp in the first place (National Research Council, 1966).

The failure of this endeavor was not due to the difficulties envisaged in the CASM I report (as discussed in previous sections of this chapter). Collaboration proved not only feasible but amiable. Rather, this was a case of promising too much too soon, of badly underrating the problems to be solved—to appreciate the complexity of those problems see cognitivist Hofstadter (1997, especially pp. 81–101 and 279–303)—and of badly underestimating the cost. In fact, this endeavor actually fulfilled the more heartening aspects of cross-disciplinary collaboration also noted in the CASM I report, namely, that efforts of this kind *"can engender research projects that have exceptional promise, both for enriching the cultures of the parent disciplines and for creating a hybrid culture that attains its own viability and establishes its own research tradition"* (Jabine et al., 1984, p. ix).

Machine translation gave birth to a new interdiscipline (computational linguistics), and it also served as a proving ground for Noam Chomsky's revolutionary theory of language (transformational grammar)—a development that was critical in enabling cognitive psychologists of the day to overturn then-prevailing behavioristic conceptions of the mind. Some former participants in the program went on to make important contributions to the information-processing substrate of cognitive psychology, and to theories of cognition based on analogies to computing systems.

Some seventeen years after the demise of the machine translation program proper, its reverberations were still being felt in fields as far removed as survey research. For example, the eminent linguist Zellig S. Harris, formerly at the helm of one of the program's nine centers, undertook to apply his method of discourse analysis to samples of the questionnaires used in the Survey of Income and Program Participation, the Panel Study of Income Dynamics, and the National Longitudinal Surveys of Labor Market Experience.

The uniqueness of this method lies in its total reliance on the occurrences, co-occurrences, and sequences of words, not on conceptions of their meanings or any other considerations contributed by the analyst. Nonetheless, the method makes it possible to code, store, and compare the information contained in sentences and whole documents (Harris and Mattick, 1988).

The method is highly computable in that it works by the application of algorithmic procedures for discovering the regularities of word combination in a given field of science, and not on the basis of subjective judgments or semantic properties that lie beyond the capacity of computers. Thus, for example, it becomes possible to compare questions *across* survey questionnaires in terms of the information imparted to respondents.

Demonstrations of the workings of the method took place at the Census Bureau, the University of Michigan's Survey Research Center, and other survey research sites. However, a full-scale project was never launched because, as it turned out, it would require extensive reprogramming of the underlying computer software ("String Analysis" programs) at a cost unacceptable to potential sources of funding.

Law and Social Sciences   Under certain circumstances, aggressive fostering of cross-disciplinary collaboration can be effective in producing the desired result, and the now well-established field of law and social sciences is a case in point. The effort to bring social scientists and legal scholars into joint research endeavors began in the late 1950's, when the National Institute of Mental Health (NIMH), convinced of the need for scientific knowledge concerning the societal effects of deviant behavior and the effectiveness of society's means of coping with deviant behavior, came up with the idea of offering *sheltered funding* for research in areas requiring an interdisciplinary or multidisciplinary approach, such as juvenile delinquency. Sheltered funding is a form of encouragement which informs researchers in new and undeveloped fields of science that their proposals will compete for support only with those of like genre rather than being placed in competition with the powerhouses of long-established disciplines.

The NIMH endeavor created a precedent that was not lost on either the legal profession, the discipline of psychology, or members of Congress interested in seeing the knowledge-base of jurisprudence expanded and, at the same time, in favor of amending the NSF Act to specifically include applied research. In 1969, NSF staff began visiting this country's law schools to prepare them for a forthcoming Law and Social Sciences Program, and in 1971 such a program was officially put in place. Since then, NSF has been a regular source of support for interdisciplinary research in this field and a subdiscipline of forensic psychology has emerged as a science and a practice.

Research Applied to National Needs (RANN)   In 1970, under its new applied research mandate, NSF launched a multidisciplinary research endeavor whose mission was to identify national problems not being addressed by existing agencies, to shorten the lead time between basic science and relevant application, and to assure the utilization of research results. By 1975, RANN was consuming approximately one-fourth of NSF's total research obligations, with social science standing at the forefront of the applied research trend. A year and a half later, NSF asked the National Academy of Sciences to evaluate RANN's performance in the area of social and behavioral science and to report accordingly. Several months later a report appeared which lambasted RANN on a number of counts, and within another year RANN was dismantled and its programs either reshuffled or terminated outright (Larsen, 1992).

According to the Academy report, RANN-engendered research evidenced no applicability to practical ends because it was based upon badly mistaken assumptions about how discoveries in the social and behavioral sciences become transformed into socially useful products. The report also criticized

RANN for wasting time, money, and effort attempting to create interdisciplinary research teams in areas where interdisciplinary research centers staffed by experienced interdisciplinary practitioners already existed (National Research Council, 1976).

Past attempts to foster interdisciplinary research endeavors embracing the social and behavioral sciences seem to suggest three key elements that help ensure success.

1. Such endeavors are most successful when structured to favor science in search of applications as well as applications in search of science. RANN illustrates the fallacy of placing the two in competition with one another.

2. Such efforts work best if the collaborators have something in common intellectually. In the case of machine translation, the linguists and computer scientists involved shared a high level of maturity in formal systems.

3. It pays to make clear at the outset that the standards of evaluation to be employed will not be the same as those applicable to the physical sciences.

3.2.3 Movements

If one views CASM as a movement, then CASM has a single predecessor insofar as social science is concerned, namely, the Social Indicators Movement (1970–1982). This previous movement is a success story with a surprise ending (Aborn, 1984).

The appellation "movement" was given Social Indicators in an article published in *Science* (Sheldon and Parke, 1975), in which the authors justified the appropriateness of the appellation as follows: The indicators enterprise has a social as well as a scientific purpose (its ultimate goal is the development of a "national social report"); it has a "cause" (the construction of measures that reflect social conditions as they really are, so that the country will never again be deceived by measures that reflect only economic conditions); it has political reality (viz., a bill in Congress to establish a Council of Social Advisors akin to the existing Council of Economic Advisors) (Congressional Record Senate, 1973); and it involves a large number of people in government and in the scientific community.

CASM's characteristics compare quite well with those set forth by Sheldon and Parke (1975) to explain why Social Indicators truly constitutes a movement. For instance, CASM has—like Social Indicators—appeared as an article in *Science* (Fienberg and Tanur, 1989). CASM too has a social as well as a scientific purpose (surveys now play a vital role in government planning, evaluating, and legislating a wide variety of health, economic, and social programs). CASM too has a "cause" (viz., to raise consciousness about the need for survey researchers and survey-takers to employ new methods for controlling heretofore uncontrollable sources of survey error). And, although small at the outset,

CASM has grown considerably in strength and number of adherents since its inception a relatively short time ago.

Although the Social Indicators Movement accomplished much research of statistical and sociological importance and strongly impacted both scientific and governmental activities during its lifespan, it ultimately lost its distinctive identity as a multidisciplinary research field because its accomplishments were readily and rapidly absorbed into the routine activities of the established disciplines. This phenomenon was termed "obliteration by incorporation" by the then-President of SSRC (Prewitt, 1983)—SSRC being the organization which helped create and was responsible for administering the Movement. The term has since become part of the jargon of social science, and is sometimes referred to by the acronym OBI. Recently, fears have been expressed concerning the CASM Movement's vulnerability to OBI as judged by, for example, the frequency with which the word *cognitive* appears in the titles of papers presented at meetings of the American Association for Public Opinion Research.

CASM may indeed be experiencing OBI, but one cannot be sure on grounds of the frequency with which the word *cognitive* appears in the titles of survey research papers. CASM may also be experiencing a certain amount of emblematism, as seems to be the case in the field of psychology. For example, in his introductory remarks to a paper titled "The Platzgeist and Cognitive Environmental Psychology," presented at a symposium on the history of psychology sponsored by the New York Academy of Sciences, Kurt Salzinger (1994) announced that the word "cognitive" appearing in the title of his paper had nothing to do with its subject, having been put there in order to assure that the paper would be published. He went on to say that the realities of publication nowadays require that the term "cognitive" be somehow associated with one's research.

3.3 THE MOVEMENT YEARS

In a paper presented at a meeting of the American Statistical Association held eight years ago, an evaluation of CASM's progress some five years into the history of CASM *as* a movement (Aborn, 1989a) was performed. Included in that evaluation was an analysis of the abstracts of forty CASM-oriented research studies appearing in a newsletter the year before (ZUMA, 1988). At the time, these studies seemed to represent a good sampling of new survey research stimulated by the Movement.

The studies were divided into three categories: method of investigation (e.g., original experiment, secondary analysis); provenance (e.g., working paper, journal article); and topic. Topic was defined by such traditional survey research variables as question wording, context effects, memory variables, comprehension, self-reported behavior, attitude–behavior relationships, and modeling of the interview process. Cognitive research tools were supposedly used in each case. Only three of the forty studies had actually been published; the rest were hopeful of the same but still in stages of preparation.

As one means of assessing the Movement's progress over the eight years subsequent to the previous evaluation, four of the major CASM-related works published during that period were selected (Jobe and Loftus, Eds., 1991; Tanur, Ed., 1992; Schwarz and Sudman, 1996; Sudman et al., 1996) and, employing the same categorizations as before, all published research studies cited therein were recorded. The results are shown in Table 3.1.

As Table 3.1 shows, there have been at least 94 CASM-relevant research studies published in 32 different journals over the past eight years. Moreover, the four major CASM-related works scanned to produce that table cited 48 books which, to one extent or another, draw together and organize the plethora of findings and counter-findings reported in the studies composing Table 3.1.

3.3.1 Impact on Survey Research

Probably the single most important contribution of the CASM Movement has been the creation of experimental laboratories within the three foremost U.S. federal statistical agencies to facilitate the application of cognitive concepts and techniques to the design and conduct of national surveys. An excellent historical review of the background and establishment of these laboratories is contained in Tanur (1992) and need not be repeated here. What does need to be promulgated here is the existence of a large fugitive literature residing in the files of these laboratories *and* now made accessible by virtue of bibliographies (Bureau of the Census, 1996; NCHS, 1996; Rope, 1995). The works referenced in these bibliographies are comprised of books and chapters in books, articles in scientific journals, papers published in convention proceedings, reports detailing recommended changes in the design of ongoing national surveys or changes actually made on the basis of cognitive research results, reports of cognitive research conducted on behalf of (and funded by) other federal agencies, and papers presented at university colloquia and other sorts of specialized scientific gatherings. All told, the three bibliographies contain a total of 547 entries (as of 1995–1996) and are currently in the process of being updated.

The entries are largely due to the work of regular members of each laboratory's staff, but include studies conducted by visiting research fellows from academic institutions. The bibliographies list entries by authorship, and include date of publication or completion of a given report, date of presentation at a given professional convention or other nongovernmental meeting, and site of professional convention or other nongovernmental meeting at which a presentation was made. Titles make it possible to identify specific topics of interest (e.g., cognitive interviewing methods) as well as specific surveys of interest (e.g., the Survey of Income and Program Participation).

Data from the three bibliographies were used to construct the graph shown in Figure 3.1, which in turn shows that the tradition of promulgating the outputs of CASM and of seeking to recruit new members to the Movement via presentations and other forms of personal contact—a tradition established soon after the occurrence of CASM I (Jabine et al., 1984, pp. 69–70)—have not only

Table 3.1   Published Research Studies Cited in Four CASM-Related Works

Journal	Number of Articles	Topic[a]
Psychological (Dedicated to Cognition)		
Applied Cognitive Psychology	21	A, C, ED, M, F, U, V
Brain and Cognition	1	M
Cognition	2	C, M
Cognitive Psychology	2	ED, M
Experimental Psychology: Learning, Memory and Cognition	4	M
Language and Cognitive Processes	1	U
Memory and Cognition	5	ED, M, X
Social Cognition	6	A, M, U
Visual Cognition	1	V
Psychological (General)		
Applied Psychology	2	C
European Journal of Social Psychology	4	Q, U
Experimental Social Psychology	2	M, S
Human Learning	1	M
Personality and Social Psychology	8	A, C, J, M, S
Personality and Social Psychology Bulletin	4	C, S
Psychiatric Research	1	I
Psychological Review	3	M, Q
Psychological Bulletin	2	A, C
Statistical		
American Statistical Association	1	I
Official Statistics	2	Q, S
Vital Health Statistics	1	ED
Mixed Readership		
Advances in Consumer Research	1	Q
Advertising Research	1	C
Consumer Research	1	Q
Marketing Research	1	S
Milbank Quarterly	1	M
Memory and Language	1	M
Political Behavior	1	M
Public Opinion Quarterly	10	A, C, Q, S
Science	1	S
Social Science Quarterly	1	S
Social Issues	1	A

[a]Topic abbreviations:
   A = Attitude–Behavior Relationship
   C = Cognitive Methodology, Modeling, Measurement
  ED = Event Dating
   I = Interviewer Bias
   M = Memory (autobiographical, recall, retrieval, long-term)
   Q = Question Wording, Order, Context Effects
   U = Understanding (comprehension)
   S = Self-Reported Behavior

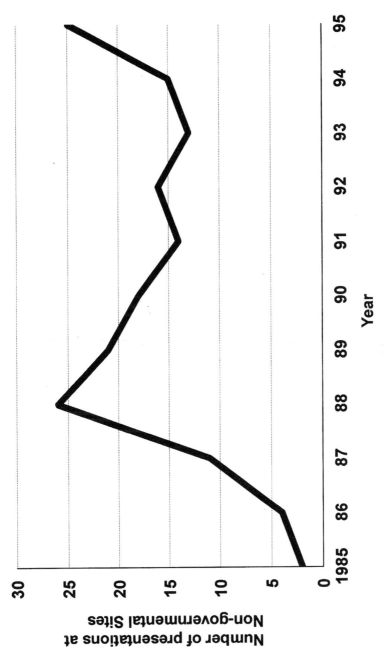

Figure 3.1 Trend in outreach activity.

33

been continued, but have grown over the years. Presentations made nationwide by staff of the NCHS, the Census Bureau, and the Bureau of Labor Statistics cognitive research laboratories have given CASM the outreach any movement needs to succeed.

Thus far, this chapter has relied upon the published and fugitive literatures as a means of assessing the Movement's impact on survey research. There is also one highly significant instance of impact—a two-day symposium sponsored by and held at the National Institutes of Health (NIH) in November 1996. Titled "The Science of Self-Report," the brochure announcing the symposium stated that its purpose was *"to present cutting-edge research on optimal methods for obtaining self-reported information for use in the evaluation of scientific hypotheses, therapeutic interventions, and prognostic indicators"* (NIH, 1996).

Participants included a number of those prominently associated with the CASM Movement, and if that in and of itself was not sufficient evidence of impact, the topics of most of the papers presented leaves no doubt (e.g., "Memory Errors and Survey Reports," "Effects of Question Wording, Interviewer Gender, and Respondent Control on Responses," and "Cognitive Laboratory Methods for Developing and Evaluating Self-Report Questionnaires").

The size of the audience was breathtaking—enough to fill a vast auditorium and leave some people standing along the sides. To add to the astonishment, those attending had to register in advance and enclose a nontrivial fee. Surprising if not exactly astonishing was the front-page press the symposium received from at least two professional media: the American Sociological Association's newsletter *Footnotes*, and the American Psychological Association's newspaper *APA Monitor* (1997). The *Monitor's* headline read: *"Poor recall mars research and treatment,"* and the article which followed noted that *"One of the most significant problems with self-report stems from the fallible nature of human memory . . ."* (ibid., p. 1). Apart from constituting a feather in CASM's cap, this symposium proves that in science, impact rarely happens overnight.

3.3.2   Impact on Cognitive Science

As noted at the outset of this chapter, the CASM Movement was launched with an agenda that called for (or perhaps more properly hoped for) the involvement of cognitive specialists in CASM-engendered research, and the building of a two-way bridge between survey research and cognitive science (Jabine et al., 1984, pp. ix–2). An evaluation of progress toward these aims made five years into the life of the Movement concluded that progress at that point in time lay somewhere between CASM's having *"a handful of studies to its credit and no lasting impact on either survey research or cognitive science,"* and CASM having become *"a whole new interdisciplinary enterprise with a theoretical core and a set of methods it can call its own"* (Aborn, 1989a, p. 15). Now, eight years after the previous evaluation, the reading on the scale of progress puts CASM much, much closer to the latter aim than to the former. This despite the conflicting evidence regarding CASM's impact on cognitive science.

The "hard" evidence can be seen in Table 3.1, which reveals the extent to which CASM has become part of literature likely to be read by cognitivists. The "soft" evidence is anecdotal; it derives from discussions with cognitivists attending various sorts of scientific gatherings, and discussions with NSF Program Directors whose programs have purview over the disciplines which compose cognitive science. These discussions revealed little or no awareness of CASM, either as a concept, an activity, or a scientific entity of any sort. Once informed of CASM's existence, there was an almost universal expression of interest in learning more about it; however, that expression may have been nothing more than collegial courtesy.

The discrepancy between the hard and the soft evidence is not difficult to explain. There are thousands of cognitivists in this country and the discussions took place with less than fifty. The sample was probably not representative even though heavy weight must be assigned to NSF Program Directors who, by their very selection and experience, are well-informed about nearly everything going on in the disciplines under their cognizance. Few scientists have time to read every article in every journal issue. And finally, there was never any reason to expect that bringing more than the existing handful of cognitive psychologists into the CASM fold, or that inducing cognitive scientists to employ surveys as vehicles for their research, would be anything but a slow and gradual process.

3.4   SUMMING UP

This chapter has not just taken a *look back* at CASM, but a *second look* at the phenomenon. In so doing, it has tried to show that CASM's origins lie not only in a long string of preceding *events*, but also in the intellectual and institutional *forces* that preceded its advent. The chapter has reexamined the remarkable agenda that launched the CASM Movement and found it prophetic with regard to the shape of things to come (apologies to H. G. Wells), but probably mistaken with regard to the importance of cross-disciplinary *collaboration*. CASM's achievements in adopting the innovative concepts and techniques of cognitive psychology to address persistent and serious survey problems demonstrates that a small number of researchers skilled in *both* parent disciplines can create interdisciplinarity without codisciplinarity.

3.5   CIRCUMSPECTION

The degree of interdisciplinarity it has attained in the short span of its existence and the favorable outlook for interdisciplinary endeavors in the immediate future tempt one to say that CASM is here to stay. However, it pays to proceed with caution. New interdisciplines arise, flourish, and then fall out of favor, and CASM is doubly vulnerable in that regard.

First, no one can tell what the public reaction will be when word gets around (as it inevitably will) that surveys have "mind-probing" experiments embedded in them—a research method one finds widely endorsed in the CASM literature. Second, there are widespread fears being expressed about the danger of a discipline becoming so much a part of cognitive science that it eventually loses its distinctive identity and disappears as an intellectual pursuit. If that seems far-fetched, consider an article which appeared in a prestigious scientific magazine a few years ago written by a neurologist who claimed that Chomsky's theory of transformational grammar makes such a perfect fit with the functioning of neural networks that before long, linguistics will become a branch of cognitive neuroscience.

Of course, not all neurologists share that belief. In his best-selling book "The Man Who Mistook His Wife for a Hat," neurologist Oliver Sacks writes: *"Our cognitive sciences are themselves suffering from an agnosia similar to the one afflicting the man who mistook his wife for a hat. The man may thus serve as a warning and parable of what happens to a science which eschews the judgmental, the particular, the personal, and becomes entirely abstract and computational"* (Sacks, 1987, p. 20).

REFERENCES

Aborn, M. (1984). The short and happy life of social indicators at the National Science Foundation. *Items, 38,2/3,* 32–41.

Aborn, M. (1989a, January). *Is CASM bridging the chasm? Evaluation of an experiment in cross-disciplinary survey research.* Paper presented at the American Statistical Association 1989 Winter Conference, San Diego, CA.

Aborn, M. (1989b). Discussion of the papers presented at the cognitive laboratories session. *Proceedings of the Section on Survey Research Methods, American Statistical Association,* 431–433.

American Psychological Association (1997). Poor recall mars research and treatment. *APA Monitor, Vol. 28, No. 1.* Washington, DC: American Psychological Association.

Bureau of the Census (1996). *Recent reports and papers related to cognitive aspects of survey methodology.* Center for Survey Methods Research, U.S. Bureau of the Census, Washington, DC: Department of Commerce.

Cantril, H. (1944). *Gauging Public Opinion.* Princeton, NJ: Princeton University Press.

Congressional Record Senate, S15314 (1973). Full opportunity and national goals and priorities.

Fienberg, S. E., and Tanur, J. M. (1989). Combining cognitive and statistical approaches to survey design. *Science, 243,* 1017–1022.

Gardner, H. (1985). *The Mind's New Science.* New York: Basic Books, Inc.

Gottfredson, L. S. (Guest ed.) (1997). *Intelligence, 24(1),* 1–320.

Harris, Z. S., and Mattick, P. (1988). Science sublanguages and the prospects for a global language of science. In M. Aborn (Ed.), *Telescience: Scientific Communication in the Information Age,* Vol. 495, pp. 73–83. The Annals of the American Academy of Political and Social Science. Beverly Hills, CA: Sage.

Herrnstein, R. J., and Murray, C. (1994). *The Bell Curve: Intelligence and Class Structure in American Life.* New York: The Free Press.

Hofstadter, D. R. (1997). *Le Ton Beau de Marot: In Praise of the Music of Language.* New York: Basic Books.

Hunt, E. (1995). The role of intelligence in modern society. *American Scientist, 83,* 356–368.

Hyman, H. H., Cobb, W. J., Feldman, J., Hart, C. W., and Stember, C. (1954). *Interviewing in Social Research.* Chicago: University of Chicago Press.

Jabine, T., Straf, M., Tanur, J., and Tourangeau, R. (Eds.) (1984). *Cognitive Aspects of Survey Methodology: Building a Bridge Between Disciplines.* Washington, DC: National Academy Press.

Jobe, J., and Loftus, E. (Eds.) (1991). Cognition and survey measurement. Special issue of *Applied Cognitive Psychology, 5,* 170–305.

Jobe, J. B., and Mingay, D. J. (1991). Cognition and survey measurement: History and overview. *Applied Cognitive Psychology, 5,* 175–191.

Larsen, O. M. (1992). *Milestones and Millstones: Social Science at the National Science Foundation, 1945–1991.* New Brunswick, NJ: Transactions Publishers.

Miller, N. E. (1995). Clinical-experimental interactions in the development of neuroscience. *American Psychologist, 50(11),* 901–911.

Mosteller, F., Hyman, H., McCarthy, P. J., Marks, E. S., and Truman, D. B. (1949). *The preelection polls of 1948.* Bulletin 60. New York: Social Science Research Council.

National Center for Health Statistics (1996). *Bibliography for the National Laboratory for Collaborative Research in Cognition and Survey Measurement,* 1984–1994. Washington, DC: Department of Health and Human Services.

National Institutes of Health, Office of Behavioral and Social Science Research (1996). *Symposia on the Science of Self-Report: Implications for Research and Practice,* Nov. 7–8, 1996. Bethesda, MD: National Institutes of Health.

National Research Council, Automatic Language Processing Advisory Committee (1966). *Language and machines.* Washington, DC: National Academy of Sciences.

National Research Council (1976). *Social and behavioral science programs in the National Science Foundation.* Report of the Simon Committee. Washington, DC: National Academy of Sciences.

National Science Foundation (1981). Survey research and opinion polls. In *Only One Science.* Twelfth Annual Report of the National Science Board. Washington, DC: U.S, Government Printing Office.

Nichols, R. W. (1997). What if? *The Sciences, 37(6),* 6.

Petroski, H. (1997). Development and research. *American Scientist, 85,* 210–213.

Prewitt, K. (1983). Council reorganizes its work in social indicators. *Items, 37(4),* 74–77.

Rope, D. (1995). *Bibliography of BLS papers.* Bureau of Labor Statistics. Washington, DC: Department of Commerce.

Sacks, O. (1987). *The Man who Mistook his Wife for a Hat.* New York: Harper & Row.

Salzinger, K. (1994). Sitzfleisch 2: The platzgeist and cognitive environmental psychology. In H. E. Adler and R. W. Rieber (Eds.), *Aspects of the History of Psychology, 1892–1992,* Vol 727, pp. 139–142. New York: Annals of the New York Academy of Sciences.

Schwarz, N., and Sudman, S. (1996). *Answering Questions: Methodology for Determining Cognitive and Communicative Processes in Survey Research.* San Francisco: Jossey-Bass.

Sheldon, E. B., and Parke, R. (1975). Social indicators. *Science, 188,* 693–699.

Sirken, M., and Herrmann, D. (1996). Relationships between cognitive psychology and survey research. *Proceedings of the Section on Survey Research Methods, American Statistical Association,* 245–249.

Sudman, S., Bradburn, N. M., and Schwarz, N. (1996). *Thinking About Answers: The Application of Cognitive Processes to Survey Methodology.* San Francisco: Jossey-Bass.

Tanur, J. M. (Ed.). (1992). *Questions About Questions: Inquiries into the Cognitive Bases of Surveys.* New York: Russell Sage Foundation.

ZUMA Newsletter on Cognition and Survey Research, No. 2 (1988).

CHAPTER 4

CASM: Successes, Failures, and Potential

Colm O'Muircheartaigh
University of Chicago

The report of CASM I (Jabine, Straf, Tanur, and Tourangeau, 1984) shows an awareness of a wide variety of ways in which survey research might be enriched by the transplantation of ideas and methods from other disciplines. Of the rich selection described in that report some have taken root and flourished; some have usurped existing methods; some have struggled to gain a foothold; and others have not, as far as I know, been attempted.

In order to evaluate the achievements of CASM, it is necessary to consider the intellectual and institutional development of survey research; it is also necessary to present a framework within which both operational and developmental work on surveys can be evaluated. In this chapter, we discuss how CASM fits into this framework, and how the success or (relative) failure of components of CASM can be seen in the context of the social and professional factors that dominate the survey community. The same considerations apply to the scope (or need) for further progress and the circumstances under which such progress might be brought about.

The discussion concentrates on the ways in which CASM has been less successful. This is not intended to denigrate its achievements, which have been considerable, but to challenge any self-satisfaction that might have arisen, and to suggest that further progress will require as much *well-directed* effort and energy as was necessary for the first wave of progress.

Cognition and Survey Research, Edited by Monroe G. Sirken, Douglas J. Herrmann, Susan Schechter, Norbert Schwarz, Judith M. Tanur, and Roger Tourangeau. ISBN 0-471-24138-5 © 1999 John Wiley & Sons, Inc.

4.1 FRAMEWORK

4.1.1 Historical Strands in Survey Research

The history of surveys (in their modern sense) goes back only 100 years, but from the outset there was a great diversity in the settings, topics, philosophies, and executing agencies involved. There are three distinct strands in the development of survey research: governmental/official statistics, social policy/social research, and commercial/advertising/market research. Each of these brought with it its own intellectual baggage, its own disciplinary perspective, and its own criteria for evaluating success and failure. Figure 4.1 gives a schematic presentation of the development of the social survey in this century. There is a complex interplay of academic discipline, area of application, and type of output.

Statistics   In statistics the evaluation of surveys was largely statistical and the survey was seen as a substitute for complete enumeration of the population. This

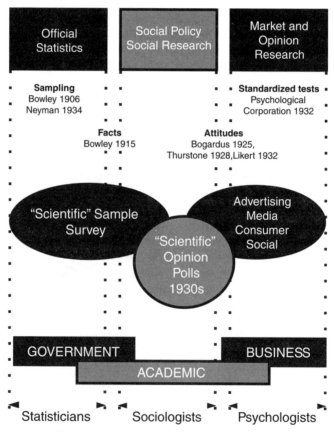

Figure 4.1   The development of survey research.

became and has remained the dominant methodology in the collection of data for government, and the government sample survey agency became an important purveyor of data both to politicians and to statesmen. Symptomatic of their genesis, these agencies tended to be located in national statistical offices, and their professional staff tended to be trained in mathematics or in statistics. Here the concept of error became synonymous with the variance of the estimator.

Social Policy/Social Research   The second major strand in the development of surveys arose from the *Social Policy* and *Social Research* movements (see for instance, Hull House Papers, 1895; Rowntree, 1902). Though not in any way a formal or organized movement, there were certain commonalities of approach and objectives across a wide range of activities. The goal of this movement was social reform, and the mechanism was community description. Here the success or failure of the activity was the effect the findings had on decision makers and politicians. In this field the disciplinary orientation was that of sociology and social psychology, with some influence from social statistics and psychometrics.

Market Research   The third strand arose from the expansion of means of communication and growth in the marketplace. From modest beginnings in the 1890's (Gale and others; see Coolsen, 1947), there was a steady increase in the extent of advertising and a development and formalization of its companion, market research. Here the effect of psychologists was particularly strong. The work of Link and others in the Psychological Corporation was influential in providing an apparently scientific basis for measurement in the market research area (see, for example, Link, 1947). For those psychologists, experimental psychology took precedence over social psychology. The terminology and the approach were redolent of science and technology. The term "error" was not used explicitly; rather there was a description of *reliability* and *validity* of instruments. This contrasts particularly with the "error" orientation of the statisticians.

   Thus, the field of survey research as it became established in the 1940's and 1950's involved three different sectors—government, the academic community, and business; it had three different disciplinary bases—statistics, sociology, and experimental psychology; and it had developed different frameworks and terminologies in each of these areas. [For an excellent and encyclopedic review of the history of survey research in the United States, see Converse (1986); for a shorter paper that emphasizes the institutional bases for survey research, see Fienberg and Tanur (1990)].

4.1.2   Evaluative Framework

In general, models of the survey process concentrate on the survey operation itself, in particular, on the data collection operation. The models may be either mathematical (presenting departures from the ideal as disturbance terms in an algebraic equation) or schematic (conceptual models describing the operational components of the data collection process).

The mathematical model starts with the variance of the sample mean of a variable measured without error and based on a simple random sample (SRS); a set of additional components may easily be added to the variance, each representing a separate source. Thus, processing, nonresponse, noncoverage, and measurement errors can all be incorporated. For generality any biases—whatever their sources—may also be added in, giving the mean squared error as the total error of the estimate. A more extensive treatment of these ideas is given in O'Muircheartaigh (1997, pp. 6–9). The model in Figure 4.2 is a development of Kish's (1965) model of survey error.

Recent advances have to some extent shifted the emphasis from a single parameter to errors in more complex models. These relate to analytic uses of survey data, and are more common in social than in government research. Mul-

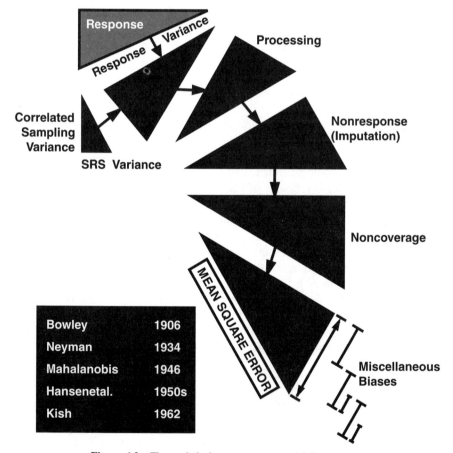

Figure 4.2   The statistical measurement model for surveys.

tilevel models permit the incorporation of hierarchical variance components directly into the analytic framework (see, for instance, Bryk and Raudenbush, 1992; Goldstein, 1995; O'Muircheartaigh and Campanelli, 1998; Wiggins, Longford, and O'Muircheartaigh, 1992).

The conceptual models focus on the interview as the core of the process. Building on the work of Hyman (1954), Kahn and Cannell (1957), Scheuch (1967) and others, Sudman and Bradburn (1974) present one of the more useful of these in their book on response effects in surveys. This (schematic) model (Figure 4.3) presents the relationship among the interviewer, the respondents, and the task in determining the outcome of the survey interview. The elaborated model identifies the potential contribution of a number of the key elements in each of these to the overall quality of the survey response.

In general, models of response errors in surveys focus on the *task*, which is constrained and structured to accomplish the research goals—in particular, to provide the data necessary for analysis. The *interviewer*, as the agent of the researcher, is seen to carry the lion's share of responsibility for the outcome of the data collection process. The *respondent* is largely disregarded, seen as an obstacle to be overcome rather than an active participant in the process.

It is clear that any model of the survey process will have to include these elements. It is not, however, sufficient to consider only these elements, as they do not take into account the context of a survey nor can they distinguish among different survey objectives. To compare the different approaches to survey research, therefore, it is necessary to provide an overarching framework that encompasses the concerns of all three major sectors.

One possible framework draws on some ideas presented by Kish (1987) in his book on statistical design for research. He suggests that there are three issues in relation to which a researcher needs to locate a research design; A slightly different nomenclature is presented here. Each of the "dimensions" is itself multidimensional; they are *representation, control*, and *realism*.

Fundamental to the argument is the belief that empirical relationships (at least among social variables) are never universal. At the level of detail (or generalization) that is necessary for comprehensible findings, the strength and even

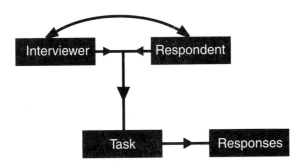

Figure 4.3   A schematic measurement model: Sudman and Bradburn, 1974.

the direction of relationships among variables are conditional. In general, we do not believe that any finding in social science will apply uniformly to all situations, in all circumstances, for all elements of the population. Indeed a good deal of social science is dedicated to understanding the ways in which differences occur across subgroups of populations or between populations.

Representation reflects this set of concerns with regard to the elements included in the investigation. In particular, it refers to the extent to which the target population is adequately mirrored in the sample of elements. In a perfectly specified model, there would be no need to be concerned about which elements from the population appeared in the sample. In the absence of complete and perfect specification of a model (with all variables with potential to influence the variables or relationship under consideration being included), the notion of representation specifically covers the appropriate representation of domains (or subclasses), the avoidance of selection bias, and the minimization of differential nonresponse. Its highest manifestation is in probability sampling, which is the approach not only in academic and (most) governmental research, but also in some media research and election polling.

Control covers issues of experimentation and control of confounding variables. Randomization is one of the methods used to avoid, or reduce the probability of, spurious correlations or misidentification of effects.

Realism arises as an issue in this context in two ways. *Realism in variables* concerns the extent to which the measured or manifest variables relate to the constructs they are meant to describe; *realism in environment* concerns the degree to which the setting of the data collection or experiment is similar to the real-life context with which the researcher is concerned. The survey context may be contrasted with observational studies in which both the variables and the environment are closer to the reality we would like to measure. These dimensions are related to the ideas of *internal validity* and *external validity* used by Campbell and Stanley (1963) and others, in describing the evaluation of social research. The validity of a comparison within the context of a particular survey is the realm of internal validity; the extent to which an internally valid conclusion can be generalized outside that particular context is the realm of external validity.

In Figure 4.4 the dimensions are defined so that the origin represents the ideal point. Thus, at the origin, there is a perfect experiment in which all factors not explicitly controlled are randomized, the experiment is carried out on a representative sample from the population, the measures used are completely realistic measures of the constructs of interest, and the setting of the experiment mirrors exactly the real environment in which the substantive variables are embedded. Clearly such a social investigation is not possible; the usefulness of the model is to locate actual investigations of different kinds in this three-dimensional space. Apart from the desirability of being aware of the limitations of our research, the model helps us to be alert to the serious problems in generalization to the social population of interest that may arise if an investigation is allowed to drift too far from the origin along any of these dimensions.

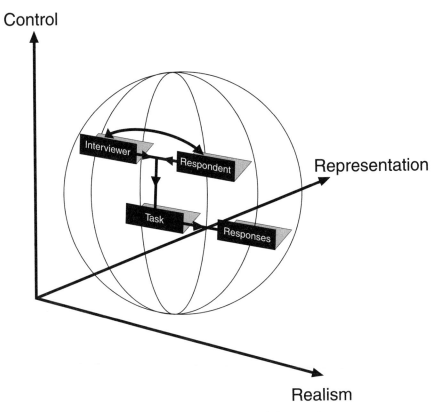

Figure 4.4 A conceptual research framework.

4.1.3 Evaluative Criteria

The concept of quality can only be defined satisfactorily in the same context as that in which the work is being done. To the extent that the context varies, and the objectives vary, the meaning of quality will also vary. It is proposed, that as a definition, we adopt the following: *work that does what it purports to do*. This redefines the problem in terms of the aims and frame of reference of the researcher rather than an arbitrary (pseudo-objective) criterion. Broadly speaking, every survey operation has an objective, an outcome, and a description of that outcome. The gaps in which quality failures (or errors) will be found are the gaps among these three elements.

In order to put the survey model in context, I suggest that we consider it to be located within the three "dimensional" space defined by *representation*, *control*, and *realism*. Ideally we would locate all our investigations at the origin of Figure 4.4, where we would have perfect representation, fully realistic environment and measures, and control over all extraneous variables. In practice, each disciplinary or subject matter approach uses a different combination of positions on these dimensions—in other words, it moves the operation to a different location in the space. A shortcoming of the formulation and evaluation of much social research is that it considers only the narrow criteria corresponding to its own position and orientation; this makes interdisciplinary collaboration and development, such as CASM, particularly difficult.

4.2   LOCATING EXAMPLES IN THREE-DIMENSIONAL SPACE

In this section we present a number of examples to illustrate the importance of taking into account the full range of considerations before accepting that the results of specific experiments will apply in a practical survey situation. The same caveats should be applied to the survey itself as a representation of the broader social reality.

Example 1: Effects of different strategies on completion rates for Census questionnaires (Dillman, Singer, Clark, and Treat, 1996) which illustrates generalizability of results from experimental situations.

The paper serving as Example 1 gives a vivid demonstration of the need for caution in generalizing results from one type of social investigation to another. Dillman et al. contrast—*inter alia*—the effectiveness of benefits appeals (emphasizing the long-run benefits to the respondent or to groups with which the respondent identifies positively of having a full response to the Census) and mandatory appeals (pointing out the legal requirement to comply with the Census) on completion rates for Census questionnaires. Qualitative research (particularly carried out through focus groups) had indicated that benefits appeals would be considerably more effective than mandatory appeals. However, a field test of the findings (self-completion of a mail questionnaire) found that only the mandatory appeals had any effect on the completion rates. There are three possible explanations: first, the social setting of the focus group did not translate into the more isolated setting in which compliance with a self-completion task is decided; second, the *meaning* and *impact* of the benefits appeal in the more leisurely and considered setting of the focus group may not have been represented in the formal (and somewhat obscure) appeal to social benefit represented by the wording on the official form in the field test;[1] and third, the samples used for the focus groups may have been unrepresentative of the population in important respects. To distinguish between the explanations would therefore require further research. In terms of the three-dimensional space in Figure 4.4, the qualitative pilot work of the focus group was carried out closer to the origin on the *control* axis but farther from the origin on the *representation* axis than the field test. The stimulus in the qualitative

---

[1]The appeal read "Why it pays to be counted in the US Census" with four bullet point reasons: "US Census counts decide: Your state and community's share of tax dollars for schools, highways, jobs, health care, services to children and the elderly, and many other programs; How many members of the U.S. Congress your state sends to Washington, D.C.; Where businesses locate, jobs are created, and housing is built; and Many other important activities."

work was also in a different location on the *realism* axis than the stimulus on the field test.

Example 2: Race/ethnicity questions on the United Kingdom Census form which illustrate a failure to consider a sufficiently broad framework.

In developing the race/ethnicity question for the 2001 Census, the United Kingdom Office for National Statistics (ONS) used as a starting point the 1991 Census question, which identified a number of categories by using skin colour or place of origin definitions (White, Black-Caribbean, Black-African, Black-Other, Indian, Pakistani, Bangladeshi, Chinese, Any other ethnic group). In 1991 there had been some complaints from the Irish community in the UK about the failure to recognize the Irish as a separate ethnic group (it is one of the largest non-British communities in the country). The pretesting of the question used various cognitive methods, including card sorts and focus groups, treating the 1991 question as a basis, and adding other ethnic groups one at a time. The research concluded that the Irish community did not wish to be identified separately. However, the set of categories offered defines the objective of the question (and therefore of the Census) as counting "problem minorities" or "people who are not white" rather than counting culturally self-identifying groups. An ancillary finding was that Irish respondents felt that having "Irish" as the only identified subcategory of "White" would cause suspicion and feelings of exclusion within the Irish community in Britain. If the initial categories had been British, French, Polish, Cypriot, Indian, Pakistani, Caribbean, and Other, the answer to the same qualitative investigation might well have been different.

Narrow use of these methods failed to address the import of a question that provides a classification of White (roughly 90 percent of the UK population) along with many subdivisions of the remaining 10 percent. What message would such a question give to the respondent about the attitudes of the researcher? Which schema or social representation of race/ethnicity is accessed by such a question? In terms of the three-dimensional space referred to earlier, this research is located in a narrowly defined area of the *realism* axis; in order to overcome the constraints of using the previous question in this inhibiting way it would have been necessary to consider a broader environment for the development of the question—*ethnography* and *social representations* (discussed later in this chapter) might provide such a framework. It is also very weak on the *representation* dimension; this deficiency could only be remedied by a large-scale field trial.

Example 3: Frequency estimates for reasonably well-defined events which illustrate implicit assumption of uniform effects and artificial examples.

A series of experiments by Schwarz and his colleagues illustrates the strengths of the cognitive approach in showing that the response alternatives presented to the respondent are treated as information by the respondent and used in constructing an answer to the question. This is illustrated particularly in experiments on reports of television watching (Schwarz, Hippler, Deutsch, and Strack, 1985, and subsequent papers). The response alternatives are shown to influence not only responses to a particular question, but also in some cases subsequent comparative judgments.

The concern here, however, is with the interpretation of the result, and with what its implications are. The elegance of the cognitive and experimental approach is that it isolates particular effects by contrasting specific conditions; it evaluates its experiment in terms of statistically significant differences, and then moves on to another topic. This has weaknesses both in substantive and in methodological terms.

The first weakness is in the (implicit) assumption of uniformity of effect. It is rare to see a consideration of gross effects in the reports of these experiments. By this we mean that where a result states that respondents "described themselves as less satisfied with . . ." under one experimental condition rather than another, this involves of course only a net difference and does not offer any information on gross impact. This is not a criticism of the authors, who present their work in an exemplary fashion, but of the recognized structure for such reports.

The second weakness relates to the first. As there is a concentration on a single uni-directional result, there is less scope for the consideration of competing explanations that apply to only part of the population; indeed, there is pressure to concentrate on just one (topic, theory, explanation) at a time.

We offer just one alternative explanation as an illustration. The response alternatives may trigger a meaning shift in the sense that there is an implication that if the numbers of hours offered are very large, what is meant by *watching* is "being in the [approximate] presence of a television set that is switched on"; if the numbers are very small, the implication may be "paying careful and concentrated attention to the program." If this were the case, the solution would lie not in the question structure, but in deciding what you want to know. This can only be done by a much broader consideration of the issue. Three possibilities come immediately to mind. First, identify the schema you want to tap into (leisure, information, background noise) and word the question and response alternatives accordingly. Second, be aware of the social representation of television watching and decide where to direct your question. Third, be aware that different parts of the population have very different social representations or schemas and establish which you are dealing with first for each respondent.

The key here is to be driven by an understanding of the substantive issue rather than taking an average effect across a particular sample as a demonstration of a general truth. We need to define what it is we want to find out, and then ask about it, while recognizing that not every respondent may be able to visualize the phenomenon in our terms. Different groups may have internally homogeneous but externally heterogeneous frameworks (schemas/social representations) for the phenomenon. Our questions need to be grounded, but the grounding may be different for different parts of our population.

The work by Schwarz and his colleagues covers a range of small-scale experimental and large-scale field studies and is therefore generally well-positioned on the representation and control dimensions. The particular aspects to which we would draw attention here are (a) the weakness on the *realism* dimension and (b) the concentration on mean differences.

Example 4: The US Current Population Survey (CPS) redesign which illustrates the need for the broadest possible context in testing and developing the survey environment.

The redesign of the CPS was a major methodological exercise undertaken jointly by the US Bureau of Labor Statistics and the US Bureau of the Census. During the redesign process a number of field and laboratory experiments were carried out to test various components of the new procedures. At the penultimate stage the new (redesigned) CPS was run in parallel with the old CPS for a period to make it possible to splice the time series together when the new version was put officially in place. This parallel operation indicated that the new version would lead to a 0.5 percent increase in the reported level of unemployment, one of the more critical measures of this survey. At the time of the official change over, the old version was continued in parallel as a further check on the splicing of the two series. To everyone's surprise, after the changeover it was the old CPS that showed the higher estimates of unemployment. Subsequent analysis (Polivka and Miller, 1995) showed a significant *parallel survey effect* on the estimates of unemployment and a nonsignificant effect for the new methodology.

The implementation of the new version and the decision to run the two versions in parallel both before and after the changeover is an example of good methodological practice. The probable explanation of the surprising outcome lies in the fact that the procedures concentrated on *representation* and *control (randomization)* in the final experiment but did not consider the issues of *realism*. It was known to all the participants which was the *actual* CPS and which was the parallel version; the procedures were necessarily different, and certainly the perception of the surveys was different. An alternative strategy that would have addressed this issue would have been to implement the new CPS in a number of states while continuing to run the old CPS in others; this might have brought to light the importance of the unmeasured factors implicit in the treatment of the real policy-relevant survey.

4.3   THE CASM APPROACH

The defining events of the CASM initiative (the CNSTAT and ZUMA conferences; see Tanur, Chapter 2, and Aborn, Chapter 3, for further discussions on the impact of these conferences) were noteworthy for two reasons. First, there was an explicit recognition of the barriers to cross-disciplinary work and some consideration of how these might be overcome. Second, there was an undertaking to generate funding for collaboration and research under the initiative; this was possible because of the *institutional* involvement of important agents in the research community, some from government and some from the academic sector.

It is widely held that the CASM initiative (or movement, as it is referred to elsewhere in this book) achieved its most noticeable progress in two methodological areas: (1) the application of the methods of cognitive psychology to issues of question form and question wording and (2) the cognitive laboratory in survey organizations.

4.3.1   The Contributions of Cognitive Psychology

The first major area of progress is in the application of the methods and theories of cognitive psychology to question formulation and questionnaire design. The change in emphasis in experimental psychology from behaviorism to cognition led to the development of testable models of memory, language comprehension, inferential reasoning, and judgment (see, for instance, Lachman, Lachman, and Butterfield, 1979). This groundwork provided legitimacy for the extension of psychologists' efforts to more naturalistic fields, such as the survey interview.

Survey research had always used split ballot methods to develop and test questionnaire forms. Muscio (1917), who describes experiments on question form, refers to research by Lippman published in 1907. The major work on survey design *Gauging Public Opinion* (Cantril, 1944) contains a large number of studies of question design and interviewing. The major breakthrough subsequent to the CASM initiative was in grounding the experiments in a theoretical framework. This framework was initially a cognitive framework, but was developed to include at least some aspects of the social nature of the survey interview. Notable among these newer inclusions is the application of Grice's (1975) four maxims of conversation—manner (intelligibility), relation (relevance), quantity (appropriate information), and quality (not dishonest or underhand).

This area of work has produced a large and distinguished literature, and has transformed the nature of academic examination of survey questioning. To give just two examples: the development of the inclusion/exclusion model of assimilation and contrast effects (Schwarz, Strack, and Mai, 1991; Strack, Schwarz, Bless, Kubler, and Wanke, 1993) has provided clarification of previously conflicting results; and Krosnick's model of the cognitive miser and satisficing (Krosnick, 1991) has considerable potential. Two other developments from a somewhat different perspective are worth mentioning: the application of Cialdini's (1988) principles to issues of nonresponse (Groves, Cialdini, and Couper, 1992), and the work of Dillman and his colleagues on self-administrative questionnaire design (Jenkins and Dillman, 1997). Furthermore, Tanur (1992) offers a valuable set of papers and Sudman, Bradburn, and Schwarz (1996) provide an excellent description of the overall field. Last, it is noteworthy that described under the CASM umbrella we now find a variety of methods not strictly cognitive; CASM has become a catch-all description in the survey literature for developments with a theoretical basis in the social sciences. Tourangeau, Chapter 8, gives a constructive overview of many of these developments in the context of attitude measurement.

4.3.2 The Cognitive Laboratory

Forsyth and Lessler (1991) provide a clear description of the methods and procedures developed and used by the major U.S. survey organizations (in particular, the Bureau of the Census, the Bureau of Labor Statistics (BLS), the National Center for Health Statistics (NCHS), the Research Triangle Institute (RTI), and Westat). The research was driven by cognitive psychology's information-processing perspective. The methods used ranged from *expert evaluation, expanded interviews*, and *targeted methods*, to *group methods*.

All of these methods had a basis in a refined theoretical framework. However, the exigencies of survey operations and in particular, the demands of large-scale survey organizations have reduced the extent to which the theoretical basis plays a role in determining the choice of method and the manner of execution of these methods.

This possibility was noted quite early in the development of CASM: ". . . insufficient attention has been given to developing theoretical models that assess the validity of various cognitive research methods . . . results from a particular implementation of a research method may be sensitive to procedural variations in implementing the method. . . . need to be sure that the methods we use to study the response process are valid indicators of that process" (Forsyth and Lessler, 1991, p. 418).

Over time, the principal activity of the cognitive laboratory has become the expanded interview. This is based on a specific contribution from cognitive psychology; the idea of *introspection*, leading to the protocol analysis proposed by Ericsson and Simon (1980, 1984) that originally informed the cognitive interview during the time of the first CASM Seminar. It would appear, however, that the methods and procedures actually in use in the cognitive laboratories have diverged substantially from those suggested by Ericsson and Simon. Indeed, we would argue that there is no longer any clear (agreed) definition of what constitutes a cognitive interview in survey research.

The function of the cognitive laboratory as it has evolved in the large agencies is not primarily the development of principles of question or questionnaire design. Rather, it is to provide a facility for pretesting questionnaires for use in ongoing surveys. It is evaluated not by its contribution to general principles or knowledge, but by its throughput of questionnaires in a documented and timely manner. As far as I can ascertain there is no systematic attempt to validate the results of these exercises. However, Willis, DeMaio, and Harris–Kojetin, propose in Chapter 9, a research program that might address these issues.

4.4 NON-CASM APPROACHES

There are powerful arguments against the standardized survey interview as a method of generating information. The artificiality of the interview situation and the performance criteria for the interviewer constitute one aspect of the problem;

the standardized questions themselves constitute the other. In this section these two issues are addressed. Although in some ways it is useful to separate their treatment, there is a common thread connecting the arguments discussed.

4.4.1 The Interview

Since surveys began there have been two poles to the interviewing role. One pole was represented by the expert interviewer who obtained information by having a "conversation" with the respondent, usually without taking notes at the time. The other was the "questionnaire" interviewer, who had a blank form and a prepared set of questions. Gradually, no doubt partly because of the increased scale of survey work, pressure to standardize interviewers' behavior grew and the leading survey organizations responded. In Public Opinion Quarterly (POQ) in 1942, Williams described the *Basic Instructions for Interviewers* for the National Opinion Research Center (NORC). Even among thoughtful methodologists, the general view of the respondent was of a relatively passive actor in the research process; see for instance, Sudman and Bradburn (1974): *"The primary demand of the respondent's role is that he answer the interviewer's questions."* The main concern was with motivating the respondent to do so. Gates and Rissland (1923) were among the first to formalize this issue.

Rapport had always been seen as a key feature of a successful interview; the term encompassed a variety of qualities that implied success on the part of the interviewer in generating satisfactory motivation for the respondent. In the recent literature the concept of rapport has been replaced with "interviewing style," with the distinction made between formal or professional style on the one hand, and informal, interpersonal, socioemotional, or personal styles on the other (see Schaeffer, 1991). Dijkstra and van der Zouwen (1982) argue that *"researchers need a theory of the processes which affect the response behaviour"*; the detailed study of the interactions that take place during the interview is one way of developing such a theory.

The more qualitative approaches to the analysis of the survey interview tend to concentrate on the role of the respondent, or at least give equal status to the respondent in considering the issue. Suchman and Jordan (1990), in an analysis of a small number of interviews, describe the various and very different images that respondents may have of the survey interview, ranging from an interrogation or test through to conversation and even therapy. There seems to be considerable promise in *conversational analysis* to contribute to our understanding of the survey response process (see Schaeffer, 1991). There is a common thread that connects the contribution of all of these sources. Overall the emphasis has shifted from the interviewer as the sole focus of attention, acting as a neutral agent, following instructions uniformly regardless of the particular respondent, to a joint consideration of the interviewer and the respondent taking part in a "speech event," a communication, where each has expectations as well as responsibilities (see Schober, Chapter 6, for further discussion on the interaction between respondents and interviewers). One pole of current thought is that of empowering the respondent [at its most extreme allowing the respondent to set the research agenda (which is not endorsed for survey research)]; the other pole is the ever more detailed specification of (uniform) permissible interactions, so that while acknowledging the importance of respondent behavior, the intention is to control it rather than to liberate it. These two poles are strikingly similar to the original situation one hundred years ago, when the expert interviewer and the questionnaire interviewer represented the two extremes of data collection.

4.4.2   The Task

Paradoxically, it would seem that the effect of introducing a more systematic consideration of cognitive issues into the survey field has been to narrow the perspective of the researcher rather than to broaden it. By adding a checklist of recognized criteria—and by introducing the cognitive laboratory as a certifying agency—there has been a further drift away from a consideration of the richness and complexity of the topics targeted by survey questions. This tendency has been exacerbated by the more formal and directive structure of most "cognitive" survey interviews, which, by their emphasis on the immediate question being answered, do not encourage the free and wide-ranging scope for discussion that might occur under less directive protocols.

There are (at least) two theoretical structures that are worth considering as alternatives. The first is *ethnography* (perhaps buttressed by *cognitive anthropology*); the second is the theory of *social representations.*

Ethnography is a methodology—dominant in anthropology but present also in sociology—that emphasizes the broad context within which a phenomenon is located and is based on an extensive and intensive knowledge of the people being studied. It emphasizes the richness and interconnectedness of phenomena. Ethnographic interviewing does not use standardized questions; by contrast, it depends on the interviewer/fieldworker not only understanding the questions but also the answers. It recognizes that a single question can draw on aspects of many schemas or scripts, and that it may draw on different schemas for different people. The anthropological view of schemas is that they contain a great deal of cultural content, that they are learned (and can therefore change), and that they are relatively robust. Gerber, Chapter 14, gives a more comprehensive view of ethnography from the perspective of anthropology.

The concept of *social representations* arises from the work of a group of French social scientists (see Moscovici, 1961) who had an interest in locating particular phenomena within the general cultural context. A social representation involves two processes—objectification and anchoring. Objectification concretizes the abstract; anchoring consists of incorporating new elements of knowledge into a network of more familiar categories. Social representations provide a structure within which different individuals occupy different positions; they are organizing principles, or dimensions of interest, along which there can be variations in the positions of different individuals. It is not just a question of consensus; only stereotypes are considered to be consensually shared within a given group or subgroup. Social representations may well result in different or even opposed positions being taken by individuals in relation to common reference points (see, for instance, Jodelet, 1991, for an excellent example of a study of social representations). Their importance is in freeing research on social attitudes and cognition from the over-emphasis on the individual's psychological organization and anchoring these attitudes and cognitions in the social field.

Both of these frameworks recognize the importance of considering a question in a broader context; their strength is in the recognition of diversity. In the case of ethnography, there is a recognition that the results of an investigation are specific to the (small) population in which it is conducted. Social representations recognize explicitly that different subpopulations may have fundamentally different organizing principles; indeed, a good deal of the interest in social representations is in identifying and describing these differences. They both suffer, however, from being difficult to translate into instruments that will satisfy the needs of large scale survey research.

4.4.3 Summary

A general ideological argument against the standardized interview is that its results lack "ecological validity" (Cicourel, 1982). The survey interview is a context in which the instrument, the interviewer, and the interviewer and respondent roles are all artificial entities that are not "grounded in the everyday." The cognitive laboratory does not address these basic concerns. Indeed, it could be argued that by its even more artificial isolation of particular questions or sets of questions it removes the enterprise even farther from the everyday; therefore, even if it does in a narrow sense improve the standardized question it fails to reduce the damage caused by the artificiality of the instrument and the setting. The basic research on cognitive aspects of question form and wording also falls short in this regard; for cognitive researchers, applications to survey questions are simply some more data points in their catalogue of experiments. Though survey interviews may be more naturalistic than laboratory conditions, neither is close to the ordinary context of information or conversational exchange—certainly not for the respondent (or laboratory subject). Part of the solution may lie in the adaptation of quite different methodologies such as ethnography (a return to the (even more) expert interviewer) and social representations (an inclusive social framework).

4.5 CONCLUSION: SCIENTIFIC KNOWLEDGE, DEVELOPMENT, AND INNOVATION

Bowley expressed, in his 1915 book on measurement, a suitably broad perspective of the issues:

"The main task . . . is to discover exactly what is the critical thing to examine, and to devise the most perfect machinery for examining it with the minimum of effort . . . we ought to realise that measurement is a means to an end; it is only the childish mind that delights in numbers for their own sake. On the one side, measurement should result in accurate and comprehensible description, that makes possible the visualization of complex phenomena; on the other, it is necessary to the practical reformer, that he may know the magnitude of the problem before him, and make his plans on an adequate scale."

This inclusion in the definition of measurement of both the substantive decisions about what to examine and the technical issues of how to examine it is a useful reminder that measurement involves three ordered components: the decision about *what* to measure (a policy decision), the decision about *how* to measure it (a methodological decision), and the determination of *how good* the measurement is (a technical issue). The second and third components are subordinate to the first; the third typically is calculated within the parameters defined by the second, although it could be argued that a quality evaluation should cover all three components. Bowley's concerns, like the social survey, locate the issue of measurement in the *social* and *policy* fields rather than in the individual cognitive framework.

We frequently fail to recognize in considering measurement issues the socially constructed nature of knowledge. The received view is based on a model of science that suggests that the objects with which we deal are all present in nature fully formed and all that we need to do is to locate them with our research. But our research does not deal in natural objects; it deals in rating scales, recall, personal and bureaucratic definitions, attitudes, lists, and other items that the investigator constructs with care. Whatever we infer about the world is constrained by this world of artifacts. The same is true for our interaction with the subjects of (respondents to) our research. We instruct others to apply our instruments with methodological zeal and this application takes place within a well-regulated *social* system—the survey interview.

What is missing from many accounts of scientific developments is any appreciation of the fundamentally social nature of scientific activity. *"Individual contributors are related by ties of loyalty, power, and conflict . . . they occupy positions in wider social structures. In this social world of science the neat distinction between the rational and irrational components of research crumbles. The fundamental issue in research is not [merely] whether the lone investigator can verify his hypothesis in the privacy of his laboratory but whether he can establish his contribution as part of the canon of scientific knowledge in his field . . . the issue is one of consensus, and consensus is not entirely a matter of logic. It involves vested interests and unexamined biases"* (Danziger, 1990).

An aspect of the CASM initiative that distinguishes it from many others is that it was conceived as an orchestrated program of research, that substantial institutional resources were committed to it ab initio, and that nontrivial funding (in both the university sector and the federal sector) was seen to be a vital part of the development. This solid foundation provided support for some developments, but may have acted as an obstacle to others. The factors that determined which components would be successful and which would fail were dominated by institutional and factional interests rather than by the intrinsic superiority of these components.

The notion that should inform our consideration of *diffusion of innovation* is resistance. There is a tendency on the part of those intimately involved with new developments to assume that all that is necessary is to produce the innovation and that its diffusion will be not only inevitable but swift. In practice, however, a number of barriers must be removed or overcome for an innovation to be accepted. Collins, Sykes, and O'Muircheartaigh (1998) discuss these issues in the context of computer assisted survey information collection; Dillman (1996) presents a discussion in the context of a large government organization.

First, we must consider the innovation itself. What societal priority is driving CASM? What has generated the agenda so far? Can a broader agenda be generated? There were a number of different routes that led to CASM—intellectual curiosity, competitive pressures, client demands. These are reflected in the output from CASM and are evident in the bifurcation between operational practice on the one hand and the academic and methodological research literature on the other.

There are both intellectual and institutional barriers to the acceptance of new ideas and methods. One of the most important of these in the operational sector of survey research is the influence of what is commonly called *Taylorism*. This stereotype of modern management consists of specialization of tasks, the idea of a single best way of carrying out a process, and perhaps most importantly, the separation of thinking from doing.

Second, we should consider functional dependency of thinking: broadly speaking this relates to the way in which in any operation we tend to ask only the questions our tools are capable of answering. This means that the advantages of new methodology will not be evident initially except insofar as the new methodology does what the old methodology does already; advances in what it is possible to do will be slower to gain acceptance. How the (presumed) improvement is measured is difficult. Without such a measure, we must ask whether the cognitive research laboratory is really necessary.

Third, innovations usually involve the redistribution and recomposition of tasks, either of which may threaten the stability or even the viability of an existing organization. Those who work with a particular methodology tend to frame their view of the activity in terms of the functions as they are currently carried out. An innovation may challenge that structure and therefore threaten to isolate or render redundant those currently in positions (and functions) of importance. For an organization there is also the crucial decision of whether to empower existing staff or to recruit new specialists—this is basically the choice between assimilation and colonization. The other CASM (computer-assisted survey methods, now CASIC) provides examples of each.

Fourth, is the innovation itself stable, or is it continuing to change? Is the original intention being met or has it been distorted? One of the most interesting aspects of any change is the way in which not only the organism or activity being changed changes over time but how the *new* concept being introduced also changes over time. Among the more interesting questions is the extent to which success is gauged by the degree to which the host organism—in this case survey research—is changed by the invasive agent—in this case cognitive methods. Would it be a success if all survey research not based on cognitive methods were to disappear?

The cognitive laboratory is a striking example of one organizational response to a radical new idea. This is essentially *Taylorization*. The innovation has been accepted, even embraced. However, it performs the classic production line function of processing a stream of "raw" materials and producing a flow of "superior" processed goods. By separating the laboratory from the rest of the organization—and in particular, from the operations side—the likelihood of a two-directional flow of ideas is minimized.

We conclude with two general points. The first relates to the importance of institutional and social norms in determining the likely success or failure of a new approach. The success of the cognitive laboratory in gaining acceptability arose in part from its usefulness, but also because it (and its name) were compatible with the culture of the organizations it was to inhabit. First, the term cognitive has comforting connotations of hard information. Second, the term laboratory carries with it the symbolism of science and scientific investigation, and borrows from the natural sciences a professionalization (or certification) of elite expertise; this makes it a suitable (and defensible) vehicle for methodological work in large accountable organizations. The name does however also constrain the practices and perspective of the work that it does. It places a premium on standardized methods, on clearly defined roles for interviewer and respondent—with a clearly asymmetric power relationship—and it reduces the likelihood that representation should be considered an important characteristic of the samples of respondents used (indeed the term subject is more likely to be considered appropriate). The respondent is likely to have an appointment, to go to an alien environment for the data collection activity, and to be in a setting that mimics the laboratory rather than the home.

If a survey organization were, instead, to set up an environment in which a more conversational, empowering, qualitative investigation of the survey and its instruments were to take place, and call it the *survey salon*, would this have corresponding acceptability?

There is still a strong *scientist* element in the attitude of quantitative survey researchers to the intrusion of less quantifiable (quantitatively judgeable) procedures. In its early years, partly because of the vagueness of the definition of a social survey and the absence of a generally recognized set of standards, the social survey was by no means held in universally high regard. Thomas (1912), in the *American Journal of Sociology*, opined that ". . . interviews in the main may be treated as a body of error to be used for purposes of comparison in future observations." Gillin (1915), in the *Journal of the American Statistical Association*, decried the tendency towards lack of quality control: ". . . *the survey is in danger of becoming a by-word and degenerating into a pleasant pastime for otherwise unoccupied people.*" The survey research community, having fought its way to respectability by the introduction of "scientific" standards, exhibits a natural reluctance to accept methodologies that do not carry similar generally accepted badges of merit. To take this narrow approach, however, is to ignore the key *social atmosphere* of the survey interview. The interview should be grounded in social norms, and the survey questions should be grounded in some social representation of reality.

The second general point relates to the need for nonuniform solutions that would be generated by a more grounded approach. The development of modern technology may provide us with at least part of the answer.

One of the major stumbling blocks in addressing the concerns of survey researchers about question wording and questionnaire design strategy has been the necessity of designing a standardized instrument that could be applied to the whole population. There were two reasons for this. The first was a physical limit on the amount of paper an interviewer could reasonably carry and the need to be able to manipulate the questionnaire during the course of the interview, often in suboptimal physical circumstances. The second was the difficulty of deciding which version of the questionnaire would be most suitable for each respondent. This constraint has colored research on question design and question wording since standardized questionnaires were first used. The arrival of powerful small computers as a means of questionnaire storage and as an aid to interview delivery has transformed the possibilities of the survey interview.

The technical problems of providing the interviewer with a wide range of manageable questionnaire versions has largely been overcome by the development of computer assisted questionnaires. (There are a number of questions about human–computer interaction that need to be addressed, of course, as the computer becomes a more important vehicle for presentation of information to respondents; Couper's Chapter 18 addresses some of these issues.) This flexibility has not been as widely used as it might be, however, because the role definitions (the interviewer as a neutral deliverer of the researcher's instrument and the perceived inertness of the respondent) have remained essentially unchanged. There is a reluctance to allow discretion to the interviewer, and the interviewer continues to be regarded (in historical terms) as a *questionnaire interviewer* rather than an *expert interviewer*.

To understand any topic, research must be carried out throughout the space designated in Figure 4.4 (reproduced in Figure 4.5). The CASM approach has, because of the pressures of institutions and of academic disciplines, been cornered; it needs to be freed from its self-imposed constraints. If there is to be a further radical shift in survey methodology, there are three key elements: the recognition of the *diversity of the population* (across categories and over time), the need to ground interview questions in some *social* realities, and the *empowerment* of the interviewer. None of these will happen, however, unless an appropriate coalition of forces evolves to support them.

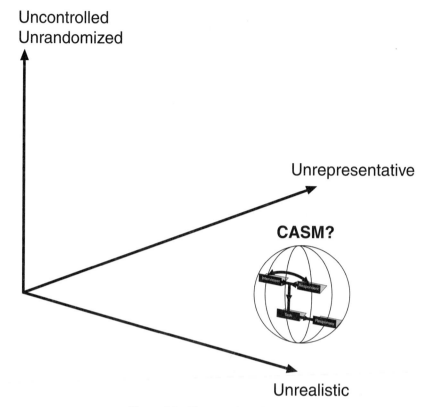

Figure 4.5 The cornering of CASM.

REFERENCES

Bowley, A. (1915; 2nd ed., 1923). *The Nature and Purpose of the Measurement of Social Phenomena.* London: PS King and Son.

Bryk, A. S., and Raudenbush, S. W. (1992). *Hierarchical Linear Models.* Newbury Park: Sage.

Campbell, D. T., and Stanley, J. C. (1963). *Experimental and Quasi-Experimental Designs for Research.* Chicago: Rand-McNally.

Cantril, H. (1944). *Gauging Public Opinion.* Princeton, NJ: Princeton University Press.

Cialdini, R. B. (1988). *Influence: Science and Practice.* Glenview, IL: Scott, Foresman.

Cicourel, A. (1982). Interviews, surveys, and the problem of ecological validity. *American Sociologist, 17,* 11–20.

Collins, M., Sykes, W., O'Muircheartaigh, C. (1998). The diffusion of technological innovation. In M. P. Couper, R. P. Baker, J. Bethlehem, C. Z. F. Clark, J. Martin, W. L. Nicholls II, and J. O'Reilly (Eds.), *Computer Assisted Survey Information Collection.* New York: Wiley.

Converse, J. M. (1986). *Survey Research in the United States: Roots and Emergence 1890–1960*. Berkeley: University of California Press.

Coolsen, F. G. (1947). Pioneers in the development of advertising. *Journal of Marketing, 12*, 80–86.

Danziger, K. (1990). *Constructing the Subject*. Cambridge, England: Cambridge University Press.

Dijkstra, W., and van der Zouwen, J. (Eds.) (1982). *Response Behaviour in the Survey-Interview*. London, England: Academic Press.

Dillman, D. (1996). Why innovation is difficult in government surveys. *Journal of Official Statistics, 12*, 113–124. Fifteen comments and a rejoinder occupy pp. 125–197 of the same issue.

Dillman, D. A., Singer, E., Clark, J. R., and Treat, J. B. (1996). Effects of benefits appeals, mandatory appeals, and variations in statements of confidentiality on completion rates for Census questionnaires. *Public Opinion Quarterly, 60*, 376–389.

Ericsson, K. A., and Simon, H. A. (1980). Verbal reports as data. *Psychological Review, 87*, 215–251.

Ericsson, K. A., and Simon, H. A. (1984). *Protocol Analysis: Verbal Reports as Data*. Cambridge, MA: MIT Press.

Fienberg, S. E., and Tanur, J. M. (1990). A historical perspective on the institutional bases for survey research in the United States. *Survey Methodology, 16*, 31–50.

Forsyth, B. H., and Lessler, J. T. (1991). Cognitive laboratory methods: A taxonomy. In P. P. Biemer, R. M. Groves, L. E. Lyberg, N. A. Mathiowetz, and S. Sudman, (Eds.), *Measurement Errors in Surveys*, pp. 393–418. New York: Wiley.

Gates, G. S., and Rissland, L. Q. (1923). The effect of encouragement and of discouragement upon performance. *Journal of Educational Psychology, 14*, 21–26.

Gillin, J. L. (1915). The social survey and its further development. *Journal of the American Statistical Association, 14*, 603–610.

Goldstein, H. (1995). *Multilevel Statistical Models*, (2nd ed. London: Edward Arnold.

Grice, H. P. (1975). Logic and conversation. In P. Cole and J. L. Morgan (Eds.), *Syntax and Semantics: 3. Speech Acts*, pp. 41–58. New York: Academic Press.

Groves, R. M., Cialdini, R. B., and Couper, M. P. (1992). Understanding the decision to participate in a survey. *Public Opinion Quarterly, 56*, 475–495.

Hull House Maps and Papers: A Presentation of Nationalities and Wages in a Congested District of Chicago, Together with Comments and Essays on Problems Growing out of the Social Conditions. Boston: Thomas Crowell and Co., 1895.

Hyman, H. H. (1954). *Interviewing in Social Research*. Chicago: Chicago University Press.

Jabine, T., Straf, M., Tanur, J., and Tourangeau, R. (Eds.) (1984). *Cognitive Aspects of Survey Methodology: Building a Bridge Between Disciplines*. Washington, DC: National Academy Press.

Jenkins, C. R., and Dillman, D. A. (1997). Towards a theory of self-administered questionnaire design. In L. Lyberg, P. Biemer, A. Collins, E. de Leeuw, C. Dippo, N. Schwarz, and F. Trewin (Eds.), *Survey Measurement and Process Quality*, pp. 165–196. New York: Wiley.

Jodelet, D. (1991). *Madness and Social Representations.* London: Harvester Wheatsheaf.

Kahn, R. L., and Cannell, C. F. (1957). *The Dynamics of Interviewing: Theory Techniques and Cases.* New York: Wiley.

Kish, L. (1965). *Survey Sampling.* New York: Wiley.

Kish, L. (1987). *Statistical Design for Research.* New York: Wiley.

Krosnick, J. A. (1991). Response strategies for coping with the cognitive demands of attitude measures in surveys. *Applied Cognitive Psychology, 5,* 213–236.

Lachman, R., Lachman, J. T., and Butterfield, E. C. (1979). *Cognitive Psychology and Information Processing.* Hillsdale, NJ: Erlbaum.

Link, H. C. (1947). Some milestones in public opinion research. *International Journal of Attitude and Opinion Research, 1,* 36–46.

Moscovici, S. (1961). *La Psychanalyse, Son Image et Son Public.* Paris: Presses Universitaires de France.

Muscio, B. (1917). The influence of the form of a question. *The British Journal of Psychology, 8,* 351–389.

O'Muircheartaigh, C. A. (1997). Measurement errors in surveys: A historical perspective. In L. Lyberg, P. Biemer, A. Collins, E. de Leeuw, C. Dippo, N. Schwarz, and F. Trewin (Eds.), *Survey Measurement and Process Quality,* pp. 1–25. New York: Wiley.

O'Muircheartaigh, C. A., and Campanelli, P. (1998). The relative impact of sampling and interviewer effect on survey variance. *Journal of the Royal Statistical Society, 161,* Part 1.

Polivka, A. E. and Miller, S. M. (1995). *The CPS redesign: Refocusing the economic lens.* Bureau of Labor Statistics Working Paper 269. Washington DC: U.S. Department of Labor.

Rowntree, B. S. (1902). *Poverty: A Study of Town Life.* London: Longmans.

Schaeffer, N. C. (1991). Conversation with a purpose—or conversation? Interaction in the standardized interview. In P. P. Biemer, R. M. Groves, L. E. Lyberg, N. A. Mathiowetz, and S. Sudman (Eds.), *Measurement Errors in Surveys* pp. 367–391. New York: Wiley.

Scheuch, E. K. (1967). Das interview in der sozialforschung. In R. Konig (Ed.), *Handbuch der Empirischen Sozialforschung,* Vol 1, 136–196. Stuttgart: F. Enke.

Schwarz, N., Hippler, H. J., Deutsch, B., and Strack, F. (1985). Response scales: Effects of category range on reported behavior and comparative judgments. *Public Opinion Quarterly, 49,* 388–395.

Schwarz, N., Strack, F., and Mai, H. P. (1991). Assimilation and contrast effects in part-whole question sequences: A conversational analysis. *Public Opinion Quarterly, 55,* 3–23.

Strack, F., Schwarz, N., Bless, H., Kubler, A., and Wanke, M. (1993). Awareness of the influence as a determinant of assimilation versus contrast. *European Journal of Social Psychology, 23,* 53–62.

Suchman, L., and Jordan, B. (1990). Interactional troubles in face-to-face survey interviews. *Journal of the American Statistical Association, 85(409),* 232–241.

Sudman, S., & Bradburn, N. (1974). *Response Effects in Surveys: A Review and Synthesis*. Chicago: Aldine Publishing Co.

Sudman, S., Bradburn, N., and Schwarz, N. (1996). *Thinking About Answers: The Application of Cognitive Processes to Survey Methodology*. San Francisco: Jossey-Bass.

Tanur, J. M. (Ed.) (1992). *Questions About Questions: Inquiries into the Cognitive Bases of Surveys*. New York: Russell Sage Foundation.

Thomas, W. I. (1912). Race psychology: Standpoint and questionnaire, with particular reference to the immigrant and the Negro. *American Journal of Sociology, 17*, 725–775.

Wiggins, R. D., Longford, N., and O'Muircheartaigh, C. A. (1992). A variance components approach to interviewer effects. In A. Westlake, R. Banks, C. Payne, and T. Orchard (Eds.), *Survey and Statistical Computing*, pp. 243–254. Amsterdam: Elsevier Science Publishers.

Williams, D. (1942). Basic instructions for interviewers. *Public Opinion Quarterly, 6*, 634–641.

SECTION B

CHAPTER 5

Cognitive Research into Survey Measurement: Its Influence on Survey Methodology and Cognitive Theory

Norbert Schwarz
University of Michigan

5.1 THE INFLUENCE OF THE CASM MOVEMENT ON COGNITIVE THEORY AND SURVEY METHODS

Survey methodology has long been characterized by rigorous theories of sampling on the one hand, and the so-called "art of asking questions" on the other hand. This state of affairs has changed over the last decade, due to an increasing collaboration of cognitive and social psychologists and survey methodologists, which was initiated by two conferences held in the United States in 1983 (See Jabine, Straf, Tanur, and Tourangeau, 1984) and in Germany in 1984 (see Hippler, Schwarz, and Sudman, 1987). In the years since, work on cognitive aspects of survey measurement has developed at a rapid pace, resulting in the publication of several books (Jobe and Loftus, 1991; Schwarz and Sudman, 1992, 1994, 1996; Tanur, 1992) and a first comprehensive monograph (Sudman, Bradburn, and Schwarz, 1996). Moreover, major survey centers, in the United States as well as in Europe, have established cognitive laboratories to help with questionnaire development and courses in cognitive psychology have entered the curricula of graduate programs in survey methodology.

Drawing on psychological theories of language comprehension, memory, and

Cognition and Survey Research, Edited by Monroe G. Sirken, Douglas J. Herrmann, Susan Schechter, Norbert Schwarz, Judith M. Tanur, and Roger Tourangeau.
ISBN 0-471-24138-5   © 1999 John Wiley & Sons, Inc.

judgment, researchers have begun to formulate explicit models of the question answering process and have tested these models in laboratory experiments and split-ballot surveys. This work links survey methodologists' expertise in the "art of asking questions" to recent developments in cognitive science, providing a useful theoretical and empirical basis for understanding the processes by which survey respondents arrive at an answer. The chapters in this section highlight what has been learned from this research. To place their discussions of theoretical and methodological developments in a broader context, it is useful to review the tasks that survey respondents face.

5.2 RESPONDENTS' TASKS

To arrive at a meaningful answer, survey respondents need to perform several tasks (see Cannell, Marquis, and Laurent, 1977; Cannell, Miller, and Oksenberg, 1981; Strack and Martin, 1987; Tourangeau, 1984). First, they need to interpret the question to understand what is meant and to determine which information they ought to provide. If the question is an attitude question, they may either retrieve a previously formed attitude judgment from memory, or they may "compute" a judgment on the spot. While survey researchers have typically hoped for the former, the latter is far more likely even when respondents have previously formed an opinion that may still be accessible in memory. This reflects that their previously formed judgment may not match the specifics of the question asked. To compute a judgment that pertains to the question asked, respondents need to retrieve relevant information from memory to form a mental representation of the target that they are to evaluate. In most cases, they will also need to retrieve or construct some standard against which the target is evaluated.

If the question is a behavioral question, respondents need to recall or reconstruct relevant instances of this behavior from memory. If the question specifies a reference period (such as "last week" or "last month"), they must also determine if these instances occurred during this reference period or not. Similarly, if the question refers to their "usual" behavior, respondents have to determine if the recalled or reconstructed instances are reasonably representative or if they reflect a deviation from their usual behavior. If they cannot recall or reconstruct specific instances of the behavior, or are not sufficiently motivated to engage in this effort, respondents may rely on their general knowledge or other salient information to compute an estimate.

Once a "private" judgment is formed in respondents' minds, they have to communicate it to the researcher. To do so, they may need to format their judgment to fit the response alternatives provided as part of the question, and in many instances the offered alternatives may both suggest and constrain what is retrieved from memory. Moreover, respondents may wish to edit their response before they communicate it, due to influences of social desirability and situational adequacy.

Accordingly, interpreting the question, generating an opinion or a representation of the relevant behavior, formatting the response, and editing the answer are the main psychological components of a process that starts with respondents' exposure to a survey question and ends with their overt report (Strack and Martin, 1987; Tourangeau, 1984). Although it is conceptually useful to present respondents' tasks in this sequence, respondents' actual performance may deviate from this ordering and they may, for example, change their interpretation of the question once they find it difficult to map their answer onto the response alternatives provided by the researcher.

5.2.1 Question Comprehension

To ensure that respondents understand the questions asked, researchers are typically advised to avoid complicated wordings and unfamiliar terms. Although this advice is sound, it misses that question comprehension involves more than an understanding of the *literal meaning* of the utterance. When asked, "What have you done today?," respondents are likely to understand the words—yet they may nevertheless not know which behaviors they are to report. Should they report that they took a shower, for example, or is the researcher not interested in this information? As this example illustrates, answering a question requires inferences about the questioner's intentions, which are at the heart of the *pragmatic meaning* of the question. In Chapter 6, Michael Schober reviews how we negotiate the meaning of questions in natural conversations and highlights how the standardization of survey interviews constrains the processes that facilitate meaningful communication. To address these problems, he recommends more flexible interviewing procedures and reports some data that indicate that flexible interviewing may improve data quality for factual reports.

5.2.2 Behavioral Reports

Many survey questions ask respondents to report on a specific event (such as one's last doctor visit) or on a class of events (such as all doctor visits during a reference period). In Chapter 7, Michael Shum and Lance Rips review how such information about one's own behavior is represented in memory. Their discussion highlights how the thematic and temporal structure of autobiographical memory determines what individuals can and cannot recall and provides helpful advice on the recall cues that may prove useful in survey research.

5.2.3 Attitude Measurement

Survey researchers have long been aware that the answers to attitude questions may be profoundly influenced by question wording and question order (see Schuman and Presser, 1981; Schwarz and Sudman, 1992; Tourangeau and Rasinski, 1988, for reviews). In Chapter 8, Roger Tourangeau reviews what has been learned about the emergence of context effects in attitude measurement and identifies the variables that determine if preceding questions result in assimilation or in contrast effects on the answers given to subsequent questions.

5.2.4 Research Methods

In addition to fostering basic research into question comprehension, autobiographical memory, and attitude measurement, the collaboration of psychologists and survey methodologists has stimulated a variety of methodological innovations designed to gain insight into respondents' mental processes. The diverse methodological developments in this area are reviewed in the contributions of Schwarz and Sudman (1996). In Chapter 9, Gordon Willis, Theresa DeMaio, and Brian Harris-Kojetin focus on the cognitive interviewing techniques that are now widely used in pretesting questionnaires in cognitive laboratories. Their discussion highlights the practical usefulness of these techniques, but also emphasizes that more attention needs to be paid to their careful evaluation (see also Groves, 1996).

5.2.5 A Case Study: Income Measurement Errors

In the final contribution to this section, Jeffrey Moore, Linda Stinson, and Edward Welniak (Chapter 10) review sources of measurement error in self-reports of income. Their discussion highlights how comprehension problems combine with the complexity of the recall task, fallible estimation strategies, and social desirability concerns. As a case study in understanding the sources of survey measurement error, it illustrates the applied implications of many of the concepts and methods introduced in the other chapters.

5.3 COGNITIVE ASPECTS OF SURVEY METHODOLOGY (CASM): A SUCCESS STORY?

The initiators of the first CASM conference hoped to build a two-way bridge between cognitive psychology and survey methods to facilitate an exchange that would advance basic research and improve survey practice. As the chapters in this section and the contributions to various books illustrate, this bridge has seen considerable traffic. Moreover, a decade and a half after the initial conferences, the developing interdisciplinary field seems well established, with cognitive laboratories in major government and academic survey centers and regular sessions at professional meetings. Yet, resting on one's laurels is a dangerous temptation in any scientific endeavor. As the collaboration of psychologists and survey methodologists moves from being a novelty to being a routine aspect of survey research, it is time to take a critical look at our practice and to identify potential pitfalls.

5.3.1 A Two-Way Bridge?

As noted repeatedly, the development of the interdisciplinary field of CASM was intended to build a two-way bridge between disciplines—yet the direction is mostly from cognitive psychology to survey methods. This is, perhaps, not surprising. Survey research offers a method, not a substantive body of theorizing about human cognition and behavior. Hence, survey research influenced cognitive theorizing mostly by providing a number of puzzling phenomena that were new to psychologists, including, for example, the emergence of telescoping effects in autobiographical recall and the impact of conversational processes on attitude judgments. These phenomena stimulated cognitive research that extends beyond the survey context (e.g., Clark and Schober, 1992; Huttenlocher, Hedges, and Bradburn, 1990; Schwarz, 1996; Schwarz and Bless, 1992; Tourangeau, 1992). Importantly, when this work found an interested audience in the cognitive sciences it did not do so because it bears on surveys, but because it told an interesting story about human cognition that transcended the survey setting.

Moreover, the criteria used in evaluating a line of research differ in important ways between disciplines, resulting in frequent experiences of mutual frustration. These differences are exaggerated by the different conditions that characterize the working environment of basic researchers in academic settings and applied researchers in the cognitive laboratories of government agencies. Cognitive and social psychologists value general theories that address cognitive processes independent of the specific content on which the process operates and they are evaluated on the basis of their ability to tell an interesting general story about human cognition. Psychologists interested in autobiographical memory may ask, for example, how people arrive at a report of how often they engage in a certain behavior under conditions where the behavior is versus is not well represented in memory (e.g., Menon, Rhagubir, and Schwarz, 1995). To address this issue, they may not only employ any behavior that conveniently matches the theoretical criteria, but will prefer behaviors that can be assessed in the undergraduate population that makes up their subject pool. And although they will choose several behaviors from each theoretically specified class to establish generality, the specific behaviors per se are of no particular interest—they are simply convenient instantiations of an unlimited number of possible operationalizations of the theoretical criteria. Moreover, the researchers trust that the principles revealed by studying these behaviors in the given population will apply to all other behaviors and populations that meet the same theoretical criteria. What they aim for is a general account of the type, "Frequency reports of poorly represented behaviors are based on estimation strategy X, whereas frequency reports of well-represented behaviors are based on a recall-and-count strategy." Any differences between specific behaviors and populations are then traced to possible differences in the quality of the underlying memory representation, which needs additional testing. The substantive behavior per se, however, is of no interest in this endeavor. Moreover, arriving at better ways to ask a frequency question is a welcome side effect of the basic research, but usually not its central goal.

The situation of applied researchers in the cognitive laboratories of government survey agencies, however, is markedly different. These researchers face the task of finding the best way to ask a frequency question about the specific behavior X and their client is not interested in finding a general principle, but solely in assessing this specific behavior. As O'Muircheartaigh (Chapter 4, p. 51) noted, "The function of the cognitive laboratory as it has evolved in the large agencies is not primarily the development of principles of question or questionnaire design. Rather, it is to provide a facility for pretesting questionnaires for use in ongoing surveys. It is evaluated not by its contributions to general principles or knowledge, but by its throughput of questionnaires in a documented and timely manner."

This focus, in combination with time and funding constraints, results in a highly pragmatic research approach: A question drafted by the client is tested in cognitive interviews (see Willis et al., Chapter 9) and "fixed" (see O'Muircheartaigh, Chapter 4). While the "fixing" may be informed by the theoretical principles discovered in basic research, the applied work rarely finds its way into the theoretical discussion. In many cases, the research design does not provide an opportunity to ask theoretically interesting questions to begin with: If we want to understand why the "fixed" question works, whereas the "nonfixed" question did not, we need to specify the characteristics of the cognitive task presented to respondents as well as the characteristics of the respective questions—and we need to test the impact of these characteristics across a range of similar questions and behaviors that meet the same criteria. Unfortunately, the daily reality of work in cognitive laboratories rarely affords researchers this luxury—with two unfortunate consequences.

On the one hand, the applied work done is rarely cumulative in nature. Since general principles are rarely identified, the "testing" and "fixing" has to start anew with each new question asked—hardly an economical use of resources. On the other hand, an enormous body of research that addresses a rich set of issues across a wide range of substantive phenomena, remains without a noticeable impact on the theoretical development of the field. In fact, the archives of the various applied laboratories may host numerous anomalies that challenge theoretical models, yet we are unlikely to learn about them. Some (but not all) laboratories make their findings available in the form of technical reports that summarize their experiences with a given question, but given the frequent lack of appropriate control conditions and theoretical discussion, it is often difficult to determine what has been learned.

As Presser and Wellens (in press) recommend, a systematic, theory-driven analysis of these archival materials promises a rich harvest, with important theoretical as well as applied benefits. At the least, we would learn which characteristics of questions result in which type of problem and which procedures are most likely to identify it. If the archival materials were available on a World Wide Web site, they would also reduce the duplication of effort in this domain, allowing researchers to see how similar items have fared in tests and which solutions proved feasible.

5.3.2 Beyond "Fixing"

In light of the extensive applied work done in cognitive laboratories, it is surprising that a systematic evaluation of the practical usefulness of cognitive laboratory procedures is still missing. Apparently, the face-validity of these procedures is sufficient to justify the resource allocations made in the absence of any hard evidence that the difference made by "cognitive pretesting" is a difference that matters in terms of actual survey data quality. Moreover, we know relatively little about the relative performance of different techniques, including verbal protocols (e.g., Ericsson and Simon, 1984), cognitive interviews based on extensive prompting (e.g., DeMaio and Rothgeb, 1996; Willis et al., Chapter 9), expert systems (e.g., Lessler and Forsyth, 1996), or behavior coding (e.g., Fowler and Cannell, 1996). In fact, many of the available techniques (for a review, see Sudman et al., 1996, chapter 2) may not be routinely employed in many laboratories, reflecting that "over time, the principal activity of the cognitive laboratory has become the expanded interview" (O'Muircheartaigh, Chapter 4), based on extensive prompting. To evaluate the efficiency of laboratory techniques, we need comparative studies that address a set of key questions specified by Groves (1996, pp. 401-402) in a related discussion:

"1. Is there evidence that the 'problem' will exist for all members of the target population? Is evidence sought that different problems exist for groups for whom the question is more/less salient, more/less threatening, more/less burdensome?

2. Do multiple measures of the same component of the question–answer technique discover the same 'problem' (convergent validity)?

3. When the 'problem' is 'fixed,' does replication of the technique show that the problem has disappeared?

4. When the 'problem' is 'fixed,' does application of other techniques discover any new 'problems?'

5. Is there evidence that the 'fix' produces a question with less measurement error than the original question?"

5.3.3 Broadening the Cognitive Agenda

As the preceding discussion and the contributions to this section indicate, the "bridge between disciplines" (Jabine et al., 1984) has mostly been a one-way street. Although psychologists picked up a number of interesting phenomena from the survey literature, this literature itself has relatively little to offer in terms of substantive theorizing about human behavior and cognition. This, of course, is not surprising: Survey research offers a methodology, not a theory of human cognition. So far, however, cognitive psychologists have not made much use of the unique opportunities that representative sample surveys afford. Working within the paradigm of general information processing models, cognitive psychologists have focused on "inside-the-head" phenomena and have trusted that any mind works pretty much like any other mind, making the use of representative samples an unnecessary luxury—in particular, when this luxury comes at the cost of procedures that offer poor control when compared to the cognitive laboratory. But as much as psychology's shift from behaviorism to information processing has rendered psychologists interesting partners for survey methodological work, more recent developments in psychology may eventually render survey methodologists interesting partners in basic psychological research. Specifically, psychologists are developing an increasing interest in the "situated" and "contextualized" nature of human cognition. A few examples may illustrate these developments.

First, there is an interest in how the immediate social setting shapes how people do their thinking. Research addressing these influences ranges from language comprehension (e.g., Schober, Chapter 6) to judgment and problem solving (e.g., Hilton, 1995; Schwarz, 1996) and collaborative memory (e.g., Dixon, 1997). Although this work does not require representative samples, many of the theoretical issues it raises can be addressed by studying the unique social setting known as the "survey interview," pursuing some of the issues noted by O'Muircheartaigh (Chapter 4).

Second, cognitive (social) psychologists are increasingly addressing the influence of individuals' social location on their subjective construction of social reality (e.g., Hardin and Higgins, 1996), an issue that has long been the domain of sociologists. Other work addresses, for example, how the historical context shapes autobiographical memory and the reconstruction of history (e.g., Schuman, Rieger, and Gaidys, 1994), or how beliefs spread in a population (e.g., Nowack, Szamrej, and Latane, 1990). Not only does this work require broader samples than can be found on a college campus, it also addresses many of the population related problems that survey methodologists found missing in CASM research.

Third, psychologists are developing an increasing awareness of cultural differences in human cognition. While much of the early evidence was compatible with the assumption that cultural differences can be traced to the same culture-neutral cognitive processes operating on different culture-specific contents, more recent work suggests that the likelihood that a given process is brought to bear to begin with differs between cultures (see Fiske, Kitayama, Markus, and Nisbett, 1998; Gerber, Chapter 14). This work is likely to foster an interest in cultural differences in cognition within a society, for which representative sample surveys again provide a useful vehicle.

As these examples illustrate, several areas of psychology are moving into research domains that may benefit from the methodological tools survey methodologists have to offer. This work may eventually broaden the range of cognitive research in ways that many survey researchers have long been asking for, including increased attention to the diversity of populations. To what extent this results in more two-way traffic on the bridge between our disciplines will ultimately depend on how intellectually interesting the emerging theoretical accounts will be.

REFERENCES

Cannell, C. F., Marquis, K. H., and Laurent, A. (1977). A summary of studies of interviewing methodology. *Vital and Health Statistics*, Series 2, No. 69. Rockville, MD: National Center for Health Statistics.

Cannell, C. F., Miller, P. V., and Oksenberg, L. (1981). Research on interviewing techniques. In S. Leinhardt (Ed.), *Sociological Methodology*, pp. 389–437. San Francisco: Jossey-Bass.

Clark, H. H., and Schober, M. F. (1992). Asking questions and influencing answers. In J. M. Tanur (Ed.), *Questions About Questions: Inquiries into the Cognitive Bases of Surveys*, pp. 15–48. New York: Russel Sage Foundation.

DeMaio, T. J., and Rothgeb, J. M. (1996). Cognitive interviewing techniques: In the lab and in the field. In N. Schwarz and S. Sudman (Eds.), *Answering Questions: Methodology for Determining Cognitive and Communicative Processes in Survey Research*, pp. 177–195. San Francisco: Jossey-Bass.

Dixon, R. A. (1997). Collaborative memory in adulthood. In D. J. Herrmann, M. K. Johnson, C. L. McEvoy, C. Hertzog, and P. Hertel (Eds.), *Basic and Applied Memory Research: Theory in Context*, pp. 359–383. Mahwah, NJ: Erlbaum.

Ericsson, K. A., and Simon, H. A. (1984). *Protocol Analysis: Verbal Reports as Data*. Cambridge, MA: MIT Press.

Fiske, A. P., Kitayama, S., Markus, H. R., and Nisbett, R. E. (1998). The cultural matrix of social psychology. In D. Gilbert, S. Fiske, and G. Lindzey (Eds.), *Handbook of Social Psychology*, 4th ed., Vol. 2, pp. 915–981. New York: Random House.

Fowler, F. J., and Cannell, C. F. (1996). Using behavioral coding to identify cognitive problems with survey questions. In N. Schwarz and S. Sudman (Eds.), *Answering Questions: Methodology for Determining Cognitive and Communicative Processes in Survey Research*, pp. 15–36. San Francisco: Jossey-Bass.

Groves, R. M. (1996). How do we know that what we think they think is really what they think? In N. Schwarz and S. Sudman (Eds.), *Answering Questions: Methodology for Determining Cognitive and Communicative Processes in Survey Research*, pp. 389–402. San Francisco: Jossey-Bass.

Hardin, C. D., and Higgins, E. T. (1996). Shared reality: How social verification makes the subjective objective. In R. M. Sorrentino and E. T. Higgins (Eds.), *Handbook of Motivation and Cognition: The Interpersonal Context*, Vol. 3, pp. 28–84. New York: Guilford.

Hilton, D. J. (1995). The social context of reasoning: Conversational inference and rational judgment. *Psychological Bulletin, 118*, 248–271.

Hippler, H. J., Schwarz, N., and Sudman, S. (Eds.) (1987). *Social Information Processing and Survey Methodology.* New York: Springer-Verlag.

Huttenlocher, J., Hedges, L. V., and Bradburn, N. M. (1990). Reports of elapsed time: Bounding and rounding processes in estimation. *Journal of Experimental Psychology: Learning, Memory, and Cognition, 16,* 196–213.

Jabine, T., Straf, M., Tanur, J., and Tourangeau, R. (Eds.) (1984). *Cognitive Aspects of Survey Methodology: Building a Bridge Between Disciplines.* Washington, DC: National Academy Press.

Jobe, J., and Loftus, E. (Eds.) (1991). Cognitive aspects of survey methodology. Special issue of *Applied Cognitive Psychology, 5,* 173–296.

Lessler, J. T., and Forsyth, B. H. (1996). A coding system for appraising questionnaires. In N. Schwarz and S. Sudman (Eds.), *Answering Questions: Methodology for Determining Cognitive and Communicative Processes in Survey Research,* pp. 259–291. San Francisco: Jossey-Bass.

Menon, G., Raghubir, P., and Schwarz, N. (1995). Behavioral frequency judgments: An accessibility-diagnosticity framework. *Journal of Consumer Research, 22,* 212–228.

Nowack, A., Szamrej, J., and Latane, B. (1990). From private attitude to public opinion: A dynamic theory of social impact. *Psychological Review, 97,* 362–376.

Presser, S., and Wellens, T. (in press). Different disciplinary perspectives on cognition in the question and answer process (Report of Working Group No. 3, CASM II Seminar). In M. Sirken, T. Jabine, G. Willis, E. Martin, and C. Tucker (Eds.), A new agenda for interdisciplinary survey research methods: *Proceedings of CASM II Seminar.* National Center for Health Statistics.

Schuman, H., and Presser, S. (1981). *Questions and Answers in Attitude Surveys: Experiments in Question Form, Wording and Context.* New York: Academic Press.

Schuman, H., Rieger, C., and Gaidys, V. (1994). Collective memories in the United States and Lithuania. In N. Schwarz and S. Sudman (Eds.), *Autobiographical Memory and the Validity of Retrospective Reports,* pp. 313–334. New York: Springer-Verlag.

Schwarz, N. (1996). *Cognition and Communication: Judgmental Biases, Research Methods, and the Logic of Conversation.* Mahwah, NJ: Erlbaum.

Schwarz, N., and Bless, H. (1992). Constructing reality and its alternatives: Assimilation and contrast effects in social judgment. In L. L. Martin and A. Tesser (Eds.), *The Construction of Social Judgment,* pp. 217–245. Hillsdale, NJ: Erlbaum.

Schwarz, N., and Sudman, S. (Eds.) (1992). *Context Effects in Social and Psychological Research.* New York: Springer-Verlag.

Schwarz, N., and Sudman, S. (1994). *Autobiographical Memory and the Validity of Retrospective Reports.* New York: Springer-Verlag.

Schwarz, N., and Sudman, S. (Eds.) (1996). *Answering Questions: Methodology for Determining Cognitive and Communicative Processes in Survey Research.* San Francisco: Jossey-Bass.

Strack, F., and Martin, L. L. (1987). Thinking, judging, and communicating: A process account of context effects in attitude surveys. In H. J. Hippler, N. Schwarz, and S. Sudman (Eds.), *Social Information Processing and Survey Methodology*, pp. 123–148. New York: Springer-Verlag.

Sudman, S., Bradburn, N., and Schwarz, N. (1996). *Thinking About Answers: The Application of Cognitive Processes to Survey Methodology*. San Francisco: Jossey-Bass.

Tanur, J. M. (Ed.) (1992). *Questions About Questions: Inquiries into the Cognitive Bases of Surveys*. New York: Russel Sage Foundation.

Tourangeau, R. (1984). Cognitive sciences and survey methods. In T. Jabine, M. Straf, J. Tanur, and R. Tourangeau (Eds.), *Cognitive Aspects of Survey Methodology: Building a Bridge Between Disciplines*, pp. 73–100. Washington, DC: National Academy Press.

Tourangeau, R. (1992). Attitudes as memory structures: Belief sampling and context effects. In N. Schwarz and S. Sudman (Eds.), *Context Effects in Social and Psychological Research*, pp. 35–47. New York: Springer-Verlag.

Tourangeau, R., and Rasinski, K. A. (1988). Cognitive processes underlying context effects in attitude measurement. *Psychological Bulletin, 103*, 299–314.

CHAPTER 6

Making Sense of Questions:
An Interactional Approach

Michael F. Schober
New School for Social Research

Imagine that you are engaged in casual conversation in your living room with a few visiting relatives. Your aunt asks you "So, how many hours a week do you work?" How do you go about making sense of the question?

Obviously there are a number of cognitive and linguistic processes involved in such an apparently ordinary encounter (see Graesser and Franklin, 1990), and these processes are worth examining for those of us interested in the cognitive aspects of survey participation. Of course, survey interviewers and respondents don't interact in the same way as your aunt interacts with you. But clarifying the similarities and differences can help us make sense of how survey respondents interpret questions.

So how do you make sense of your aunt's question? You need to parse the continuous stream of sound in your aunt's utterance into words, and this requires a great deal of phonological knowledge. You need to know the grammatical structure of English well enough to figure out that your aunt's utterance had the syntactic form of a question, and you need to parse the utterance into its grammatical parts—nouns, verbs, subjects, objects, etc. You need to know the conventional meanings of the words in the utterance, like *hours*, *week*, and *work*. You need to access those meanings and combine them, using your knowledge of grammatical structure, to create a coherent sense of what the utterance means. You need to know enough about social situations to understand that when a question is asked, the questioner wants an answer.

Beyond these cognitive and linguistic processes, you also need to figure out

Cognition and Survey Research, Edited by Monroe G. Sirken, Douglas J. Herrmann, Susan Schechter, Norbert Schwarz, Judith M. Tanur, and Roger Tourangeau. ISBN 0-471-24138-5 © 1999 John Wiley & Sons, Inc.

how the conventional meanings apply in the current context. This isn't always straightforward. You need to know whether the "you" in the question refers to yourself alone, or you and your spouse together. You need to know whether your aunt is asking for a ballpark estimate or an absolutely precise number of hours. You need to know whether your aunt wants you to think of work broadly or narrowly: Do the hours you spend socializing with the boss count as work time? If you are wary about your family, you need to know your aunt's agenda in asking the question. Is she genuinely curious or is she comparing you unfavorably to your more industrious cousin?

If you are uncertain about what she meant by the question, you are likely to ask: "Do you mean both of us?" or "Do you mean everything I do that's related to work?" or (most dangerously) "What are you getting at?" Alternatively, you can answer the question assuming one interpretation: "Well, *I* work 45 hours a week" or "*Officially* I'm paid for 40 hours of work per week" or "I just won the Employee of the Month Award." You do this on the presumption that if your aunt is not satisfied with your answer she will ask a follow-up question.

This example illustrates that making sense of questions involves more than just individual processes like computing syntactic structure or accessing conventional word meanings. Making sense of questions also involves an *interactive* element: People make inferences about the questioner's intentions by relying on assumptions about how the social world works (Grice, 1975) and they rely on the questioner to help interpret questions in subsequent dialogue. I propose that we can only understand some things people do as they make sense of questions if we consider them to be engaged in interactive processes, not just individual cognitive processes (see Clark, 1992, 1996; Schober, 1998; Schober and Conrad, in press).

6.1   MAKING SENSE OF QUESTIONS IN SPONTANEOUS CONVERSATION

In ordinary conversation, addressees make sense of speakers' questions by relying on at least two resources that could be called interactive. First, to arrive at an initial interpretation, addressees presume that questioners have followed a principle of *audience design*, basing the wording and framing of their questions on the knowledge, beliefs, and assumptions that both participants share. Second, addressees rely on *grounding* procedures to make sure they have understood the question. Because addressees' initial interpretations of questions aren't guaranteed to match speakers' intentions, conversational participants can engage in additional conversational turns to reach agreement that a question has been understood as intended.

6.1.1   Audience Design

When you hear your aunt address her question to you, you know that she has designed it with (at least) you in mind, and this affects how you interpret what she says. For the moment you have been cast in the listener role of being her designated *addressee*, rather than in any of the other listener roles you might play in a conversation. Your relatives on the sofa who have been taking part in the conversation, but aren't currently being addressed, have been cast in the role of *side participants*. Your spouse in the kitchen, who hasn't been taking part in the conversation but can overhear it, is playing the role of *bystander*. Your neighbors who, unbeknownst to you, are listening through the paper-thin walls are taking the role of another kind of overhearer: *eavesdroppers* (see Clark, 1992, 1996; Clark and Carlson, 1982; Goffman, 1981; Schober and Clark, 1989, for discussion of listener roles).

How does your listener role affect the way you make sense of her question? As an addressee, you can assume that your aunt believes that you will be able to interpret her through the words she utters (Grice, 1957), against the background premises and knowledge that you and she share. The side participants have to make a slightly different assumption: Your aunt probably wants them to understand the question she is addressing to you, but her intentions toward them are different than her intentions toward the addressee (you). She is probably doing something like informing them of her intentions toward you (see Clark and Carlson, 1982). She could also be doing something more complicated, like trying to let the side participants know something about her attitude toward you without your knowing it. Bystanders and eavesdroppers make sense of your aunt's question on different bases still. Bystanders must understand that your aunt may not even be attempting to inform them of her intentions toward you; eavesdroppers are certain that she is not.

As you can see, inferences about a speaker's intentions for even a simple question can become complicated. For any particular listener, such inferences rest on an assessment of what the parties involved think about each other—what knowledge, beliefs, or attitudes they believe are shared. This has been called their mutual knowledge (Lewis, 1969; Schiffer, 1972) or *common ground* (Clark and Marshall, 1981). For a belief or piece of information to be a part of two people's common ground, both parties must hold a vast (in principle, infinite) set of reciprocal beliefs: Each person must believe that the other person holds the belief; each must believe that the other believes they hold the belief; each must believe that the other believes that they believe the other holds the belief, etc. (see Clark and Marshall, 1981; Schiffer, 1972).

So, as an addressee you infer your aunt's intentions against the common ground you assume you share with her, along with your assumptions about the common ground she shares with the side participants (the other relatives present). The side participants infer your aunt's intentions against what they assume your common ground with her is, along with their knowledge about their own common ground with her. Bystanders make similar inferences, but with less certainty that the aunt's utterance has been designed with them in mind; eavesdroppers know that the aunt's utterance has not been designed with them in mind. Side participants, bystanders, and eavesdroppers have all been shown to understand the references in utterances less accurately than addressees

(Schober and Clark, 1989; Wilkes-Gibbs and Clark, 1992), and this results in part from the fact that a speaker's utterances are designed most particularly for the addressee.

People can assume common ground with each other on two main bases—cultural and personal. At the cultural level, people can assume that they share relevant kinds of mutual knowledge with other members of the many communities they belong to—English speakers, women, U.S. citizens, parents, New Yorkers, theater aficionados, woodworkers, musicians, etc. At the personal level, people can assume that the physical environment they are currently in is mutually known to both parties. They can also assume that experiences they have shared with the other person are mutually known, including what has been said in current and previous conversations.

Both cultural and personal common ground come into play as you interpret your aunt's question. You interpret your aunt's question knowing that her social role is as a visiting relative, and not as your boss or the IRS; on the basis of your knowledge of social roles, you can assume that her purposes in asking the question don't include using your answer for employing you or levying taxes. If, through earlier personal experience with your aunt, you know that she finds office socializing a burden, you may assume she thinks of "work" in a broadly inclusive way, and you may thus include your lunches with the boss as work hours. If every other question your aunt has pursued thus far during the visit has been about you and your spouse together, you may rely on this personal common ground to infer that her current question probably refers to both of you, and not just yourself.

6.1.2   Grounding

Personal common ground accumulates during a conversation through an interactive process that has been called *grounding* (Clark and Brennan, 1991; Clark and Schaefer, 1987, 1989; Clark and Wilkes-Gibbs, 1986). The idea is simple: an utterance can only be said to have entered common ground if both conversational participants ratify that it has been understood. If your aunt tells you that she just returned from a trip to Singapore, she can assume that this is now mutually known—that this utterance is grounded—only if you give her evidence that you understood what she told you. Her utterance alone does not guarantee grounding; you might not have heard her at all, you might not have been paying close enough attention to understand what she said, or you might have misheard her. Only once you provide her with evidence of understanding (or at least no evidence of misunderstanding) can she presume that you have understood her. And only once your aunt presents you with evidence that she has accepted *your* evidence of understanding can you both truly believe that the utterance has entered your common ground.

You can provide your aunt with evidence of understanding in various ways (Clark and Schaefer, 1989), some more explicit than others. You can explicitly acknowledge comprehension by nodding, saying "uh-huh" or "okay," or saying

"I understand." You can repeat your aunt's utterance verbatim: "So you went to Singapore!" You can demonstrate that you have understood by continuing the conversation with another utterance that is relevant. Or you can show her implicitly that you understood by continuing to pay attention to her, on the assumption that if you hadn't understood her you would have asked for clarification.

You can also show your aunt that you *haven't* understood her utterance by providing explicit or implicit evidence. If you know she said something but you didn't hear it, you can say "Huh?" or "What was that?" or "I didn't hear you." If you only heard part of what she said you can say "You went where?" or "What was that about Singapore?" You can gaze at her uncomprehendingly, implicitly requesting her to repeat or reframe her utterance.

The point is that only once the two of you have agreed that her initial utterance was understood can the utterance be considered grounded. This can take several conversational turns or several clarification sequences. You might provide your aunt with evidence that you have interpreted her utterance in a way that she finds unacceptable ("Oh, your visa expired and you were thrown out?"), and she can work with you to make sure that you understand her statement as she intended it ("No, I didn't feel like staying any longer").

It isn't only statements like "I just got back from Singapore" that must be grounded, but also questions like "So, how many hours a week do you work?" The question is only grounded once both of you agree that it has been understood as intended. If you answer your aunt's question for both yourself and your spouse ("Probably about 100 hours") when she only intended to be asking about you, she can persist ("No, I mean just you") until you finally do interpret her question her way.

Of course, grounding does not guarantee absolutely accurate understanding; it only guarantees that people understand each other as well as they want to for current purposes. In some situations—say, casual cocktail party conversations—people may not care if they understand precise references and exact underlying intentions. In other situations—say, a telephone conversation with someone giving you technical help on using your computer—you may want to be more certain that you have understood. In such cases you will probably use explicit grounding procedures to their fullest extent.

6.2 MAKING SENSE OF SURVEY QUESTIONS

Now suppose that you are asked exactly the same question—"How many hours a week do you work?"—in a telephone survey interview administered by a government agency like the Census Bureau or the Bureau of Labor Statistics. Do you make sense of the question in the same way as you do when your aunt asks the question informally? As a respondent in an official survey, you probably know intuitively that you are not engaged in the same kind of conversation as the one with your visiting relatives, even though you may never have thought about this explicitly. There are clearly different rules of the game at work in the survey interview.

In particular, conversation in standardized surveys is restricted in various ways (see, e.g., Clark and Schober, 1992; Schaeffer, 1991; Suchman and Jordan, 1990, 1992), for all sorts of legitimate historical reasons like concerns about interviewer bias and generalizability (see Beatty, 1995, for a discussion of what has led to current prescriptions for interviewer behavior). Because of these restrictions, audience design and grounding operate differently in standardized surveys than they do in spontaneous conversation.

6.2.1 Audience Design

In standardized surveys, interviewers read questions that other people (the survey designers) have scripted at another time and place. The questions are supposed to be read verbatim, with absolutely no deviation. Rather than designing questions on the fly for particular addressees, as your aunt does in her informal conversation with you, survey interviewers are *intermediaries* for the survey designers (see Clark and Schober, 1992). Unlike your aunt, who can freely change the course of her questioning, interviewers are required to follow the script under all circumstances. They must do this even if following the script seems absurd, as when a respondent's elaboration on an answer to one question would ordinarily make the entire set of follow-up questions irrelevant.

Questions in survey interviews aren't designed with particular addressees in mind. Rather, they are scripted (and pretested) to be appropriate for a generic, nonspecific member of the culture. In this respect, audience design in survey interviews resembles the kind of community-based audience design that authors or journalists rely on when they write for the general public, or the kind of audience design that a lecturer or broadcaster uses for speaking to a large audience.

So, if you are savvy about how survey interviews work, you can't presume that the question "How many hours a week do you work?" was designed with exactly your circumstances in mind. It was designed with the average respondent in mind, and to answer the question appropriately you need to imagine how the question could have been intended to be interpretable for the average respondent. If your work situation doesn't match what you guess is a usual situation—you have wildly varying hours, or you perform various activities that may or may not be classifiable as work—you need to figure out how your circumstances map onto the average situation.

You also attribute different motives to the survey designers who created the question than to your aunt, and you have different notions about what your answer will be used for. If you know that the survey you are taking part in is an official government survey that measures unemployment rates, you might assume (correctly or incorrectly) that "work" and "hours" are intended to have particularly stringent interpretations. If you have a fearful streak, you may imagine that your responses will be checked against IRS records for accuracy, and that your answer could get you in trouble for underreported income. If the survey is a public opinion poll commissioned by a labor advocacy group, or a market research survey, you may assume that the purposes in asking the question are quite different, and you may interpret words in the questions quite differently.

6.2.2 Grounding

The grounding procedures available to you and the survey interviewer are severely restricted. As you will notice if you try to ask for clarification ("What exactly do you mean by work?"), standardized interviewers are trained to take a neutral stance in order to avoid suggesting answers or biasing responses in any way. This translates into some very specific procedures for interviewers to adopt (Fowler and Mangione, 1990).

Consider Fowler and Mangione's prescriptions for what an interviewer should do if you ask for clarification on the work question. For questions like this that demand a numerical response, the interviewer must probe to get you to present one number, but the probe must not bias your response in any way. The permitted methods for doing this include repeating the question ("The question asks: How many hours a week do you work?"), describing what kind of response is needed ("I need a number of hours") or explicitly saying that your interpretation is required ("What would be *your* best estimate?" or "We need *your* interpretation" or "Whatever it means to you").

Note how different this is from what your aunt would do. If you ask your aunt "What exactly do you mean by work?" she will probably feel some obligation to tell you what she means, or at least to find out why you want to know and to handle your concern—to ground your understanding. Officially, the survey interviewer is supposed to *avoid* grounding your understanding of the question.

Fowler and Mangione's other prescriptions also differ from what your aunt is likely to do. For questions that provide response alternatives ("How would you rate your schools—very good, good, fair, or poor?"), if the respondent doesn't answer with exactly one of these alternatives (as in "The schools around here are not very good"), the interviewer must repeat the question and/or repeat *all* the response alternatives ("Would that be very good, good, fair, or poor?"). Any deviation from this procedure, like only re-presenting some of the response alternatives ("Well, would you say fair or poor?"), would be considered a directive probe (Fowler and Mangione, 1990, pp. 39–40).

Your aunt probably wouldn't ask your opinion about schools by providing a list of response alternatives. But if she did ask you a question with response alternatives, like "How do you like your oatmeal, thick or runny?" she would be unlikely to re-present all the response alternatives if your answer did not match the ones she had proposed. If you answered, "Oh, fairly thick," your aunt probably would not say "Would you say thick or runny?" Instead, she would tailor any further inquiries to what you had just said: "Just how thick do you mean?" or "Do you mean really thick or just medium thick?" In other words, unlike an interviewer required to follow standardized procedures, she would adjust her follow-up question to show that she had understood your answer.

If your aunt were to ask you your opinion about local schools, rather than presenting you with response alternatives, she would probably ask the question in an open-ended way: "How would you rate your schools?" Here is what standardized survey interviewers are licensed to do when responses to open-ended questions are inadequate (Fowler and Mangione, 1990, p. 42). They can:

1. Repeat the question.
2. Probe saying, "How do you mean (that)?"
3. Probe saying, "Could you tell me more about (that)?"
4. Ask, "Is there anything else?"

Your aunt might use some of these interviewer behaviors as you and she ground your understanding of her question, although some of them would sound odd or stilted or interview-like in an informal interaction. But it is highly unlikely that she would restrict herself to *only* these behaviors or that she would try to adopt the neutral stance that they attempt to embody. Rather, she would try to ground understanding by adapting whatever she said to what you said.

In sum, interviewers deviate from ordinary grounding techniques by using some licensed probing techniques. This is especially true for interviewer behaviors that adopt a "neutral" stance, like the "whatever it means to you" probes. On the other hand, interviewers sometimes use ordinary grounding techniques, as when they accept responses by saying "okay" or "uh-huh," when they repeat a question that the respondent didn't hear, and when they give explicit or implicit evidence that they have not understood an answer (see Schober and Conrad, 1997).

Note that grounding in standardized interviews is *asymmetrical*. Respondents can't expect a substantive answer if they ask interviewers what a question means; interviewers aren't supposed to help ground respondents' understanding. But interviewers are licensed to ask respondents what they mean by an answer, and the rules of the game allow—in fact, require—respondents to tell them.

6.3 IMPLICATIONS FOR SURVEY RESEARCH

At some level, respondents usually understand that a survey interview is not like an ordinary conversation. Ideally, they would understand the implications of the fact that audience design and grounding only partially operate as they ordinarily do. But I don't believe they always do. I propose that the similarities between grounding techniques in ordinary conversation and survey interviews may sometimes fool respondents into acting as if they were real addressees, rather than participants in an unusual conversation with an intermediary (the interviewer) for an absent third party (the survey designer or designers). Of course, respondents in survey interviews *aren't* like real addressees, although it is an open question whether they should be considered more like side participants, bystanders, or eavesdroppers. Following Gerrig's (1994) reasoning on the status of readers of fiction, survey respondents are like side participants because questions are written for an audience toward whom the writer has at least *some* intentions. But one could argue that survey respondents are like bystanders and eavesdroppers because, unlike side participants, they cannot interact with the speaker to resolve misunderstandings.

In other words, there are both superficial and substantive ways in which survey interactions appear to be like ordinary conversations (see also Schaeffer, 1991). Superficially, survey interviews and spontaneous conversations both involve two parties exchanging information in a series of turns. More substantively, some of the techniques interviewers use to get answers they can use—saying "uh-huh," repeating a question the respondent hasn't heard, etc.—overlap with ordinary grounding techniques. Of course, interviewers aren't really supposed to ground meaning; they are supposed to probe neutrally in order to obtain codable, usable answers (see Houtkoop-Steenstra, 1996), ideally without providing any interpretations for the respondent.

This overlap creates an ambiguous situation for survey respondents. Are they real addressees, conversing with a partner who is presenting them with evidence of understanding, or not? When the interviewer says "uh-huh" or "okay," going on to the next question, does this mean the interviewer has really understood the respondent—with all the concomitant assumptions to be made about what has been grounded—or not?

Such ambiguity, I propose, is at the heart of interviewer-respondent interaction in standardized surveys. It is part of what leads both interviewers and respondents to deviate from official scripts (see Houtkoop-Steenstra, 1996; Schober and Conrad, in press; Suchman and Jordan, 1990, 1992). And it is part of what creates some of the mysterious "response effects" in survey interviews, where question wording, response alternatives, and question ordering can drastically affect the respondents' answers (see also Clark and Schober, 1992; Schwarz, 1994, 1996; Strack and Schwarz, 1992).

6.3.1 Response Effects

As a growing body of evidence shows (see, e.g., Bless, Strack, and Schwarz, 1993; Schwarz, 1996, 1998; Strack and Schwarz, 1992; Sudman, Bradburn, and Schwarz, 1996), survey respondents don't necessarily switch off their ordinary conversational reasoning. Although survey respondents shouldn't depend on their ordinary assumptions about audience design and their ordinary procedures for grounding as they make sense of survey questions, it seems that sometimes they do.

An exhaustive overview is beyond the scope of this chapter, but a few examples can help make the point. In each case, the oddness of audience design and grounding in the survey situation helps explain the seemingly anomalous survey results.

Example 1: Response Alternatives   Several studies have shown that the response alternatives presented as part of a question influence respondents' answers. In one study (Schwarz, Hippler, Deutsch, and Strack, 1985), German respondents were asked to report how much television they watched per day. One set of respondents were presented with response alternatives in a low range, from "up to a half hour" to "more than two and a half hours." The other set of respondents were presented with the same number of response alternatives in a high range, from "up to two and a half hours" to "more than four and a half hours." When asked the question with the low range of response alternatives, 16.2 percent of the respondents reported watching more than two and a half hours of television per week. In contrast, 37.5 percent of the respondents given the high range of response alternatives reported watching that much television.

How does audience design help explain this? Respondents assume—perhaps mistakenly—that the question wording (in this case, the response scale intervals) is informative about what the survey researchers intended. The survey researchers who designed the questions presumably knew something about national TV-watching habits, and so the response scales probably represent what is normal within the population. Indeed, later questioning showed that respondents with the high range estimated that Germans watch more television (3.2 hours per week) than respondents given the low range (2.7 hours per week). Respondents mistakenly took the distribution of the response scales as based on normal TV-watching behavior, because they presumed the survey designers were knowledgeable.

Note that this audience design operates through cultural and not personal common ground. That is, these respondents aren't assuming that the response alternatives were designed for them personally based on what the researchers know about them, or based on what they have told the interviewer thus far. Instead, respondents are judging what the interviewer must have meant through the wording of the question itself.

In ordinary conversation, addressees regularly make sense of questions through response alternatives. If your aunt asks "Do you prefer red or white wine?" you will probably assume that these are your only choices (possibly this is all she has in the house), and that rosé is not among the alternatives. This is because the response alternatives she provided have restricted the *domain of inquiry* (see Clark and Schober, 1992) and set the presuppositions for asking the question.

So it isn't surprising that response scales affect more than just frequency estimates in surveys. They also affect respondents' reports of their subjective experience (e.g., Schwarz, Strack, Müller, and Chassein, 1988). Other differences in response alternatives also affect people's answers, as when a set of response alternatives includes or excludes "don't know" as an option (see Clark and Schober, 1992, for a review).

It is also not surprising that effects of response alternatives can be reduced or eliminated entirely. This can be done if researchers explicitly block the basis for respondents' assuming that the response alternatives are informative. For example, as Schwarz (1996) reports, when respondents are told that they are participating in the study of German TV-watching so that the researchers can determine what the right response alternatives are, the effects of the response alternatives disappear.

Example 2: Question Ordering   Even though survey researchers often intend each question in a survey to be answered independently of all the others, respondents sometimes provide different answers depending on the order the questions appear in (see, for example, Hippler and Schwarz, 1987; Hyman and Sheatsley, 1950; Rugg and Cantril, 1944; Schuman and Presser, 1981; Strack, Martin, and Schwarz, 1988). Here is one particularly clear example (Schwarz, Strack, and Mai, 1991).

In the study, respondents were asked several questions (in German) to be answered on an 11-point rating scale, from very dissatisfied to very satisfied. One question (A) was about general life satisfaction: "How satisfied are you currently with your life as a whole?" Another question (B) was about relationship satisfaction: "Please think about your relationship to your partner (spouse or date). How satisfied are you currently with your relationship?" When question A preceded question B, the answers were significantly less highly correlated ($r = 0.32$) than when question A followed question B ($r = 0.67$).

How might audience design and grounding help account for this? In ordinary conversation, addressees presume that questions are designed based not only on cultural common ground but also on personal common ground. When your aunt asks you a question after you have just answered an earlier question, you assume that the content of your answer was grounded. The mere fact that she asked you another question on a different topic provides evidence that she understood your answer well enough for her current purposes, and that your answer entered your personal common ground with her. So her next question, you presume, is being asked *in light of* your answer to the first question, and you will interpret the question accordingly.

So if your aunt asks you how satisfied you are with your current relationship and then asks you about your happiness with life in general, you will assume that her intention is something like this: "Given what you just told me about your current relationship, how happy are you with life in general?" In other words, you will assume that the second question is asking for elaboration on your first response, and that your aunt assumes the two areas of your life are connected. On the other hand, if your aunt asks you about your general happiness and then asks about your satisfaction in your current relationship, you will assume her intention is something like this: "You just told me how happy you are in general. Now I'm asking about a more particular aspect of your life." In this case, you are less likely to assume that the second question builds on the first, because your evaluation of general life happiness may include many other aspects of life.

This interpretation is supported by some additional results. Schwarz, Strack, and Mai reworded the questions to explicitly suggest the inclusive interpretation, by starting off with "Including the life-domain that you already told us about, how satisfied are you currently with your life as a whole?" This led the two answers to be highly correlated again ($r = 0.61$), just about as highly as they had been correlated when they had simply been presented in that order. Schwarz et al. also altered the wording to make explicit that the two questions should not be interpreted as related ("We would first like to ask you to report on two aspects of your life, which may be relevant to people's overall well-being: (a) relationship satisfaction; (b) satisfaction with life as a whole."). This led the two answers to be far less correlated ($r = 0.18$), which is as low as the two answers were correlated when the instructions specifically stated that the second question was about a different area entirely ($r = 0.20$).

So unless they are explicitly informed that they shouldn't, respondents interpret subsequent questions in light of previous ones, just as they do in ordinary conversation. Now, there are also cognitive processes like priming involved in question order effects (see, e.g., Sudman et al., 1996, for a discussion), and these may not depend on interactive processes like audience design or grounding. But the point is that interactive processes can be involved: respondents don't shut off their ordinary assumptions about what has been grounded, even though they might be aware that interviewers are following a script written at an earlier time by survey designers who aren't there to ground understanding.

Example 3: Differing Interpretations of Questions   Different respondents can interpret exactly the same questions in radically different ways. We know this from comparisons of survey responses to official records, like police reports, bank statements, and hospital records (see Wentland, 1993, for a review). We also know it from subsequent questioning of respondents, when they are asked, after they have participated in a survey, why exactly they answered the questions as they did.

To take an extreme (and controversial) example of a finding using this second technique, Belson (1981, 1986) reported that a notable percentage of respondents interpreted even ordinary words and phrases in survey questions—words like "weekend," "you," "children," and "generally"—differently than the question designers had intended. For example, when respondents were asked "For how many hours do you usually watch television on a weekday? This includes evening viewing," 15 percent of the respondents interpreted "you" as including other people too, 33 percent included times the television was on but they hadn't been paying attention to it, and 61 percent included days other than the five weekdays intended or excluded some weekdays for other reasons (see Belson, 1981, pp. 127–137).

Why might this be? Part of the problem is that alternate interpretations are always available for words in ordinary conversation. Even though addressees presume that speakers have used audience design in selecting their words, this doesn't guarantee perfect understanding. When your aunt asks about your work hours, you have to make a judgment about who exactly "you" includes. Since you have the ability to ground understanding, you can ask your aunt exactly what she means by "you." You also know that if your interpretation of "you" turns out to have been different from your aunt's intention, the mistake can be corrected, if you and your aunt so desire.

In standardized survey interviews, respondents are not licensed to ground understanding of words and phrases in the questions. Respondents must answer questions with a *presumption of interpretability* (Clark and Schober, 1992)—they have to assume that their best guess as to an interpretation must be the one the survey designers intended. And their best guess can easily be wrong.

Evidence for this explanation—that the inability to ground understanding is one major factor leading to inaccurate interpretations—can be found in some studies that Fred Conrad and I have carried out. In two studies, one laboratory study (Schober and Conrad, 1997) and one field study using a national telephone sample (Schober and Conrad, 1998), we have compared respondents' answers to the same questions administered by interviewers who either (a) follow strict standardized interviewing procedures or (b) follow more conversationally flexible procedures. Respondents answered fact-based questions from major government surveys about housing (e.g., "How many bedrooms are there in your house or apartment?"), jobs (e.g., "Last week, did you have more than one job, including part-time, evening, and weekend work?"), and purchases (e.g., "In the past year, did you purchase or have expenses for household furniture?").

In both studies, we trained one group of experienced telephone interviewers to adhere to Fowler and Mangione's (1990) prescriptions for standardization. They were to take a neutral stance, never providing any definitions or interpretations of questions for respondents. We trained another group to help the respondents understand the questions in the way the survey designers had intended—that is, according to the official definitions for key words and phrases in the questions that the survey organizations had provided. Specifically, this second group of interviewers were allowed to rephrase questions, answer requests for clarification, and provide respondents with additional information or clarification even if they had not directly asked for it.

In the laboratory study, respondents answered questions on the basis of fictional scenarios we had devised—floor plans of houses, descriptions of work situations, or purchase receipts. Thus, we knew what the facts were and could code response accuracy—that is, whether responses matched official definitions. In the field study, we couldn't measure response accuracy directly, because we didn't know people's living and work situations. So we relied on indirect measures of response accuracy: how often their responses in a second interview (flexible or standardized) differed from their responses to the same question in an initial standardized interview, and how often the purchases they listed when they answered "yes" to a purchase question conformed to the official definitions.

In both studies, flexible interviewing techniques led respondents to produce answers that were more consistent with the official definitions. This was especially the case when the mapping between respondents' circumstances and the official definition wasn't obvious to the respondents—for example, when respondents weren't sure whether to classify a television purchase as a household furniture purchase or not (the official government definition excludes televisions as furniture).

What these results show is that the match between survey designers' intentions and respondents' interpretations of questions can be quite poor. Standardized interviewing techniques, which do not allow respondents to ground their understanding of questions, can lead to poor response accuracy, exactly because this match is poor. More flexible interviews, which allow interviewers and respondents to ground the respondent's understanding, can improve the match, thus leading to more accurate responses.

But the improvement in response accuracy resulting from licensed grounding comes at a real cost: increased interview duration. These results should not be seen as promoting large-scale adoption of flexible interviewing techniques; far too much remains to be seen about the potential costs and benefits of the technique. The point here is that if we conceive of survey interviews as interactional phenomena, in which some unusual versions of audience design and grounding are taking place, we can study new theoretically-based ways to improve response accuracy.

6.4 CONCLUSIONS

Respondents make sense of survey questions in ways that are related to the ways they make sense of questions as addressees in ordinary conversation. They make inferences about the intentions and agendas underlying the questions—how they presume the question designers implemented *audience design*. This amounts, in part, to judgments about the common ground, both cultural and personal, that the question designers may have taken into account as they designed the survey.

Beyond these individual judgments that respondents make, they also rely on interactive *grounding* procedures to make sense of questions, although the procedures available to them in surveys only overlap partially with the procedures available in ordinary conversation. Respondents sometimes seem to treat surveys as if grounding is proceeding in the ordinary way; this can help account for certain question order effects in surveys. The inability of survey respondents to fully ground their understanding of questions can lead to wide variability in how respondents interpret particular words and phrases in survey questions.

I should note that standardized practice varies from survey to survey, and that not everyone follows Fowler and Mangione's (1990) prescriptions to the letter. It remains to be seen what the precise effects of particular deviations are. But in examining such effects, an interactional analysis of the type proposed here is a good place to start.

ACKNOWLEDGMENTS

This work was supported in part by NSF grant IRI-94-02167. Many thanks to Herbert Clark, Fred Conrad, and Norbert Schwarz for comments on earlier versions of this paper. Any opinions, findings, and conclusions or recommendations expressed in this material are those of the author and do not necessarily reflect the views of the National Science Foundation.

REFERENCES

Beatty, P. (1995). Understanding the standardized/non-standardized interviewing controversy. *Journal of Official Statistics, 11*, 147–160.

Belson, W. A. (1981). *The Design and Understanding of Survey Questions.* Aldershot, England: Gower.

Belson, W. A. (1986). *Validity in Survey Research.* Brookfield, VT: Gower.

Bless, H., Strack, F., and Schwarz, N. (1993). The informative functions of research procedures: Bias and the logic of conversation. *European Journal of Social Psychology, 23*, 149–165.

Clark, H. H. (1992). *Arenas of Language Use.* Chicago: University of Chicago Press.

Clark, H. H. (1996). *Using Language.* Cambridge, UK: Cambridge University Press.

Clark, H. H., and Brennan, S. E. (1991). Grounding in communication. In L. B. Resnick, J. M. Levine, and S. D. Teasley (Eds.), *Perspectives on Socially Shared Cognition,* pp. 127–149. Washington, DC: APA.

Clark, H. H., and Carlson, T. B. (1982). Hearers and speech acts. *Language, 58,* 332–373.

Clark, H. H., and Marshall, C. R. (1981). Definite reference and mutual knowledge. In A. K. Joshi, B. Webber, and I. Sag (Eds.), *Elements of Discourse Understanding,* pp. 10–63. Cambridge, UK: Cambridge University Press.

Clark, H. H., and Schaefer, E. F. (1987). Collaborating on contributions to conversations. *Language and Cognitive Processes, 2,* 19–41.

Clark, H. H., and Schaefer, E. F. (1989). Contributing to discourse. *Cognitive Science, 13,* 259–294.

Clark, H. H., and Schober, M. F. (1992). Asking questions and influencing answers. In J. M. Tanur (Ed.), *Questions About Questions: Inquiries into the Cognitive Bases of Surveys,* pp. 15–48. New York: Russell Sage Foundation.

Clark, H. H., and Wilkes-Gibbs, D. (1986). Referring as a collaborative process. *Cognition, 22,* 1–39.

Fowler, F. J., and Mangione, T. W. (1990). *Standardized Survey Interviewing: Minimizing Interviewer-Related Error*. Newbury Park, CA: Sage.

Gerrig, R. (1994). *Experiencing Narrative Worlds*. New Haven: Yale University Press.

Goffman, E. (1981). *Forms of Talk*. Philadelphia: University of Pennsylvania Press.

Graesser, A. C., and Franklin, S. P. (1990). QUEST: A cognitive model of question-answering. *Discourse Processes, 13*, 279–303.

Grice, H. P. (1957). Meaning. *Philosophical Review, 66*, 377–388.

Grice, H. P. (1975). Logic and conversation. In P. Cole and J. L. Morgan (Eds.), *Syntax and Semantics: 3. Speech Acts*, pp. 41–58. New York: Academic Press.

Hippler, H. J., and Schwarz, N. (1987). Response effects in surveys. In H. J. Hippler, N. Schwarz, and S. Sudman (Eds.), *Social Information Processing and Survey Methodology*, pp. 102–122. New York: Springer-Verlag.

Houtkoop-Steenstra, H. (1996). Probing behaviour of interviewers in the standardised semi-open research interview. *Quality and Quantity, 30*, 205–230.

Hyman, H. H., and Sheatsley, P. B. (1950). The current status of American public opinion. In J. C. Payne (Ed.), *The Teaching of Contemporary Affairs: Twenty-First Yearbook of the National Council of Social Studies*, pp. 11–34. New York: National Education Association.

Lewis, D. (1969). *Convention: A Philosophical Study*. Cambridge, MA: Harvard University Press.

Rugg, D., and Cantril, H. (1944). The wording of questions. In H. Cantril (Ed.), *Gauging Public Opinion*, pp. 23–50. Princeton, NJ: Princeton University Press.

Schaeffer, N. C. (1991). Conversation with a purpose—or conversation? Interaction in the standardized interview. In P. P. Biemer, R. M. Groves, L. E. Lyberg, N. A. Mathiowetz, and S. Sudman (Eds.), *Measurement Errors in Surveys*, pp. 367–391. New York: Wiley.

Schiffer, S. (1972). *Meaning*. Oxford, UK: Clarendon Press.

Schober, M. F. (1998). Different kinds of conversational perspective-taking. In S. R. Fussell and R. J. Kreuz (Eds.), *Social and Cognitive Approaches to Interpersonal Communication*, pp. 145–174. Mahwah, NJ: Erlbaum.

Schober, M. F., and Clark, H. H. (1989). Understanding by addressees and overhearers. *Cognitive Psychology, 21*, 211–232.

Schober, M. F., and Conrad, F. G. (in press). A collaborative view of standardized surveys. To appear in J. van der Zouwen, N. C. Schaeffer, D. Maynard, and H. Houtkoop (Eds.), *Interviewer-Respondent Interaction in the Standardized Survey Interview*.

Schober, M. F., and Conrad, F. G. (1997). Does conversational interviewing reduce survey measurement error? *Public Opinion Quarterly, 60*, 576–602.

Schober, M. F., and Conrad, F. G. (1998). Does conversational interviewing improve survey data quality beyond the laboratory? *1997 Proceedings of the Section on Survey Research Methods, American Statistical Association*, pp. 910–915.

Schuman, H., and Presser, S. (1981). *Questions and Answers in Attitude Surveys: Experiments in Question Form, Wording and Context*. New York: Academic Press.

Schwarz, N. (1994). Judgment in a social context: Biases, shortcomings, and the logic of conversation. In M. Zanna (Ed.), *Advances in Experimental Social Psychology*, Vol. 26, pp. 123–162. San Diego, CA: Academic Press.

Schwarz, N. (1996). *Cognition and Communication: Judgmental Biases, Research Methods, and the Logic of Conversation*. Mahwah, NJ: Erlbaum.

Schwarz, N. (1998). Communication in standardized research situations: A Gricean perspective. In S. R. Fussell and R. J. Kreuz (Eds.), *Social and Cognitive Approaches to Interpersonal Communication*, pp. 39–66. Mahwah, NJ: Erlbaum.

Schwarz, N., Hippler, H. J., Deutsch, B., and Strack, F. (1985). Response scales: Effects of category range on reported behavior and comparative judgments. *Public Opinion Quarterly, 49,* 388–395.

Schwarz, N., Strack, F., and Mai, H. P. (1991). Assimilation and contrast effects in part-whole question sequences: A conversational logic analysis. *Public Opinion Quarterly, 55,* 3–23.

Schwarz, N., Strack, F., Müller, G., and Chassein, B. (1988). The range of response alternatives may determine the meaning of the question: Further evidence on informative functions of response alternatives. *Social Cognition, 6(2),* 107–117.

Strack, F., Martin, L. L., and Schwarz, N. (1988). Priming and communication: The social determinants of information use in judgments of life-satisfaction. *European Journal of Social Psychology, 18,* 429–442.

Strack, F., and Schwarz, N. (1992). Communicative influences in standardized question situations: The case of implicit collaboration. In G. R. Semin and K. Fiedler (Eds.), *Language, Interaction and Social Cognition*, pp. 173–193. London: Sage.

Suchman, L., and Jordan, B. (1990). Interactional troubles in face-to-face survey interviews. *Journal of the American Statistical Association, 85(409),* 232–241.

Suchman, L., and Jordan, B. (1992). Validity and the collaborative construction of meaning in face-to-face surveys. In J. M. Tanur (Ed.), *Questions About Questions: Inquiries into the Cognitive Bases of Surveys*, pp. 241–267. New York: Russell Sage Foundation.

Sudman, S., Bradburn, N., and Schwarz, N. (1996). *Thinking About Answers: The Application of Cognitive Processes to Survey Methodology*. San Francisco: Jossey-Bass.

Wentland, E. J. (1993). *Survey Responses: An Evaluation of Their Validity*. San Diego, CA: Academic Press, Inc.

Wilkes-Gibbs, D., and Clark, H. H. (1992). Coordinating beliefs in conversation. *Journal of Memory and Language, 31,* 183–194.

CHAPTER 7

The Respondent's Confession: Autobiographical Memory in the Context of Surveys

Michael S. Shum and Lance J. Rips
Northwestern University

We might almost define autobiographical memory as what nonprofessionals falsely believe the study of human memory is about. By contrast, professionals know that the study of human memory is about stimuli—memory for material that a researcher has invented or selected for presentation to subjects in laboratory experiments. Stimuli can be natural items, such as photos or stories, or unnatural items, such as nonsense syllables or visual illusions, but stimuli are not events that are part of a person's history outside the experimental lab. They do not include landmark events, such as college graduation, weddings, births of children, and deaths of friends or relatives, nor do they include everyday events, such as going to class, driving to work, and buying groceries. These events are not what the field of laboratory memory is about, but they *are* what this paper is about.

At the time of the first CASM Conference in the early 1980's (Jabine, Straf, Tanur, and Tourangeau, 1984), the standard texts and journals on human memory were almost solely about memory for "stimuli." The same is largely true today. During this interval, though, cognitive scientists have produced a large body of work on autobiographical memory, including three monographs (Conway, 1990; Kolodner, 1984; Thompson, Skowronski, Larsen, and Betz, 1996) and six collections of original papers (e.g., Conway, Rubin, Spinnler, and Wagenaar, 1992; Neisser and Winograd, 1988; Rubin, 1986, 1996; Schwarz and Sud-

Cognition and Survey Research, Edited by Monroe G. Sirken, Douglas J. Herrmann, Susan Schechter, Norbert Schwarz, Judith M. Tanur, and Roger Tourangeau.
ISBN 0-471-24138-5   © 1999 John Wiley & Sons, Inc.

man, 1994; Winograd and Neisser, 1992). Some of this research has even found its way into the standard journals, although most appears in newer outlets, such as *Applied Cognitive Psychology* and *Memory*. The research has also invaded a textbook or two (Cohen, 1989; Searleman and Herrmann, 1994).

Survey methodology has been one force responsible for this acceleration of research on autobiographical memory. Survey questions about factual matters obviously require respondents to recall and to report information about their own lives, including incidents in which they received medical treatments, bought a product, took a job, moved to a new location, and many others. Distorted memories about these matters usually imply response errors in survey data (Bradburn, Rips, and Shevell, 1987).

This is not to say, however, that memory is the only factor that determines respondents' answers to factual questions. Sometimes respondents use heuristics, inferences, and estimates rather than retrieving explicit information about separate events (e.g., Burton and Blair, 1991; Menon, 1994). When memory information is unavailable or inaccessible, these processes dominate responses. But although respondents may sidestep exhaustive retrieval, fragmentary facts from memory may still play a role by providing a premise for an inference or an initial value for a heuristic.

Our goal in this chapter is to describe recent work on autobiographical memory as it informs and is informed by survey methodology. In this limited space, there is no hope of fully reviewing the mass of new work on autobiographical memory that we have just cited. Instead, we focus in the first section on some common assumptions that arise from this research that may be important to survey designers, especially assumptions about how people retrieve events. We then relate this consensus view of retrieval to some data of our own in which students recall events from the last few months of their lives. The final section draws some implications about the role of memory in answering questions about our past.

7.1 RETRIEVAL FROM AUTOBIOGRAPHICAL MEMORY

Rival theories of autobiographical memory differ mainly in the organization they impose on personal incidents. Episodes from your life might exist in memory in independent, minimally connected units (Tulving, 1983), in hierarchies organized by their distinctive properties (Kolodner, 1984), or in thematically and chronologically structured histories or streams (Barsalou, 1988; Conway, 1996). Although these structural differences are important, we concentrate in this section on some of the assumptions that current cognitive theories share. We think a fairly consistent picture of autobiographical memory emerges from these assumptions, and it is helpful to bring it into the open, since it provides a framework for much of the research in this area.

The central ideas of these theories are that (a) autobiographical memory is memory for *representations* of personal events, and (b) people retrieve these event representations by describing a sufficient number of their parts (for the notion of memory descriptions, see Norman and Bobrow, 1979). For example, the memory of a mugging that happened to you might consist of a memory record of what took place (from your point of view), where the event happened, who was involved, and other details. You might attempt to retrieve this information in turn by describing what took place or other parts of the representation.

In thinking about theories and data in this area, we can distinguish two kinds of memory tasks, both of which are relevant to surveys. Sometimes what we want from autobiographical memory is the representation of a prespecified individual event. For example, you may know (on independent grounds) about one specific occasion on which you were mugged; you then search memory for the details of that mugging ("describe the mugging that occurred to you on April 19"). Sometimes, however, we search memory for the representation of *any* event that meets a general description (or as many events as possible that meet that description). An example is retrieving the memory of (any) mugging (or all muggings) in which you were a victim in the last six months. As we will see, conclusions about retrieval depend on this distinction between tasks that demand memory for *individual events* and tasks that demand memory for *event classes*.[1]

7.1.1 Autobiographical Memory Is Memory for Event Representations

Most theories of autobiographical memory assume that the units in which people store and later retrieve information about their past are individual event representations.[2] Although this event-centered way of representing personal facts might not be inevitable, these theories accord with the view that thought and language organize information around individual events that have properties of their own (e.g., being fast or slow, blameworthy or meritorious) and that interact with other events in causal ways (e.g., Davidson, 1967; Parsons, 1990).

---

[1]We mean this individual/class distinction to apply to memory *tasks* rather than to memory *representations*. In particular, we are not claiming that people store representations corresponding to arbitrary event descriptions (e.g., events in which you felt frustrated, events in which you were stared at). You can, however, be *asked* about arbitrary event descriptions, and some survey questions may do just that. We also note that for some purposes it may be useful to distinguish between tasks that require retrieval of all events in a class and those that require (only) some.

[2]Survey questions also target personal information that are properties or states of respondents, rather than events, and that fall outside the domain of current theories of autobiographical memory. Surprisingly little research has been directed at memory for mundane enduring properties, such as a person's address, political affiliation, or birthplace, and this information does not fit comfortably in traditional ways of taxonomizing memory, such as the semantic/episodic distinction (Tulving, 1983). Most research on memory for properties of the self has focused on personality traits, such as friendliness, rather than on these more factual matters (see Kihlstrom and Klein, 1994, for a review).

Some studies of very long-term memory have examined names of high school classmates (Bahrick, Bahrick, and Wittlinger, 1975; Williams and Hollan, 1981), street names of a town you lived in (Bahrick, 1983), and exams or grades (Bahrick, Hall, and Dunlosky, 1993; Strube and Neubauer, 1988). These items are neither enduring properties nor events, but perhaps we remember them as parts of events (agents, locations, or outcomes of autobiographical incidents).

The importance of these causal links in our personal history may account for this assumption.

On these theories, then, remembering our past means recalling information about individual occurrences. However, the recalled information is not necessarily integrated and complete. The information that you retrieve about an event may have to be assembled on each occasion into a coherent retelling (Conway, 1996). Moreover, the information you retain and recall about an incident is merely one of many possible representations of that event. This is partly a psychological point, because the selectivity of perception and memory makes it impossible to remember all that occurs in even fairly simple everyday events like washing the dishes. But it is also a logical point, because there are an infinite number of representations of any event. It might be possible to describe a mugging that happened to you as, say, the immediate cause of your calling police sergeant Jones, the one and only crime committed by Elmer Smith's favorite nephew, and many others. The moral is that "remembering an event"—retrieving a representation of the event—does not entail being able to answer all possible questions about it.

7.1.2 Retrieval from Autobiographical Memory Depends on Uniqueness, Specificity, and Faithfulness

According to most theories of long-term memory, people retrieve information by specifying parts of the to-be-remembered trace. Fragmentary clues from prior thought or from the environment evoke the more complete memory record that these clues match. Different memory models carry out this retrieval process in different ways. According to some theories, retrieval occurs when activation from the cues converges on the memory trace (Anderson, 1983; Gillund and Shiffrin, 1984); in others, activation from the cues reinstates a global condition of the memory system that embodies the old information (McClelland and Rumelhart, 1986; Smith, Chapter 16 of this book); in still others, the cues trigger mental rules that contain the trace (Newell, 1990), or they provide routes through an indexing system that lead to the appropriate memory (Kolodner, 1984). All these theories are similar, however, in assuming that retrieval is a probabilistic matter that depends on how close the cue is to the stored information. In most theories, retrieval can also be incremental or cyclical rather than all or none: Initial cues can produce new ones through memory or inference; the new cues then aid in retrieving further cues, until the person retrieves the target memory or gives up trying.

These general retrieval theories also apply to autobiographical incidents (Conway, 1996; Williams and Hollan, 1981). In retrieving the memory of an event, you use cues that comprise a partial description, based on information currently available from external or mental sources. Your success in recovering the episode will depend not on how well this partial description fits the original event, but on how well it fits the stored representation of the event. A hospital's report of a medical treatment may be of no help to you in remembering the treatment if the report fails to coincide with the way you registered the incident in the first place.

We ought to expect that the likelihood of retrieving an individual event will depend jointly on the *uniqueness* of the original memory representation, the *specificity* of the cue description, and the *faithfulness* of the cue to the memory. Thus, the more distinctive (unique) the memory, the more detailed the partial description in the cue, and the more faithful the match to the stored representation, the more successful retrieval will be.

Uniqueness of Memory Descriptions　Clearly, no cue will be able to single out an individual event if the stored representation of that event is the same as many others. Lack of uniqueness can come about as people forget identifying details of otherwise similar occurrences, but it can also happen because of vague or context-sensitive elements in the encoding. If you encode an event as one in which you sent in the final draft of your manuscript, then that event representation may be impossible to retrieve later when the draft turns out to be only one of a series of "final" versions (Linton, 1975).

Of course, if the goal in searching memory is to retrieve an event class rather than an individual event, then lack of uniqueness is not necessarily an impediment. On one hand, as long as the events within the class have separate representations, we can at least count the events in answering a question like "How many times were you the victim of a crime in the last six months?" On the other, it is not clear that the memory system preserves separate representations of events that are no longer individuated internally. Kolodner's (1984) model, for example, prohibits retrieval unless a cue provides information that distinguishes a target event from others. In line with this, evidence from introspection (Linton, 1975) and free recall (Barsalou, 1988) suggests that events within a class sometimes coalesce into a single summarized representation (e.g., "I played tennis a lot last year," "I watched a lot of TV this week"). Such events are no longer unique, even in the sense of being separately countable.

Faithfulness and Specificity of Cues　If the original representation is unique, then retrieval of an individual event will depend on the number of details in the cue and on how well the details match the representation. Faithfulness and specificity trade off, however. It is possible to stay perfectly faithful to the remembered representation by being vague and equally possible to be highly specific by being fanciful; however, neither type of cue is likely to retrieve a designated event. A cue of the form "an incident that happened to me in the past" is not likely to make contact with the memory of a specific mugging or medical treatment, even though it is perfectly faithful to the stored representation. But neither will a novel-length cue if it fails to match anything in your experience. What matters are the number of predicates of the cue that are faithful to those of the stored representation.

Questions about event classes have to sacrifice specificity in order to cover the entire class. A question about all criminal incidents has to be general enough to cover the entire set. Nevertheless, it may be the case that people cannot recall particular incidents within the set unless the cue contains some threshold level of specificity. "Any criminal incident" may not suffice for retrieval, and the respondents may have to elaborate the cue with further detail (robberies, burglaries, incidents when police were present) before it is useful. Unpacking cues ("Were you the victim of a burglary, robbery, or other type of crime?") increases the subjective frequency of events in the class (Tversky and Koehler, 1994). Likewise, a series of questions about specific income sources produces a larger total estimate than a global question about income (Moeller and Mathiowetz, 1994; U.S. Bureau of the Census, 1979). Unpacking may also increase memory accuracy (e.g., Cohen and Java, 1995), although evidence on the latter point is mixed (see Jobe, Tourangeau, and Smith, 1993, for a review).

Cue Strength and Cue Validity   It is possible that some individual cues are better able to make contact with an autobiographical event. Certain cues may have more inherent ability to retrieve a memory than other cues, even when uniqueness, faithfulness, and specificity are equal. "Recount an incident when you were mugged" may be a more successful probe than "Recount an incident that happened to you on April 19, 1991 at 11 P.M.," even if the respondent's one and only mugging took place on April 19, 1991 at 11 P.M. Current theories allow for this possibility: Activation-based approaches can send more mental current from some descriptors than from others. Index and search approaches also accommodate these differences. People must begin their retrieval at the top level of the index before exploring lower levels, and some index paths may be more successful than others.

Researchers have used two methods to determine which individual cues are most effective. In one paradigm, the investigator gives subjects a cue and asks them to retrieve as quickly as possible any personal memory that the cue describes—any item within an event class (Barsalou, 1988; Conway and Bekerian, 1987; Reiser, Black, and Abelson, 1985). Response time to retrieve a memory is the dependent variable. In the second type of experiment, subjects record personal experiences at the time they happen in a diary or photograph, usually over a period of weeks or months. After the subjects complete their diaries, the investigator tests them by presenting a cue from an entry and seeing whether they can report the full incident—a task requiring retrieval of an individual event (Brewer, 1988; Burt, Mitchell, Raggatt, Jones, and Cowan, 1995; Wagenaar, 1986).

Results on event classes are equivocal, in part, because the definition of action, location, or time cues varies across experiments. The small amount of available evidence suggests an advantage for cues specifying the participants in an event (e.g., events involving your sister) rather than ones specifying an action (watching TV), a time (evening), or a location (a motel room); see Barsalou (1988, Table 8.4).[3] For retrieval over long stretches of time, lifetime periods, such as the time you spent in college or the year you spent on sabbatical in Japan, may also be effective cues (Conway and Bekerian, 1987). By contrast, individual-event studies (e.g., the diary studies cited above) suggest an advantage for actions over locations and times. This result seems to hold both in situations where the diary keeper picks the events to be remembered (Burt et al., 1995; Wagenaar, 1986) and in those where the diary keeper records events that occur when a beeper goes off at random intervals (Brewer, 1988).

The results of both types of experiment, however, probably depend on the number of stored events that match each cue. In the event-class experiments, the larger the number of matching events, the faster the response, other things being equal. The more events you did with your sister, for example, the easier it should be to bring any one of them to mind. The results of the individual-event experiments may also depend on the number of matching events, but in the opposite direction. In these experiments, the larger the number of events that match the cue, the less likely it is that any one of those events will be the one that the experimenter targets. If your diary (and your memory) contains 20 events in which your sister participated, then the chance is only 0.05 of selecting the right event when the experimenter prompts with the cue "your sister." The opposite effect of event frequency may thus explain discrepancies in the results of the two types of experiments. At the same time, frequency variations make it difficult to say anything about possible differences in the inherent strength of the cue. We simply do not know whether actions or participants are better able to retrieve events in the absence of confounding factors.

7.2 MEMORY FOR A YEAR'S EVENTS

The current perspective on autobiographical memory suggests that there is no retrieval without cues—partial descriptions of an event. When we ask people questions about their past, the questions provide cues to start their memory search. It is often the case, however, that these cues are not specific enough to make contact with a relevant memory (e.g., "recall some events from the past year"), and in these situations people must supplement the question with additional cues of their own. What cues do they use in these circumstances? In this section, we present research suggesting that individuals' life roles, especially the calendars or schedules they live under, can aid them in retrieving personal memories.

---

[3]Reiser et al. (1985) found faster retrieval in response to activities (e.g., *had dinner in a restaurant*) than to what they termed "generalized actions" (e.g., *did not get what you wanted*). However, this may be due less to the superiority of the activities than to the inferiority of generalized actions as cues (see Barsalou, 1988; Conway and Bekerian, 1987; Felcher, 1992).

7.2.1   The Calendar Effect

Several investigators have asked college students to recall events that happened to them in the last year or two and then to date these events (Kurbat, Shevell, and Rips, 1998; Pillemer, Rhinehart, and White, 1986; Robinson, 1986). The students can recall any personal event they can remember as long as it is unique and takes place within the span of a day. When plotted as a time series, the frequency of the recalled events displays a characteristic pattern that is apparent in Figure 7.1 (from Kurbat et al., 1998): Students recall more events at boundary points between school terms and vacations than at other times of the year.

The data in Figure 7.1 come from an experiment that was conducted in February 1989 (Kurbat et al., 1998), and the students' task was to recall 20 events that happened to them in the previous year (1988). The students were enrolled at the University of Chicago (panels 1 and 3) and at Cornell College in Iowa (panels 2 and 4), and the groups included both freshmen (panels 1 and 2) and upperclassmen (panels 3 and 4). The x-axis gives the week of the year in which the event occurred, with week 1 corresponding to January 1–7, week 2 to January 8–14, and so on. The vertical lines in the figure show the beginnings and the ends of the academic terms for these students and bring out the radical difference in their schedules. Cornell College has nine month-long terms during the academic year, during each of which students take a single course; the University of Chicago has a standard quarter system. We show the term boundaries for freshmen during only the last part of the year, since these students entered college in September or October.

Despite differences in the schedules, students from both schools produced similar distributions of memories. Peaks in the distributions occur at the beginning of the school year, around Christmas vacation, around spring vacation (for Chicago upperclassmen), and at the end of the school year. Most adjacent terms at Cornell College are separated by only a four-day interval, and these minor term boundaries do not seem to produce many memories. The boundaries that matter are those separating school from longer vacations. Note, too, that the peaks around summer vacation shift to the right at Chicago relative to Cornell, in line with Chicago's later school year. Thus, the recall peaks are probably due to the school calendar itself, rather than to nonschool holidays or seasonal changes that are the same at the two colleges.

7.2.2   What Accounts for Recall Peaks?

One explanation of the cusps in Figure 7.1 is that the students were using their knowledge of the academic calendar as cues for the year's memories. The calendar provides students a faithful and specific account of what must have occurred to them. As students at Cornell or Chicago, they must have found a way back to school in the fall, gone through the usual beginning-of-school activities, and found their way to classes. At the end of the school year, they must have taken finals or turned in papers, packed their belongings, moved out of their apart-

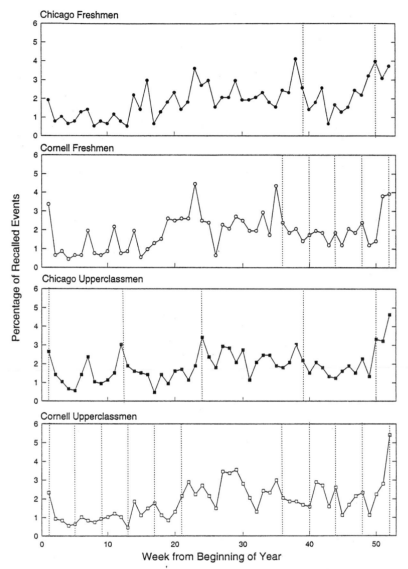

Figure 7.1 Percentage of recalled events by week for freshmen and upperclassmen at the University of Chicago and Cornell College. X-axis shows weeks in 1988, beginning with January 1. (From Kurbat et al., 1998.)

ments or dorms, said goodbye to friends, and found transportation to their summer destinations. This framework is richly predictive (or post-dictive) in a way that could make contact with specific events. Once a student realizes that she must have moved into her dorm in October, she may be able to recover the fact that she stayed at Margaret's apartment the night she arrived in Chicago, until the dorm officially opened the next day. Kurbat et al.'s instructions did not mention the academic year or academic calendar; students were simply told to list events from 1988. Still, the academic calendar is an obviously important source of information about what the students had done during the year, and they could draw on it freely to retrieve events. We refer to this explanation as the *cuing hypothesis* in what follows.

Evidence for Cuing   The best evidence for this cuing hypothesis comes from a separate experiment at Chicago (Kurbat et al., 1998, Experiment 3) in which students recalled events on the pages of a calendar. In one condition, we indicated at the appropriate place on the calendar the beginning and the end of each quarter (e.g., "beginning of fall quarter," "end of fall quarter"); in a second condition, we marked the position of six holidays (e.g., Thanksgiving, Valentine's Day, Independence Day) whose dates do not coincide with the academic term boundaries. Students were free to recall events from any day of the year, and no mention was made in the instructions about the highlighted dates. Nevertheless, students in the term boundaries condition produced frequency distributions that closely resembled those of Figure 7.1, whereas students in the holiday condition produced distributions that increased memories around the designated holidays and decreased memories around term boundaries. Explicit cues, then, alter the shape of the distribution, although they do not completely eliminate the effect of the academic calendar.

There is an obvious alternative hypothesis, however, to our cuing explanation. Events that take place at the beginning and the end of academic terms may be more important to students on average than events that take place during the middle of terms or vacations. If so, then recall peaks in Figure 7.1 might be due to the ease with which people can recall important information. One difficulty with the importance hypothesis, however, is that it does not explain the results of the holiday/academic term experiment that we just described. Also, in some of the experiments in question (Kurbat et al., 1998; Robinson, 1986), students rated the importance and the distinctiveness of the events after they recalled them, and no clear differences in the ratings emerged between events around term boundaries and events at other time points. It is possible to claim, however, that what matters is not the importance of the recalled events but the importance of *all* events that happened to the students. Suppose students recall all events that exceed some threshold level of importance. If more of these important events occur near term boundaries, then importance could still be driving the calendar effect, even though students rate the recalled events equal in importance.

Test of the Importance Hypothesis   To check the importance hypothesis, we carried out a pair of studies in which we asked students to rate the importance of events at the time the events occurred and to recall them some weeks later. We modeled the procedure for the two experiments after Brewer (1988): At the start of the experiments students received watches with programmable weekly alarms, and they wore them between 10 A.M. and 10 P.M. every day for nine to ten weeks. We programed the watches to go off at random times in this period: four times a day in Experiment 1 and three times a day in Experiment 2. (The random schedule was different each day.) Every time the alarm sounded, the students wrote down in a response booklet a description of an event that was taking place and rated the event for personal importance. At the end of each day, the students also described and rated the "most memorable" event of that day.[4] The experimenter met with the students each week to collect response booklets from the previous week, and to program the watches for the upcoming week. The students were Northwestern University undergraduates: six students in Experiment 1 and seven students in Experiment 2. We selected the response periods of the experiments so that they would span a term boundary (the end of the school year in Experiment 1 and Christmas vacation in Experiment 2), allowing us to look at the calendar effect in this context.

After the end of the response period, the students returned for a final session, and we gave them a free recall task for all the events they had described.[5] The methods of Experiments 1 and 2 differed, however, after the recall phase. In Experiment 1, we asked the students to date the recalled events. In Experiment 2, students received a list of all the events they had written down during the response period, and they dated them. In this way, we could ascertain dating accuracy and importance for both recalled and nonrecalled events.

The recall test in both these experiments produced a calendar effect: The number of events the students recalled was reliably greater in weeks adjacent to the term boundaries (*critical weeks*) than in other weeks.[6] There was also a trend in the data for the students to rate events from the critical weeks as more important than other weeks' events at the time the events occurred. This accords with the importance hypothesis, but the effect was quite small: The difference is less than 0.3 scale points on the 1-to-7 rating scale, marginally significant in Experiment 1 and nonsignificant in Experiment 2.

The central issue, then, is whether we can attribute the calendar effect to this difference in importance. To find out, we performed regression analyses on the week-to-week percentage of recalled events, using as predictors the rated importance of the events and whether the event came from a critical week or a noncritical week. The analysis showed that with critical versus noncritical week as a predictor, there is no significant residual effect of importance. By contrast, when importance is the predictor, the difference between critical and noncritical weeks remains reliable. Thus, although events that occur at the beginning or the

---

[4]Subjects later recalled the same proportion of memorable events as of random events, so we consider these two sets together in reporting the results.

[5]In Experiment 1 the recall test followed the response period by just one week, and there was a substantial recency effect in these data. To remove the effect of recency, we have omitted the last two weeks of the response period in the data that we discuss below. In Experiment 2 the recall test followed the response period by one month, reducing the recency effect. We therefore included all data from Experiment 2 in the analyses.

[6]The dates we used to assess the calendar effect were the true dates of the events, rather than the self-reported dates that we had used in earlier experiments (Kurbat et al., 1998). These results, therefore, eliminate the possibility that the calendar effect is due to students' bias in dating events.

end of the quarter may be slightly more important than other events, there is no evidence that importance is responsible for the calendar effect. Examination of the events students rate as important suggests that these events are usually social–personal episodes, such as breaking up with a boyfriend or girlfriend, events that are only incidentally tied to the school calendar.

7.3 SUMMARY

Most theories of autobiographical memory assume that people store their personal histories as representations of past events. They likewise assume that people retrieve these traces by building a cue—a partial description that matches the stored representation. We have tried to show that certain consequences follow from this picture that apply to the design of survey questions. In the survey context, it is the question that provides the initial cue, although respondents may elaborate the cue to bring it in line with their own situation. Thus, if the point of the question is to get the respondent to retrieve information about a particular, previously designated event ("Tell me about the mugging that happened to you on April 19"), then specific and faithful questions will ease the respondent's task. Of course, success in retrieval will also depend on properties of the original representation. Retrieval will be more accurate if the respondent encoded the event uniquely and if the cue matches only that unique item. The situation is more complicated for questions that ask the respondent to retrieve any event (or all events) within some class ("Tell me about an event in which you were mugged," "Tell me about all events in which you were mugged"). These questions cannot be more specific than the event class itself; however, unpacking the question ("Tell me about all the times you were mugged on city streets, in a public building, or in other areas") may increase specificity without excluding events.

Our studies of the calendar effect substantiate this picture of autobiographical memory. Even when people have no overt cues, they search memory by relying on schedules and calendars that provide partial information about what must have transpired. Altering the cues they use changes the shape of the distribution of memories. Although events' importance or distinctiveness may make them more memorable, our results suggest that importance is not sufficient to explain the pattern of events people retrieve. Instead, recalling personal events appears to be a kind of archeological process. The more accurate and detailed your hypotheses about past events—based, perhaps, on records, schedules, and other evidence—the more likely you are to recover them.

ACKNOWLEDGMENTS

The research in this chapter was partly supported by NSF grant SBR-9514491.

REFERENCES

Anderson, J. R. (1983). *Architecture of Cognition*. Cambridge, MA: Harvard University Press.

Bahrick, H. P. (1983). The cognitive map of a city: Fifty years of learning and memory. In G. H. Bower (Ed.), *The Psychology of Learning and Motivation*, Vol. 17, pp. 125–163. Orlando: Academic Press.

Bahrick, H. P., Bahrick, P. O., and Wittlinger, R. P. (1975). Fifty years of memory for names and faces: A cross sectional approach. *Journal of Experimental Psychology: General, 104*, 54–75.

Bahrick, H. P., Hall, L. K., and Dunlosky, J. (1993). Reconstructive processing of memory content for high versus low test scores and grades. *Applied Cognitive Psychology, 7*, 1–10.

Barsalou, L. W. (1988). The content and organization of autobiographical memories. In U. Neisser and E. Winograd (Eds.), *Remembering Reconsidered: Ecological and Traditional Approaches to the Study of Memory*, pp. 193–243. Cambridge, England: Cambridge University.

Bradburn, N. M., Rips, L. J., and Shevell, S. K. (1987). Answering autobiographical questions: The impact of memory and inference on surveys. *Science, 236*, 151–167.

Brewer, W. F. (1988). Memory for randomly sampled autobiographical events. In U. Neisser and E. Winograd (Eds.), *Remembering Reconsidered: Ecological and Traditional Approaches to the Study of Memory*, pp. 21–90. Cambridge, England: Cambridge University.

Burt, C. D. B., Mitchell, D. A., Raggatt, P. T. F., Jones, C. A., and Cowan, T. M. (1995). A snapshot of autobiographical retrieval characteristics. *Applied Cognitive Psychology, 9*, 61–74.

Burton, S., and Blair, E. (1991). Task conditions, response formulation processes, and response accuracy for behavioral frequency questions in surveys. *Public Opinion Quarterly, 55*, 50–79.

Cohen, G. (1989). *Memory in the Real World*. Hillsdale, NJ: Erlbaum.

Cohen, G., and Java, R. (1995). Memory for medical history. *Applied Cognitive Psychology, 9*, 273–288.

Conway, M. A. (1990). *Autobiographical Memory: An Introduction*. Milton Keynes, England: Open University Press.

Conway, M. A. (1996). Autobiographical knowledge and autobiographical memories. In D. C. Rubin (Ed.), *Remembering Our Past: Studies in Autobiographical Memory*, pp. 67–93. Cambridge, England: Cambridge University Press.

Conway, M. A., and Bekerian, D. A. (1987). Organization in autobiographical memory. *Memory and Cognition, 15*, 119–132.

Conway, M. A., Rubin, D. C., Spinnler, H., and Wagenaar, W. A. (Eds.) (1992). *Theoretical Perspectives on Autobiographical Memory*. Dordrect: Kluwer Academic.

Davidson, D. (1967). The logical form of action sentences. In N. Rescher (Ed.), *The Logic of Decision and Action*, pp. 81–120. Pittsburgh: The University of Pittsburgh Press.

Felcher, E. M. (1992). *Estimating the response frequency of autobiographical events in response to survey questions.* Unpublished doctoral dissertation, Northwestern University.

Gillund, G., and Shiffrin, R. M. (1984). A retrieval model for both recognition and recall. *Psychological Review, 91,* 1–67.

Jabine, T., Straf, M., Tanur, J., and Tourangeau, R. (Eds.) (1984). *Cognitive Aspects of Survey Methodology: Building a Bridge Between Disciplines.* Washington, DC: National Academy Press.

Jobe, J. B., Tourangeau, R., and Smith, A. F. (1993). Contributions of survey research to the understanding of memory. *Applied Cognitive Psychology, 7,* 567–584.

Kihlstrom, J. F., and Klein, S. F. (1994). The self as a knowledge structure. In R. S. Wyer, Jr., and T. K. Srull (Eds.), *Handbook of Social Cognition,* 2nd ed., Vol. 1, pp. 153–208. Hillsdale, NJ: Erlbaum.

Kolodner, J. L. (1984). *Retrieval and Organizational Strategies in Conceptual Memory.* Hillsdale, NJ: Erlbaum.

Kurbat, M. A., Shevell, S. K., and Rips, L. J. (1998). A year's memories: The calendar effect in autobiographical recall. *Memory and Cognition, 26,* 532–552.

Linton, M. (1975). Memory for real-world events. In D. A. Norman and D. E. Rumelhart (Eds.), *Explorations in Cognition,* pp. 376–404. San Francisco: Freeman.

McClelland, J. L., and Rumelhart, D. E. (1986). A distributed model of human learning and memory. In J. L. McClelland, D. E. Rumelhart, and PDP Research Group (Eds.), *Parallel Distributed Processing: Explorations in the Microstructure of Cognition,* Vol. 2, pp. 170–215. Cambridge, MA: MIT Press.

Menon, G. (1994). Judgments of behavioral frequencies: Memory search and retrieval strategies. In N. Schwarz and S. Sudman (Eds.), *Autobiographical Memory and the Validity of Retrospective Reports,* pp. 161–172. New York: Springer-Verlag.

Moeller, J. F., and Mathiowetz, N. A. (1994). Problems of screening for poverty status. *Journal of Official Statistics, 10,* 327–337.

Neisser, U., and Winograd, E. (Eds.) (1988). *Remembering Reconsidered: Ecological and Traditional Approaches to the Study of Memory.* Cambridge, England: Cambridge University Press.

Newell, A. (1990). *Unified Theories of Cognition.* Cambridge, MA: Harvard University Press.

Norman, D. A., and Bobrow, D. G. (1979). Descriptions: An intermediate stage in memory retrieval. *Cognitive Psychology, 11,* 107–123.

Parsons, T. (1990). *Events in the Semantics of English.* Cambridge, MA: MIT Press.

Pillemer, D. B., Rhinehart, E. D., and White, S. H. (1986). Memory of life transitions: The first year in college. *Human Learning, 5,* 109–123.

Reiser, B. J., Black, J. B., and Abelson, R. P. (1985). Knowledge structures in the organization and retrieval of autobiographical memories. *Cognitive Psychology, 17,* 89–137.

Robinson, J. A. (1986). Temporal reference systems and autobiographical memory. In D. C. Rubin (Ed.), *Autobiographical Memory,* pp. 159–188. Cambridge, England: Cambridge University Press.

Rubin, D. C. (Ed.) (1986). *Autobiographical Memory*. Cambridge, England: Cambridge University Press.

Rubin, D. C. (Ed.) (1996). *Remembering our Past: Studies in Autobiographical Memory*. Cambridge, England: Cambridge University Press.

Schwarz, N., and Sudman, S. (Eds.) (1994). *Autobiographical Memory and the Validity of Retrospective Reports*. New York: Springer-Verlag.

Searleman, A., and Herrmann, D. (1994). *Memory from a Broader Perspective*. New York: McGraw-Hill.

Strube, G., and Neubauer, S. (1988). Remember that exam? In M. M. Gruneberg, P. E. Morris, and R. N. Sykes (Eds.), *Practical Aspects of Memory: Current Research and Issues*, pp. 247-252. New York: Wiley.

Thompson, C. P., Skowronski, J. J., Larsen, S. F., and Betz, A. L. (1996). *Autobiographical Memory: Remembering What and Remembering When*. Mahwah, NJ: Erlbaum.

Tulving, E. (1983). *Elements of Episodic Memory*. Oxford, England: Oxford University Press.

Tversky, A., and Koehler, D. (1994). Support theory: A nonextensional representation of subjective probability. *Psychological Review, 101*, 547–567.

U.S. Bureau of the Census (1979). *Vocational School Experience: October, 1976* (Current Population Reports Series P-20, No. 343). Washington, D.C.: U.S. Government Printing Office.

Wagenaar, W. A. (1986). My memory: A study of autobiographical memory over six years. *Cognitive Psychology, 18*, 225–252.

Williams, D. M., and Hollan, J. D. (1981). The process of retrieval from very long-term memory. *Cognitive Science, 5*, 87–119.

Winograd, E., and Neisser, U. (Eds.) (1992). *Affect and Accuracy in Recall*. Cambridge, England: Cambridge University Press.

CHAPTER 8

Context Effects on Answers
to Attitude Questions

Roger Tourangeau
The Gallup Organization

If there is one area where psychological theory and survey methodology have merged easily, it is the area of context effects. The survey literature is full of intriguing examples in which the "same" question produces different answers depending on the context in which it is asked (Bradburn, 1982; Schuman and Presser, 1981). The psychology literature offers many parallel demonstrations of order and context effects on social judgments. As early as 1946, Asch had demonstrated that the same information had more impact on impressions of another person when it was presented first in a description than when it was presented later on. Interest in order effects within psychology reached a peak during the 1970's and 80's, when several studies demonstrated that carrying out one task could "prime" concepts—that is, make them highly accessible—which would then be used in ostensibly unrelated judgments later on (e.g., Higgins, Rholes, and Jones, 1977). The work in psychology inspired several theories that applied equally well to the survey findings (Strack and Martin, 1987; Sudman, Bradburn, and Schwarz, 1996; Tourangeau and Rasinski, 1988).

8.1 THE PROCESS OF ANSWERING ATTITUDE QUESTIONS

Most efforts to explain survey context effects share some key assumptions about how respondents answer attitude questions. Perhaps the most basic assumption is that attitude judgments are often "temporary constructions" (Wilson and

Cognition and Survey Research, Edited by Monroe G. Sirken, Douglas J. Herrmann, Susan Schechter, Norbert Schwarz, Judith M. Tanur, and Roger Tourangeau.
ISBN 0-471-24138-5 © 1999 John Wiley & Sons, Inc.

Hodges, 1992); although they may rest on a base of long-standing evaluations and beliefs, the judgments themselves must often be created anew. A corollary is that the answers to an attitude question often reflect whatever information about a topic happens to be accessible when the question is asked; the answers are based on a quick-and-dirty sampling of what the respondent knows or feels about an issue (Tourangeau, 1992; Tourangeau and Rasinski, 1988). Many studies suggest that differences in how an issue is construed—in the information sampled on different occasions—are a major source of context effects.

Another assumption is that attitude judgments are rarely absolute; rather, they implicitly or explicitly involve comparisons with some standard. Judging someone's height in feet and inches may involve an absolute judgment, but judging whether someone is tall is inherently comparative—tall is a relative concept. Because attitude judgments are almost always relative, they will vary with the standard that is salient when the question is asked.

If the answers to attitude questions are the product of a sampling of beliefs about a topic and if they represent comparative judgments, the stability of the answers across contexts will depend on such factors as the degree of homogeneity within the pool of beliefs from which the samples are drawn, the amount of overlap in the beliefs sampled on different occasions, and whether the same standards are applied each time. Context can affect all three—altering the beliefs seen as relevant, the ones tapped in making the judgment, and the standards used in evaluating them.

8.2   MECHANISMS PRODUCING CONTEXT EFFECTS

Researchers have cited a range of principles in explaining context effects. In their classic review, Schuman and Presser (1981) distinguish seven categories of context effects, based mainly on the relative generality of the context and target items and on the overall direction of the impact of context. The level of generality variable refers to whether the context item(s) concerns a broader issue that encompasses the one raised by the target item, a narrower issue, or an issue at the same level. For example, a question about free speech is broader than one on free speech for extremist groups, which is broader in turn than an item on free speech for Ku Klux Klan members (Ottati, Riggle, Wyer, Schwarz, and Kuklinski, 1989).

Schuman and Presser also distinguished between *consistency* and *contrast* effects, based on whether responses to the target item move toward or away from answers to the context items. Both terms would seem to imply changes in the correlation between responses to the context and target items. Consistency suggests a higher-than-normal correlation, contrast a lower-than-normal (or even negative) correlation between the context and target items. But most of the examples discussed by Schuman and Presser (and by later researchers as well) involve directional shifts in answers to the target item. Correlational effects are infrequently observed in practice because the context items are typically chosen to produce more or less uniform reactions. If everyone answers the context items the same way, their target answers will all be pushed in the same direction, changing the mean (or marginals) of the responses but not the target–context correlation. Most researchers have retained this distinction, but use the terms *assimilation* and *contrast* to distinguish the two types of directional shift.

Context effects can also be classified according to the component of the response process affected by the question context. For example, one influential model distinguishes four major components—comprehension, retrieval, judgment, and reporting (Tourangeau, 1984). Context effects are, in this framework, traced to the impact of context on the interpretation of the target question, the information retrieved in answering it, the use of that information in judging the target issue, or the reporting of the judgment.

8.2.1 Context and Question Comprehension

Assimilation Effects   With survey questions, as with most language, the interpretation often depends on context. One component of the comprehension process involves identifying the higher level structures into which the current question fits—the schema or category of which it is a part (Abelson, 1981; Bransford and Johnson, 1972) or the speech act it is intended to carry out (Clark, 1985). The context is often crucial in determining which structure is relevant. Contextual clues to the interpretation of an attitude question are particularly important when the question concerns an ambiguous or unfamiliar issue.

Studies by Tourangeau and Rasinski (1986) and Strack, Schwarz, and Wänke (1991) illustrate the effects of context on responses to items about such issues. Rasinski and I examined responses to an item about the Monetary Control Bill, an obscure, though real, piece of proposed legislation. Reported support for the bill increased when the question about the Monetary Control Bill followed a series of items on inflation. Under this context, many respondents apparently inferred that the Monetary Control Bill was an anti-inflation measure. When the preceding inflation items were scattered among unrelated items, they had no impact on responses to the question on the Monetary Control Bill; evidently, when the questions skipped from one topic to the next, they no longer carried implications about the meaning of the later questions. The study by Strack, Schwarz, and Wänke (1991) asked students at German universities about an undefined "educational contribution"; that item followed an item about college tuition in the United States or one about government financial support for students in Sweden. Support for the educational contribution was higher in the latter condition. Respondents given the United States context presumably inferred that the educational contribution was to be taken *from* them as tuition while those given the Swedish context inferred that it was to be given *to* them as financial aid.

Contrast Effects   Inferences about the relation between adjacent items can also produce contrast effects. Such effects have been observed when a specific item precedes a more general question on the same topic. The specific item may ask about respondents' happiness with their marriages and the general item about their overall happiness (Schuman and Presser, 1981); or the items may ask about young drivers and drivers in general (Kalton, Collins, and Brook, 1978) or about the local and state economy (Mason, Carlson, and Tourangeau, 1995). In most cases, the result is a contrast effect on responses to the general item (see Smith, 1982; Turner and Martin, 1984, for exceptions).

The general item in these studies asks for an evaluation of some domain and the specific item for an evaluation of a salient subdomain. When the general item comes second, it poses an interpretive problem. It seems to ask for a summary ("Taken altogether, how would you say things are these days?"), but respondents have made only one prior point. The item seems to violate Grice's (1975) maxim that contributions to a conversation should avoid redundancy; a summary of one item is unlikely to contain much new information. To resolve the difficulty, respondents may infer that the general item excludes the subdomain covered in the preceding question.

Two studies have tested this account (Schwarz, Strack, and Mai, 1991; Tourangeau, Rasinski, and Bradburn, 1991). Both examined versions of the general happiness item that specified what to include or leave out. For example, in the study by my colleagues and me, the inclusion version read, "Taking things all together, including your marriage and other important aspects of your life, how would you say things are these days?" The study by Schwarz and his co-workers used similar wording, although the specific item concerned respondents' current romantic relationships. As predicted, the correlations were high when the general item came first (indicating that respondents spontaneously considered their marriages/relationships in evaluating their overall happiness) and when the general item explicitly instructed respondents to include their marriage/relationship. They were low when the general item came second (indicating that respondents exclude their marriages/relationships under this order) and when the item explicitly called for exclusion. The first two columns of Table 8.1 shows these results.

Table 8.1   Correlations Between Responses to General and Specific Questions

| | | Schwarz et al. (1991) | |
Condition	Tourangeau et al. (1991)	One Specific Item	Three Specific Items
General/specific	0.54 (60)	—	0.32 (50)
Specific/general—no introduction	—	0.67 (50)	0.46 (50)
Specific/general—joint lead-in	0.28 (53)	0.18 (56)	0.48 (56)
Specific/general—exclusion wording	0.27 (54)	0.20 (50)	0.11 (50)
Specific/general—inclusion wording	0.52 (59)	0.61 (50)	0.53 (50)

Note: Parenthetical entries are sample sizes.

Clearly, respondents do not always exclude the most salient subdomain from their overall evaluation of a domain. They will include it when the general question follows items on several subdomains, since a summary that covers many particulars is no longer uninformative. The last column in Table 8.1 shows what happened in the study by Schwarz and his colleagues when the questionnaire asked about three subdomains rather than one. Unless the item instructed respondents to exclude their relationships, the correlation between the ratings of romantic relationships and overall happiness remained high under both question orders.

In addition, respondents seem to understand that when questions shift from one topic to the next—as they often do in surveys—the Gricean conversational maxims do not apply in the usual way. They will not reinterpret the general item as excluding the subdomain covered in the preceding item unless they see the two as forming a connected sequence. The study by Schwarz and his colleagues varied whether the happiness questions were administered with or without a lead-in that emphasized the connection between them. Without the introduction, the correlation between the two items remains high under both question orders. For the subdomain to be excluded, then, the respondents must infer that the Gricean maxims *do* apply.

8.2.2 Context and Retrieval Processes

If respondents tap only a portion of their relevant beliefs in answering an attitude question, those that happen to be accessible when the question is asked may dominate the responses. By affecting the accessibility of beliefs related to a later question, earlier questions can affect the response process for later ones. Answering one question about an issue enables respondents to answer a second one more quickly (Judd, Drake, Downing, and Krosnick, 1991; Tourangeau, Rasinski, and D'Andrade, 1991). Similar accessibility changes can affect the direction of the answers as well as their speed if the beliefs made accessible by the earlier items point to a different answer from the beliefs that would otherwise have been considered.

Retrieval-Based Assimilation Effects   My colleagues and I conducted two studies that illustrate these retrieval-based assimilation effects (Tourangeau, Rasinski, Bradburn, and D'Andrade, 1989a,b). The context items in our studies were selected to increase the accessibility of considerations supporting one side of an issue. For example, some respondents received several questions about the government's responsibility to the poor before they answered a question on welfare; others received questions about economic individualism before the welfare question. In seven of ten comparisons, respondents who answered different context questions gave significantly different answers to the target item; the differences were all in the direction of assimilation. In both studies, respondents who answered the questions on government's responsibility to the poor were more likely to support welfare spending than those who had answered the items on economic individualism.

Many other studies have demonstrated the impact of accessible material on responses to survey questions (Bishop, 1987; Rasinski and Tourangeau, 1991; Schwarz and Bless, 1992a, 1992b), judgments of life satisfaction (Schwarz and Clore, 1983; Strack, Martin, and Schwarz, 1988; Strack, Schwarz, and Gschneidinger, 1985), or other judgments (Higgins et al., 1977; Srull and Wyer, 1979).

Disregarding Accessible Material   Highly accessible material is not necessarily used and, when it is used, it is not necessarily included in the sample of material on which the judgment is based. Instead, contextually salient material may be used to construct a standard with which the target is compared. Other members from the same category as the target—especially, extreme ones—may serve as standards in this way. In addition, sometimes material that is irrelevant, invalid, or already used will be disregarded.

A study by Schwarz and Bless (1992a) illustrates the exclusion of accessible material whose relevance is called into question. In their study, German college students evaluated the two major German parties—the Christian Democratic Union (CDU) and Social Democratic Party (SDP). Evaluations of the CDU were affected by earlier questions about Richard von Weizsäcker, the President of the Federal Republic of Germany at the time, and a respected politician and long-term CDU member. Views about von Weizsäcker were, thus, potentially quite relevant to evaluations of the CDU. At the same time, the President was prohibited by law from participating in partisan politics; it was, therefore, also reasonable for respondents to disregard their views about von Weizsäcker in evaluating the CDU. Respondents who answered a prior item about von Weizsäcker's party membership ("Do you happen to know which party Richard von Weizsäcker has been a member of for more than twenty years?") gave higher evaluations of the CDU than respondents who answered an item about his office ("Do you happen to know which office Richard von Weizsäcker holds, setting him aside from party politics?"). Respondents who did not get either question gave evaluations in between those of other two groups (see Table 8.2). When the context brought von Weizsäcker's relevance into question, respondents disregarded their views about him in evaluating the CDU (to the CDU's detriment). Because he had no connection to the SDP, the questions on von Weizsäcker did not influence evaluations of that party.

Respondents may also discount information that they regard as unrepresentative in some way. In a study by Wilson, Hodges, and La Fleur (1995), respondents who had recalled an unrepresentative set of characteristics of a target person took this bias into account in rating the person. Similarly, in a study by Schwarz, Bless, Strack, Klumpp, Rittenauer-Schatka, and Simons (1991) respondents who recalled 12 instances of assertive behavior rated themselves as *less* assertive than respondents who recalled six instances; the difficulty of recalling the additional examples apparently suggested to the respondents how unrepresentative these examples must be.

Table 8.2   Mean Ratings of German Political Parties by Prior Items on von Weizsäcker

Party Evaluated	Prior Item on von Weizsäcker		
	Question on Party	None	Question on Office
Christian Democrats (CDU)	6.5	5.2	3.4
Social Democrats (SDP)	6.3	6.3	6.2

Note: Data from Schwarz & Bless, 1992a; scale ranged from 1 to 11, with higher numbers indicating more favorable evaluations.

Respondents may, finally, disregard accessible information simply because they feel the task or question calls for something new (e.g., Martin, 1986). We have already seen how Grice's maxim of quantity can lead to the inference that a general question is meant to exclude material used in answering an earlier, more specific question (Schwarz, Strack, and Mai, 1991; Tourangeau, Rasinski, and Bradburn, 1991); in such cases, it may seem repetitious to reuse accessible material and so it is not considered.

The exclusion of accessible material will produce contrast only if respondents end up discounting considerations they would otherwise have taken into account. For instance, in the Schwarz and Bless study, respondents who received the item on von Weizsäcker's office gave the CDU lower ratings than those who were not asked about von Weizsäcker at all; apparently, some respondents spontaneously considered von Weizsäcker in evaluating the CDU.

8.2.3   Context and Judgment

It may not be obvious how to use a specific consideration in answering an attitude question or how to combine them when several considerations come to mind. Anderson's (1981) information integration theory suggests that respondents first assign scale values to each consideration, reflecting its implications for the judgment; they may also assign weights to the different considerations. Then, depending on the nature of the question, they apply some integration rule to derive a final judgment. Each of these processes—scaling considerations, assigning weights, and applying rules—can be context dependent.

Assimilation Effects on Judgment   One of the clearest cases of context affecting the rule applied in making a judgment involves two items reported by Hyman and Sheatsley (1950). One item asks whether "the United States should let Communist reporters come in here and send back to their papers the news as they see it?" The other asks whether "a Communist country like Russia should let American newspaper reporters come in and send back to their papers the news as they see it?" Support for free access for the Communist reporters varies sharply according to whether that item precedes or follows the item on American reporters (see Table 8.3). The original result may be the largest context

Table 8.3   Proportion Endorsing Free Access for Communist Reporters, by Question Order

	Study	
Question Order	Hyman and Sheatsley (1950)	Schuman and Presser (1981)
Communist item first	36%	55%
Communist item second	73%	75%

effect in the survey literature, with the two contexts producing a shift of 37 percentage points in responses to the Communist reporters item. Schuman and Presser (1981) confirmed the original finding and follow-up studies by Schuman and Ludwig (1983) demonstrated similar results when the items involve imports to the United States from Japan and exports from the United States to Japan.

Most researchers accept the account of these findings offered by Schuman and his collaborators. When the Communist item comes first, responses to the item often reflect attitudes toward Communism; when it comes after the item on American reporters, responses reflect the norm of evenhandedness, which requires that both parties receive similar treatment.

Judgmental Contrast   If attitude judgments are comparative, then the standard of comparison will have a major impact on the judgment. One's assessment of one's current life may be quite positive when one thinks about the worst times of the past, but quite negative when one thinks about the best times (Strack et al., 1985). Such contrast effects may partly reflect how respondents use the response scale, but they also reflect how respondents evaluate the specific considerations retrieved in assessing their lives. One's life may actually *seem* better or worse depending on the standard adopted.

What standards of comparison are likely to be used in judging a stimulus? The standard may involve the average value for the category from which the stimulus is drawn, some prototypical category member, or an extreme member. Assimilation is more likely when the standard is seen as representing the typical value, contrast when it is seen as representing an extreme value. Moreover, contrast often seems to result when respondents judge several things of the same type on a single dimension—when they judge their satisfaction with both recent and past events (Strack et al., 1985), their favorability toward several politicians (Schwarz and Bless, 1992b), the heights of many college students (Manis, Biernat, and Nelson, 1991), and the attractiveness of faces depicted in photographs (Wedell, Parducci, and Geiselman, 1987).

With judgments of causality, a special standard of comparison may be applied, involving a set of counterfactual cases that contrast with the actual outcome. The preferred causal explanation specifies what made the difference between what actually happened and what might have happened instead (Chang and Novick, 1990; Hilton, 1990; Wells and Gavanski, 1989). When the question wording or context alters the background cases to which the actual outcome is compared, the explanation changes as well (McGill, 1989). It is one thing to explain why *Bill* chose this major, another to explain why he chose *this major*. In one case, Bill is compared with other students; in the other, the major he chose is compared with other majors.

8.2.4 Context and Reporting

Context can alter how respondents map their judgments onto the response scale and how they edit their answers before reporting them. Once again, both assimilation and contrast effects can result.

Consistency and Response Editing When attitude items are placed next to each other in a questionnaire, the responses to them may become more consistent (McGuire, 1960). By highlighting their relationship, juxtaposing the items increases the pressure to answer them consistently (see also Wyer and Rosen, 1972). McGuire's formulation of the "Socratic effect" assumed that the items involved were logically related. However, Smith (1983) found that placing related items together in a questionnaire increases their correlation, even if the relationship is not strictly logical. Respondents try to avoid inconsistent answers and grouping related items makes inconsistencies more obvious.

Contrast and the Mapping of Responses According to Parducci's (1965, 1974) range-frequency model, the mapping of judgments onto response categories depends on the range of stimuli being judged; the most extreme stimuli are mapped onto the end points of the scale, thereby defining the entire scale (see also Ostrom and Upshaw, 1968). Extreme stimuli can also serve as standards of comparison, affecting the judgment itself and not merely how it is reported. In either case, the result is a contrast effect. Although it may be possible in principle to distinguish judgmental contrast effects from the effects of anchors on reporting, it is difficult in practice to determine whether a prior item served as a standard of comparison, an anchor for the response scale, or both.

8.2.5 Variables Affecting the Likelihood and Direction of Context Effects

Context effects are the outcome of several processes. The ordering of items may carry the implication that an item about an unfamiliar issue is about the same issue as the previous items or that a general item excludes an important subdomain. An item can increase the accessibility of considerations that may then be used in making later judgments—or that may be disregarded if they are seen as tainted in some way. Placing items next to each other can highlight the need to treat both the same way, or it can encourage comparisons between them. Juxtaposing related questions can make their relationship more salient, promoting consistency, or the earlier items can anchor the response scale, producing contrast. The array of potential processes makes it difficult to predict what will happen in any specific case.

Variables Affecting the Likelihood of Context Effects   Are there general conditions or characteristics of attitudes that promote or inhibit context effects? One basic hypothesis concerns the relationship between the context and target items.

Relations Among the Items   The mechanisms thought to produce context effects all presuppose some close topical or logical connection between the context and target items. For instance, when questions shift from one topic to the next without warning, respondents are unlikely to see the earlier items as carrying implications about the meaning of later ones. In fact, as Table 8.2 illustrates, it may be necessary to emphasize the relation between two items—for example, by an introduction—for certain context effects to occur. Even with psychophysical judgments, one stimulus affects the judgments of later ones only when it is seen as part of the same sequence (Brown, 1953). The relationship between the context and target items may involve logical implication, class inclusion, shared category membership, or association in long-term memory, but *some* relationship is needed and the mere existence of a relationship may not be enough; the presentation of the questions may have to emphasize that relationship for context effects to occur.

When the context of a question varies, but the variations involve items unrelated to the target question, nothing much happens. The National Opinion Research Center's General Social Survey employs a design in which batteries of items are rotated across three different versions of the questionnaire. Smith (1988) examined 358 questions that varied in their context across different versions; the answers differed significantly by version only for about 4 percent of the items—slightly fewer than would be expected by chance. When the contextual variations consist of differences in questions on subjects unrelated to the target question, context effects seem rare.

Characteristics of the Attitude or Attitude Issue   It seems plausible that accessible, well-formed attitudes would be relatively invulnerable to the effects of question context. With accessible attitudes, one would expect relevant questions to trigger the automatic retrieval of an existing evaluation that would provide the basis for an answer (Fazio, 1989). If so, the item context should have little or no impact (Basili, 1993). Similarly, central attitudes are more stable over time than noncentral attitudes (Judd and Krosnick, 1982); it seems logical that they would be more stable across contexts as well.

Neither hypothesis receives much support from the literature. A review by Krosnick and Schuman (1988) found little relation between measures of attitude intensity and importance, on the one hand, and susceptibility to item wording and context effects, on the other. The large number of context effects involving judgments of life satisfaction also suggests that context effects can occur with very familiar topics about which respondents presumably have strong opinions.

If there is one topic that people care strongly about, surely it is the quality of their lives.

Some researchers have argued for the opposite hypothesis—that well-developed attitudes may be a prerequisite for certain context effects. Bickart (1992), for example, points out that experts may be more affected by contextually salient information than novices are since only the experts have the background knowledge needed to recognize its implications. Both empirical results and theoretical arguments, then, suggest that no type of attitude is immune to the effects of context.

Variables Affecting Specific Components of the Response Process Context affects each component of the response process, and different variables govern whether context effects will arise during each one. Consider the comprehension of a question. Assimilation effects are more likely when the question concerns an unfamiliar issue than when it concerns a familiar one, and when the question appears to be on the same topic as the prior questions. Contrast effects are more likely when a general question follows similar items on one or two important subdomains, the wording does not specify the scope of the general question, and the context emphasizes connection between the general and specific items. But there is no reason to believe that these variables also determine whether context affects retrieval or judgment. Instead, the governing variables tend to be specific to a single component of the response process.

8.2.6 Variables Affecting the Direction of Context Effects

Numerous studies have also attempted to pinpoint variables that determine the direction of the effect of context, given that context has an effect at all. Table 8.4 lists 14 such variables. Most findings in the table follow two general principles. The first is that deeper processing is necessary for contrast than for assimilation effects. For instance, respondents in one study showed assimilation effects when they were distracted, but contrast when they were not (Martin, Seta, and Crelia, 1990). Contrast effects also emerged when respondent motivation was high (Martin et al., 1990, Experiments 2 and 3) or when an explicit cue called attention to the context (Strack, 1992), but assimilation effects emerged when motivation was low or when no cue made the context salient. When respondents are willing and able to process the questions deeply (or when explicit cues eliminate the need for deep processing), contrast effects are the rule. When processing is superficial, assimilation effects predominate.

The mechanisms responsible for assimilation and contrast effects help explain these findings. Assimilation effects are largely the product of automatic processes that occur without effort and outside of awareness. Unless respondents are motivated (or cued) to question these outputs, they generally accept them uncritically. Contrast effects, on the other hand, result from controlled processes, in which respondents consciously detect some bias in the considerations that come to mind and try to compensate for it.

Table 8.4 Variables Affecting the Direction of Context Effects

Study	Task	Variable	Result
Herr, Sherman, and Fazio (1983)	Rate ferocity, size of animals	Extremity of context items	Assimilation with moderate context items; contrast with extreme ones
		Real vs. unreal targets	Assimilation with made-up target animals; contrast with real targets
Manis, Biernat, and Nelson (1991)	Judge height of models	Scale for judgment	Assimilation when judgment in feet and inches; contrast when judgment on 7-point scale
Martin (1986)	Form impression from description	Interruption during context task	Assimilation when prior task interrupted; contrast when task completed
Martin, Seta, and Crelia (1990)	Form impression from description	Distraction	Assimilation with distraction; contrast without distraction
		Motivation	Assimilation with low motivation; contrast with high motivation
McMullen (1997)	Imagine alternative outcome for negative event	Focus on actual or counterfactual outcome	Assimilation with counterfactual focus; contrast with focus on actual

Study	Dependent measure	Variable	Results
Ottati, Riggle, Wyer, Schwarz, and Kuklinski (1989)	Rate agreement with statements	Separation between target and context items	Assimilation when context separated from target; contrast when adjacent
Schwarz, Strack, and Mai (1991)	Rate life satisfaction	Introduction to target and context items	Assimilation with no lead-in; contrast with lead-in
		Number of context items	Assimilation with three context items; contrast with one
Strack (1992)	Rate likability of target	Awareness of context	Assimilation with no reminder of prior context; contrast with reminder
Strack, Schwarz, and Gscheidinger (1985)	Rate life satisfaction	Recall past vs. present events	Assimilation with present events; contrast with past events
		Vividness of recall	Assimilation with vivid recall; contrast with pallid recall
Wedell, Parducci, and Geiselman (1987)	Rate attractiveness of faces	Simultaneous vs. successive presentation	Assimilation when context and target presented at same time; contrast when presented successively

The second principle linking the findings is that conditions fostering comparisons between the target and context stimuli promote contrast effects. Vague rating scales (Manis et al., 1991), tasks requiring respondents to focus on both the target and context stimuli (McMullen, 1997), and context items with extreme values on the dimension of judgment (Herr, Sherman, and Fazio, 1983) encourage comparisons and tend to produce contrast. Absolute response scales, tasks in which the focus is mainly on the context stimuli, and context items with typical values discourage comparisons, and assimilation tends to result. In addition, it is easier to compare things that are familiar; when the target item is unfamiliar or completely novel, it will make comparison difficult and encourage assimilation.

Differences in the level of generality of the context and target items can also affect whether respondents compare them. When the context item concerns an issue or category that encompasses the target issue, respondents may make inferences about the target based on what they know about the broader issue raised by the context item. Just as we make inferences about a species of birds based on our knowledge of birds in general, we may make inferences about an issue based on our views about the more general issue of which it is a part (see Figure 8.1). Respondents are less likely to fill in gaps in their knowledge in this way when the target issue is a familiar one; instead, inferences from the general to the specific are likely to involve novel issues like the Monetary Control Bill.

We also make inductive inferences about a general category based on the instances we encounter. Respondents are willing to generalize about their overall happiness based on the recent experiences they recall (Strack et al., 1985) or about a political party based on a prominent member (Schwarz and Bless, 1992a). Such inferences are more likely if the category members presented are numerous, prototypical, and cover the entire category of interest (Osherson, Smith, and Shafir, 1986). We make inferences in both directions—from the members to their category and from the category to its members. By affecting which members are accessible, context can affect judgments regarding the category; and, by affecting what information about the category is accessible, context can affect judgments of the members. In either case, assimilation effects will be the typical result.

Members of the same category can also affect judgments about each other. Members of a category may be compared, and when the member serving as the standard of comparison is extreme or simply quite different from the target on the dimension of judgment, contrast will result. In addition, by creating expectations about the class as a whole, other category members can indirectly affect judgments about the target (cf. Manis et al., 1991). When the contextually-salient category members are extreme (and therefore unrepresentative of the category), they will serve as standards of comparison and contrast effects will predominate; when they are numerous or are prototypical, expectation-based assimilation effects will predominate (see Figure 8.1).

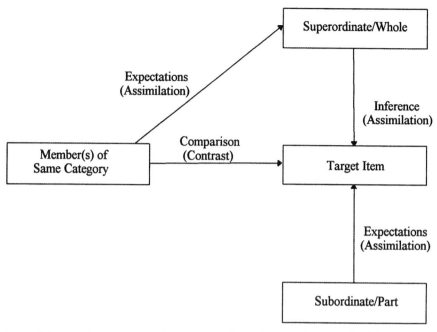

Figure 8.1   Relations between context and target items. The middle box on the right represents the target item; the remaining boxes represent context items at higher, the same, or lower levels of generality than the target. The paths involving assimilation assume that information is not consciously excluded.

8.2.7   General Models of Context Effects

The Inclusion/Exclusion Model   The most successful attempt to tie findings like those in Table 8.4 into a single model is the inclusion/exclusion model proposed by Schwarz and Bless (1992a). Their model assumes that, in judging a target stimulus, respondents form a representation of the target and of a standard with which the target is compared. Both representations rely on information accessible when the judgment is being formed. According to Schwarz and Bless, information made accessible by context can be used in three ways. First, it may be incorporated into the representation of the target, producing an assimilation effect. Second, it may be excluded from the representation of the target, producing a subtraction-based contrast effect. Finally, it may be incorporated into the representation of the standard, producing a comparison-based contrast effect.

The impact of including or excluding contextually accessible material will depend, according to the model, on what would have been included in the representations of the target and standard in the absence of the context items. Larger assimilation effects will result as the representation of the target draws more heavily on temporarily accessible information or when this information has more extreme implications for the judgment; smaller effects will result as the representation draws more on chronically accessible or less extreme information. In addition, when context increases the accessibility of several items, it will reduce the impact of any one of them. Similar arguments apply to the impact of *exclusions* on the representation of the target; as more information or information with more extreme implications is excluded from the representation of the target, subtraction-based contrast effects will grow larger.

One of the most appealing features of the inclusion/exclusion model is its delineation of the conditions that foster exclusion of accessible material from the representation of a target. According to Schwarz and Bless (1992a) (see also Sudman et al., 1996, Figure 5.1), material will be excluded from the representation of the target when it fails to meet any of four tests:

- Does the material bear on the question?
- Did it come to mind because of relevant influences?
- Does it fit the target?
- Is the respondent intended to use it?

These tests filter out irrelevant, biased, false or inapplicable, and redundant information. Information left out of the representation of the target that nonetheless bears on the judgment may be used to construct a standard of comparison.

The Belief-Sampling Model   The assumptions of the inclusion/exclusion model are quite similar to those of the belief-sampling model, which has guided the discussion here. The belief-sampling model assumes that respondents sample a few beliefs in making an attitude judgment, using a form of averaging to integrate these beliefs. (The sample of beliefs corresponds to the representation of the target.) Combining beliefs into a judgment requires that the beliefs be scaled, and the scale values will reflect salient anchors or standards. Like the inclusion/exclusion model, the belief-sampling model traces assimilation effects to a change in the underlying pool of beliefs from which the sample is drawn (as when context changes the interpretation of the target item) or to a change in the accessibility of specific beliefs, which affect the likelihood of their being included in the sample. Similarly, it traces contrast effects to the exclusion of specific beliefs from the sample on which the judgment is based (as in the subtraction-based contrast effects of the inclusion/exclusion model) or to the impact of contextual standards on the scale values assigned to the beliefs (as in comparison-based contrast effects).

Although the differences between the two models are not very striking, they differ in emphasis. The inclusion/exclusion model is more explicit about the factors that determine whether a given consideration is excluded from the sample on which the target judgment rests. The belief-sampling model is more explicit about how considerations that enter the sample affect the answer and about the variables that promote consistent answers to the same question on different occasions.

The belief-sampling model asserts that the correlation between responses to an item presented on two occasions (or in two different contexts) depends on three parameters:

- the level of consistency in the scale value assigned to the same belief on different occasions;
- the internal consistency of the pool of relevant beliefs;
- the degree that the specific beliefs sampled on the two occasions overlap.

This analysis implies that context will affect the responses when it changes the beliefs on which they are based or the scale values assigned to those beliefs. The beliefs will change when the context alters either the interpretation of the item or the specific beliefs included in (or excluded from) the sample on which the judgment is based. To the extent that these beliefs change across contexts, there will be reduced overlap and less consistency among the beliefs taken into account. The scale values assigned to a given consideration will change when the standard of comparison changes across contexts. Changes of standard will alter the evaluation given to the same consideration under the two contexts.

In summary, both models emphasize the flexible and constructive nature of attitude judgments. These judgments are context-sensitive because attitude objects can be construed in many different ways, and context helps determine what is included and what is left out. In addition, the judgments themselves are inherently comparative and, as the context changes the standards, it alters the judgments. If these models are right, then context effects are not so much a measurement artifact but an essential by-product of the process of asking and answering attitude questions.

REFERENCES

Abelson, R. P. (1981). Psychological status of the script concept. *American Psychologist, 36*, 715–729.

Anderson, N. (1981). *Foundations of Information Integration Theory.* New York: Academic Press.

Asch, S. (1946). Forming impressions of personalities. *Journal of Abnormal and Social Psychology, 41*, 258–290.

Basili, J. N. (1993). Response latency versus certainty as indexes of the strength of voting intentions in a CATI survey. *Public Opinion Quarterly, 57*, 54–61.

Bickart, B. A. (1992). Question order effects and brand evaluations: The moderating role of consumer knowledge. In N. Schwarz and S. Sudman (Eds.), *Context Effects in Social and Psychological Research*, pp. 63–79. New York: Springer-Verlag.

Bishop, G. (1987). Context effects in self-perceptions of interest in government and public affairs. In H. Hippler, N. Schwarz, and S. Sudman (Eds.), *Social Information Processing and Survey Methodology*, pp. 179–199. New York: Springer-Verlag.

Bradburn, N. M. (1982). Question-wording effects in surveys. In R. Hogarth (Ed.), *Question Framing and Response Consistency*, pp. 65–76. San Francisco: Jossey-Bass.

Bransford, J., and Johnson, M. (1972). Contextual prerequisites for understanding: Some investigations of comprehension and recall. *Journal of Verbal Learning and Verbal Behavior, 11*, 717–726.

Brown, D. R. (1953). Stimulus similarity and the anchoring of subjective scale. *American Journal of Psychology, 66*, 199–214.

Chang, P., and Novick, L. (1990). A probabilistic contrast model of causal induction. *Journal of Personality and Social Psychology, 58*, 545–567.

Clark, H. (1985). Language use and language users. In G. Lindzey and E. Aronson (Eds.), *The Handbook of Social Psychology: Vol. 2. Special Fields and Applications*, 3rd ed., pp. 179–231. New York: Random House.

Fazio, R. (1989). On the power and functionality of attitudes: The role of attitude accessibility. In A. Pratkanis, S. Breckler, and A. Greenwald (Eds.), *Attitude Structure and Function*, pp. 153–179. Hillsdale, NJ: Erlbaum.

Grice, H. P. (1975). Logic and conversation. In P. Cole and J. L. Morgan (Eds.), *Syntax and Semantics: 3. Speech Acts*, pp. 41–58. New York: Academic Press.

Herr, P., Sherman, S., and Fazio, R. (1983). On the consequences of priming: Assimilation and contrast effects. *Journal of Experimental Social Psychology, 19*, 323–340.

Higgins, E., Rholes, W., and Jones, C. (1977). Category accessibility and impression formation. *Journal of Experimental Social Psychology, 13*, 141–154.

Hilton, D. (1990). Conversational processes and causal reasoning. *Psychological Bulletin, 107*, 65–81.

Hyman, H. H., and Sheatsley, P. B. (1950). The current status of American public opinion. In J. Payne (Ed.), *The Teaching of Contemporary Affairs: Twenty-First Yearbook of the National Council for the Social Studies*, pp. 11–34. New York: National Education Association.

Judd, C., Drake, R., Downing, J., and Krosnick, J. (1991). Some dynamic properties of attitude structures: Context-induced response facilitation and polarization. *Journal of Personality and Social Psychology, 60*, 193–202.

Judd, C., and Krosnick, J. (1982). Attitude centrality, organization, and measurement. *Journal of Personality and Social Psychology, 42*, 436–447.

Kalton, G., Collins, M., and Brook, L. (1978). Experiments in wording opinion questions. *Journal of the Royal Statistical Society (Series C), 27*, 149–161.

Krosnick, J. A., and Schuman, H. (1988). Attitude intensity, importance, certainty, and susceptibility to response effects. *Journal of Personality and Social Psychology, 54*, 940–952.

Manis, M., Biernat, M., and Nelson, T. (1991). Comparison and expectancy processes in human judgment. *Journal of Personality and Social Psychology, 61*, 203–211.

Martin, L. L. (1986). Set/reset: Use and disuse of concepts in impression formation. *Journal of Personality and Social Psychology, 51*, 493–504.

Martin, L. L., Seta, J. J., and Crelia, R. A. (1990). Assimilation and contrast as a function of people's willingness and ability to expend effort in forming an impression. *Journal of Personality and Social Psychology, 59,* 27–37.

Mason, R., Carlson, J., and Tourangeau, R. (1995). Contrast effects and substraction in part-whole questions. *Public Opinion Quarterly, 58,* 569–578.

McGill, A. (1989). Context effects in judgments of causation. *Journal of Personality and Social Psychology, 57,* 189–200.

McGuire, W. (1960). A syllogistic analysis of cognitive relationships. In M. Rosenberg, C. Hovland, W. McGuire, R. Abelson, and J. Brehm (Eds.), *Attitude Organization and Change,* pp. 65–111. New Haven, CT: Yale University Press.

McMullen, M. (1997). Affective contrast and assimilation in counterfactual thinking. *Journal of Experimental Social Psychology, 33,* 77–100.

Osherson, D., Smith, E., and Shafir, E. (1986). Some origins of belief. *Cognition, 24,* 197–224.

Ostrom, T. M., and Upshaw, H. S. (1968). Psychological perspective and attitude change. In A. C. Greenwald, T. C. Brock, and T. M. Ostrom (Eds.), *Psychological Foundations of Attitudes,* pp. 65–111. New York: Academic Press.

Ottati, V. C., Riggle, E. J., Wyer, R. S., Schwarz, N., and Kuklinski, J. (1989). Cognitive and affective bases of opinion survey responses. *Journal of Personality and Social Psychology, 57,* 404–415.

Parducci, A. (1965). Category judgment: A range-frequency model. *Psychological Review, 72,* 407–418.

Parducci, A. (1974). Contextual effects: A range-frequency analysis. In E. Carterette and M. Friedman (Eds.), *Handbook of Perception: Psychophysical Judgment and Measurement,* Vol. II: pp. 127–141. New York: Academic Press.

Rasinski, K., and Tourangeau, R. (1991). Psychological aspects of judgments about the economy. *Political Psychology, 12,* 27–40.

Schuman, H., and Ludwig, J. (1983). The norm of evenhandedness in surveys as in life. *American Sociological Review, 48,* 112–120.

Schuman, H., and Presser, S. (1981). *Questions and Answers in Attitude Surveys: Experiments in Question Form, Wording, and Context.* New York: Academic Press.

Schwarz, N., and Bless, H. (1992a). Constructing reality and its alternatives: Assimilation and contrasts effects in social judgment. In L. L. Martin and A. Tesser (Eds.), *The Construction of Social Judgment,* pp. 217–245. Hillsdale, NJ: Erlbaum.

Schwarz, N., and Bless, H. (1992b). Scandals and public trust in politicians: Assimilation and contrast effects. *Personality and Social Psychology Bulletin, 18,* 574–579.

Schwarz, N., Bless, H., Strack, F., Klumpp, G., Rittenauer-Schatka, H., and Simons, A. (1991). Ease of retrieval as information: Another look at the availability heuristic. *Journal of Personality and Social Psychology, 61,* 195–202.

Schwarz, N., and Clore, G. L. (1983). Mood, misattribution, and judgments of well-being: Informative and directive functions of affective states. *Journal of Personality and Social Psychology, 45,* 513–523.

Schwarz, N., Strack, F., and Mai, H. (1991). Assimilation and contrast effects in part-whole question sequences: A conversational logic analysis. *Public Opinion Quarterly, 55,* 3–23.

Smith, T. (1982). *Conditional order effects.* GSS Technical Report No. 33. Chicago: NORC.

Smith, T. (1983). An experimental comparison between clustered and scattered scale items. *Social Psychology Quarterly, 46,* 163–168.

Smith, T. (1988). *Ballot position: An analysis of context effects related to rotation design* (GSS Methodological Report No. 55). Chicago: NORC.

Srull, T., and Wyer, R. (1979). The role of category accessibility in the interpretation of information about persons: Some determinants and implications. *Journal of Personality and Social Psychology, 37,* 1660–1672.

Strack, F. (1992). Order effects in survey research: Activative and informative functions of preceding questions. In N. Schwarz and S. Sudman (Eds.), *Context Effects in Social and Psychological Research,* pp. 23–34. New York: Springer-Verlag.

Strack, F., and Martin, L. L. (1987). Thinking, judging, and communicating: A process account of context effects in attitude surveys. In H. Hippler, N. Schwarz, and S. Sudman (Eds.), *Social Information Processing and Survey Methodology,* pp. 123–148. New York: Springer-Verlag.

Strack, F., Martin, L. L., and Schwarz, N. (1988). Priming and communication: The social determinants of information use in judgments of life satisfaction. *European Journal of Social Psychology, 18,* 429–442.

Strack, F., Schwarz, N., and Gschneidinger, E. (1985). Happiness and reminiscing: The role of time perspective, affect, and mode of thinking. *Journal of Personality and Social Psychology, 47,* 1460–1469.

Strack, F., Schwarz, N., and Wänke, M. (1991). Semantic and pragmatic aspects of context effects in social and psychological research. *Social Cognition, 9,* 111–125.

Sudman, S., Bradburn, N., and Schwarz, N. (1996). *Thinking About Answers: The Application of Cognitive Processes to Survey Methodology.* San Francisco: Jossey-Bass.

Tourangeau, R. (1984). Cognitive sciences and survey methods. In T. Jabine, M. Straf, J. Tanur, and R. Tourangeau (Eds.), *Cognitive Aspects of Survey Methodology: Building a Bridge Between Disciplines,* pp. 73–100. Washington, DC: National Academy Press.

Tourangeau, R. (1992). Context effects on attitude responses: The role of retrieval and memory structures. In N. Schwarz and S. Sudman (Eds.), *Context Effects in Social and Psychological Research,* pp. 35–47. New York: Springer-Verlag.

Tourangeau, R., and Rasinski, K. A. (1986). *Context Effects in Attitude Surveys.* Unpublished manuscript.

Tourangeau, R., and Rasinski, K. A. (1988). Cognitive processes underlying context effects in attitude measurement. *Psychological Bulletin, 103,* 229–314.

Tourangeau, R., Rasinski, K. A., and Bradburn, N. M. (1991). Measuring happiness in surveys: A test of the subtraction hypothesis. *Public Opinion Quarterly, 55,* 255–266.

Tourangeau, R., Rasinski, K. A., Bradburn, N. M., and D'Andrade, R. (1989a). Carryover effects in attitude surveys. *Public Opinion Quarterly, 53,* 495–524.

Tourangeau, R., Rasinski, K. A., Bradburn, N. M., and D'Andrade, R. (1989b). Belief accessibility and context effects in attitude measurement. *Journal of Experimental Social Psychology, 25,* 401–421.

Tourangeau, R., Rasinski, K. A., and D'Andrade, R. (1991). Attitude structure and belief accessibility. *Journal of Experimental Social Psychology, 27,* 48–75.

Turner, C. F., and Martin, E. (1984). *Surveying Subjective Phenomena.* New York: Russell Sage Foundation.

Wedell, D. H., Parducci, A., and Geiselman, R. E. (1987). A formal analysis of ratings of physical attractiveness: Successive contrast and simultaneous assimilation. *Journal of Experimental Social Psychology, 23,* 230–249.

Wells, G., and Gavanski, I. (1989). Mental simulation of causality. *Journal of Personality and Social Psychology, 56,* 161–169.

Wilson, T. D., and Hodges, S. D. (1992). Attitudes as temporary constructions. In L. L. Martin and A. Tesser (Eds.), *The Construction of Social Judgments,* pp. 37–66. Hillsdale, NJ: Erlbaum.

Wilson, T. D., Hodges, S. D., and LaFleur, S. J. (1995). Effects of introspecting about reasons: Inferring attitudes from accessible thoughts. *Journal of Personality and Social Psychology, 69,* 16–28.

Wyer, R. S., and Rosen, N. (1972). Some further evidence for the Socratic effect using a subjective probability model of cognitive organization. *Journal of Personality and Social Psychology, 24,* 420–424.

CHAPTER 9

Is the Bandwagon Headed to the Methodological Promised Land? Evaluating the Validity of Cognitive Interviewing Techniques

Gordon B. Willis
Research Triangle Institute

Theresa J. DeMaio
U.S. Bureau of the Census

Brian Harris-Kojetin
The Arbitron Company

There is wide agreement in the field of cognition and survey methodology (CASM) that the purpose of studying the survey response process is to understand sources of response error in survey questions, and to reduce this error. However, over the past fifteen years, two divergent viewpoints have emerged concerning the issue of how this is best achieved. First is the view that the major purpose of CASM investigations is to understand the *general* nature of the survey response, and that insights from our investigations should lead to cognitive principles that can be applied across a wide range of survey questionnaires. In this view, the results of our experimental investigations consist of findings pertaining to cognitive *process*, and the application of these findings will allow us to design survey questions which exhibit less error. For example, Loftus (1984) suggested that cognitive techniques could be used to determine whether

Cognition and Survey Research, Edited by Monroe G. Sirken, Douglas J. Herrmann,
Susan Schechter, Norbert Schwarz, Judith M. Tanur, and Roger Tourangeau.
ISBN 0-471-24138-5   © 1999 John Wiley & Sons, Inc.

survey respondents typically recall events in a forward or backward temporal order, and that such knowledge would allow survey designers to develop questions that match respondent tendencies. As a further example, Schwarz and his colleagues have conducted research to study the behavior of survey respondents who are administered behavioral-frequency questions containing closed-ended response categories (Schwarz and Bienias, 1990; Schwarz, Hippler, Deutsch, and Strack, 1985). These studies reveal that respondents use those categories to infer norms which they then use as a basis for reporting their own behavior.

The second main approach to CASM research, which has come into use especially by questionnaire designers at U.S. Federal agencies, focuses less on determination of general cognitive principles, and more on the *pretesting* of survey questions (Beatty, Willis, and Schechter, 1997; Campanelli, 1997; Jobe and Mingay, 1991; Tanur and Fienberg, 1992). Here, a variety of cognitive techniques, representing a "toolbag" of methods, are utilized in an interview with a small number of subjects, generally in a cognitive laboratory.[1] The basic purpose is usually to detect faults in a specific set of survey questions, and to obtain information useful in improving these questions. These methods rely on a very general cognitive model of the survey response process, usually the Tourangeau (1984) scheme emphasizing encoding, retrieval, decision, and response processes (see Jobe and Herrmann, 1996, for a review of cognitive models). Because the focus of these investigations tends to be the "clinical" application of CASM techniques to the questionnaire, there has been limited emphasis on augmenting the more general body of knowledge concerning the survey response process.

CASM investigations which emphasize the second of these approaches—question pretesting—have tended to rely on a set of procedures commonly referred to as cognitive interviewing techniques. The focus of this paper is a review of major issues related to the evaluation of these techniques, as they relate to question pretesting. In this chapter, we explicate a number of challenges to evaluation, discuss studies that have adopted varying approaches to overcoming these challenges, and make recommendations concerning the types of evaluation designs that may be promising. First, we discuss some preliminary issues in evaluating cognitive interviewing methods, especially in terms of the definition of the methods used and the results reported. Next, we consider whether the use of cognitive methods can be shown to be responsible for improvements in survey question quality. Finally, we discuss cognitive interviewing in the context of other pretesting methods, and consider how the evaluation of multiple methods can be accomplished. For each section, we offer recommendations and considerations for research, using as examples evaluation studies that have been conducted.

---

[1]Typically, the term "subject" refers to a volunteer who is administered the questionnaire during laboratory pretesting, and "respondent" denotes an individual administered the question in the field environment.

9.1 DEFINITION AND DESCRIPTION OF COGNITIVE INTERVIEWING

An initial issue in conducting an evaluation of the effectiveness of any procedure is whether the method itself has been adequately defined and described. In fact, several types of procedures have been labeled as "cognitive interviewing," and these have divergent features, as well as different historical backgrounds. In particular, we distinguish the cognitive interview as conducted by survey researchers from the "Cognitive Interview" developed by Fisher and Geiselman (1992) for the purpose of facilitating retrieval from memory of information by witnesses to crimes. We are concerned here only with the variety of cognitive interviewing that is applied by survey researchers to study the general cognitive processes (including, but not limited to retrieval) that are assumed to be used naturally by survey respondents as they answer questions. Beyond this distinction, we consider two relevant questions: (a) Does survey-oriented "cognitive interviewing" constitute a generally well-defined activity? and (b) Is this activity practiced similarly by different researchers?

9.1.1 Specifying the Cognitive Interview: "Think-Aloud" versus "Verbal Probing"

Even within the relevant rubric of survey-oriented cognitive interviewing, we note considerable inconsistency in the terminology and defining approach that characterize the key method. Although the general technique has often been termed simply as "think-aloud" interviewing, this is a poor label, because a number of other procedures are commonly used, even by those explicitly describing their method as "think-aloud" (Forsyth and Lessler, 1991). The "think-aloud" procedure developed out of experimental psychology, especially in the context of problem solving (Ericsson and Simon, 1980; 1984), and was advocated by Loftus (1984) as a potential means for understanding the survey response process. In a true think-aloud interview, the subject verbalizes his or her thoughts while engaged in a cognitive activity, with little interjection by the interviewer other than encouragement to "tell me what you are thinking."

However, an alternative basic procedure currently used by survey researchers involves more extensive *verbal probing*. Specially trained cognitive interviewers make use of specific directive probes, such as requesting that subjects paraphrase survey questions, explain their understanding of key terms, and report their level of confidence in the answers they provide to the questions. These methods were introduced well before the advent of CASM, and were derived from a survey research tradition, rather than from experimental psychology (Cantril, 1944). The use of similar probing methods has been advocated by several authors since that time (Belson, 1981; DeMaio, Mathiowetz, Rothgeb, Beach, and Durant, 1993; Selltiz, Jahoda, Deutsch, and Cook, 1959; Willis, Royston, and Bercini, 1989). Interviewing techniques involving both think-aloud and verbal probing are described in detail by Willis (1994).

A number of researchers have considered the relative advantages of think-aloud and verbal probing methods, and especially the extent to which these techniques are useful with different population groups, or the manner in which they may bias the cognitive processing of survey respondents (Biehal and Chakravarti, 1989; Davis and DeMaio, 1993; DeMaio and Rothgeb, 1996; Foddy, 1996; Forsyth and Lessler, 1991; Gerber and Wellens, 1997; Jobe, Keller, and Smith, 1996; Sudman, Bradburn, and Schwarz, 1996; Willis, 1994; Willis, Royston, and Bercini, 1991; Von Thurn and Moore, 1994). To this point, there has been no definitive conclusion concerning which technique is more effective, or most likely to exhibit problems related to either artificiality or to what Sudman et al. (1996) term as Heisenberg Uncertainty Effects. Bolton (1993), Bolton and Bronkhorst (1996), and Bickart and Felcher (1996) have advocated the use of the think-aloud technique. Willis and Schechter (1997), on the other hand, focused on the use of a combined method of verbal probing and think-aloud, but with verbal probing as the focus. Foddy (1996) evaluated a range of these techniques, and has concluded that the overall effectiveness of probes increases with their level of specificity, with "think-aloud" being the least effective.

Our review of a number of reports of cognitive interviewing studies indicates that most researchers at this point tend to approach cognitive interviewing as a combination of think-aloud and verbal probing, but with a much greater emphasis on the use of specific verbal probes than was envisioned initially. In fact, Blair and Presser (1993) and Gerber and Wellens (1997) have concluded that there is little evidence that the "think-aloud" technique in the form proposed by Ericsson and Simon (1980; 1984) is practiced by many survey researchers when conducting cognitive interviews. Therefore, we believe that the continued use of the term "think-aloud" to refer generally to survey-based cognitive interviewing techniques has caused considerable failure of communication between researchers, and we propose that this term be generally reserved for situations in which subjects are asked to engage in pure think-aloud activity. Instead, we follow DeMaio et al. (1993) and Schechter, Blair, and Vande Hey (1996) in suggesting that the more frequently applied techniques involving a combination of think-aloud and verbal probing be referred to under the more general umbrella term of *cognitive interviewing techniques*. We emphasize however, that, to the extent that cognitive interviewing in practice subsumes a continuum of techniques, from the very general (think-aloud) to the very specific (use of targeted probes), evaluation studies must take into account this variability.

9.1.2   Description of How Cognitive Interviewing Techniques Are Practiced

Even if it is generally the case that practitioners currently emphasize a combination of think-aloud and verbal probing, there remain a number of parameters that can be varied, and that may have a major impact on the effectiveness of the procedure (Tucker, 1997). In essence, there exists no single "cognitive interviewing" technique; variables include the use of probes that are "scripted" prior to interviewing as opposed to those produced spontaneously by the interviewer during the interview, the use of concurrent versus retrospective probing,[2] the level of education and experience of interviewers, number of interviews conducted per iterative "round" of testing, type of practice questions used, and nature of the summarization of the information gathered (see DeMaio and Rothgeb, 1996; Forsyth and Lessler, 1991; Willis, 1994).

Blair and Presser (1993) have noted that little agreement exists among researchers concerning the levels at which these variables should be set. Worse, these factors are not usually well-described in research studies in a way that allows clear comparison. Forsyth and Lessler (1991) have proposed a typology of the specific procedures that may be used within the cognitive interview (confidence judgments, paraphrasing, etc.), but specification at this level of detail is often absent from written reports on cognitive interviewing studies. Further, although Esposito, Rothgeb, and Campanelli (1994) have proposed a systematic scheme for description of the major parameters in studies involving behavior coding techniques (the analysis of interviewer-respondent behavioral interactions; see Cannell, Fowler, and Marquis, 1968; Fowler and Cannell, 1996), no such scheme exists for use in cognitive interviewing research.

Note that it is of course feasible, and to some extent even vital, to conduct evaluation studies simply to determine the relative merits of approaches that vary these multiple procedural parameters, rather than attempting a global determination of whether cognitive interviewing in general is valid. Such studies would assess both *repeatability* of results within a laboratory, and *reproducibility* across laboratories (Tucker, 1997). In any event, as emphasized by Presser and Blair (1994), a key point is that the results of a study evaluating only one variety of cognitive interviewing have limited applicability to cognitive interviewing that is conducted in a different manner.

9.1.3 Recommendations for Evaluation

Clearly, cognitive interviewing consists of a variety of activities, and these activities may be practiced quite differently. Our initial recommendation, then, is that researchers put more effort into documenting "what goes on" in the cognitive interview. However, doing so entails more than simply describing the methods used, even when using a systematic descriptive scheme. Cognitive interviewers have sometimes described precisely "what we do" in the cognitive laboratory (e.g., Willis, 1994), but these descriptions are more accurately conceptualized as the authors' *perceptions* of what they do in the laboratory. Espe-

---

[2]The use of terminology related to concurrent versus retrospective probing has been inconsistent. Although retrospective probing has typically been viewed as probing done *after* the completion of the interview, Sudman et al. (1996) describe such probing as occurring immediately after the presentation of each question, to distinguish this from concurrent think-aloud. We prefer to use three somewhat self-explanatory terms to describe the relevant procedures: (a) concurrent think-aloud, (b) immediate retrospective probing, and (c) delayed retrospective probing.

cially for procedures in which interviewer-based verbal probing is frequent, it may be most productive to apply an objective coding system in order to assess key variables such as the relative degree to which think-aloud and verbal probing are used by interviewers, and the extent to which specific probes are balanced versus biased in nature.

As an example of this type of work, a study by Beatty, Schechter, and Whitaker (1996) has examined the ways in which interviewers used different categories of probes, and the degree to which interviewer probing in turn affected subject behavior. In particular, they found that the use of *elaborative* probes, such as "Tell me more about this," tended to reduce the level of precision in respondents' answers to the targeted survey questions, relative to the use of *reorienting* probes which refocus attention on answering the question (e.g., "Can you give me an answer in terms of number of days your health was not good?"). To the extent that level of precision (and therefore codability) of responses is often seen as an indicator of question problems, it is likely that the type of probing applied by interviewers, and its resultant effects on subject precision, will influence conclusions concerning whether a particular survey question is problematic.

9.2   DESCRIBING THE OUTCOMES OF COGNITIVE INTERVIEWING

Beyond the question of whether the *practice* of cognitive interviewing is well-defined, a second basic issue is the adequacy of description of interview *outcomes*, or findings, in a way that enables critical evaluation. Evaluation is essentially a quantitative and statistical science. However, findings from cognitive interviewing have typically been qualitative in nature (Tucker, 1997). Normally, these results are presented in terms of a summary written after each tested question, detailing the overall findings pertinent to that question (see Willis, 1994), or as overall written summaries of each interview (see DeMaio et al., 1993). However, qualitative written summaries have been difficult to analyze or to present succinctly, and Tucker has suggested that these may reflect interviewer bias. Clearly, there are subjective elements to the qualitative results of cognitive interviewing, and whether or not this subjectivity constitutes bias, or is overall a positive or negative feature, are unresolved questions (Bercini, 1992; Willis, 1994). In any event, the relevant point is that whether or not qualitative descriptions are useful, they are only rarely accompanied by objective, quantified data.

9.2.1   Recommendations for Evaluation

A coding system which produces systematic quantitative summaries of interview outcomes would be very useful for purposes of evaluation. Although Sudman et al. (1996) have suggested that it may not be possible to produce a general scheme for coding the results of cognitive interviews, several researchers have made such attempts. Willis et al. (1989, 1991), Bolton (1993), Blixt and Dykema (1993), Presser and Blair (1994), and Conrad and Blair (1996) have proposed the use of coding systems that summarize the results of cognitive interviews, using a limited number of major categories of outcomes (generally between two and nine). For example, Conrad and Blair (1996) categorized the general problems found in survey questions as *lexical, temporal, logical, computational,* and *omission/inclusion,* whereas Willis et al. (1991) simply distinguished between *cognitive* and *logical* problems. We suggest that continued development of such schemes will greatly facilitate evaluation efforts. For example, Willis and colleagues have begun an experiment in which cognitive interviews on a single target questionnaire are conducted across two laboratories, and a common coding scheme is used to assess similarity of interviewing outcomes.

9.3 THE RELEVANCE OF COGNITIVE INTERVIEWING TO "PROBLEMS" IN SURVEY QUESTIONS

In the discussion above, we have dealt mainly with preliminary issues related to the description of cognitive interviewing, in terms of both process and product. The next sections of the chapter consider what are perhaps the key issues of interest to many survey researchers: Does the cognitive interview serve in its intended capacity as a means for finding flawed survey questions? Further, does the use of these techniques lead to any measurable improvement in these questions?

Finding Flawed Questions: Verification versus Discovery To determine the efficacy of cognitive interviewing in detecting flawed questions, we must first specify the exact means by which we expect this detection to occur. Even though the purpose of cognitive testing is widely accepted to be "finding problems," the meaning of this statement is itself somewhat vague. This phrase could mean either that, upon testing: (a) Problems that are suspected to exist are "found" (*verification*), or (b) problems that are unanticipated are "found" (*discovery*) (Willis and Schechter, 1997). For example, it might be anticipated that parents will be unable to recall from memory detailed information concerning their children's immunization history, and cognitive testing may be used to provide *verification* of this expectation. Alternately, it may also be *discovered,* although not anticipated, that memory for certain shots is particularly poor.

The distinction between verification and discovery is a basic one, because the intended purpose of the cognitive interview, with respect to these, has implications concerning whether such interviewing is deemed a worthwhile practice. We might believe that verification of problems already anticipated is useful, as this reinforces our expectations. Alternately, such a finding might be criticized as failing to provide any additional diagnostic information. Discovery, rather than verification, might therefore be viewed as the desirable testing outcome.

Improving Questions: Finding versus Fixing   To further complicate matters, several authors have pointed out that a critical function of cognitive interviewing is not simply to find (to either verify or discover) problems, but additionally, to suggest ways in which the problems detected may be rectified (DeMaio et al., 1993; Schechter and Herrmann, 1997; Willis and Schechter, 1997). For example, one may determine through cognitive testing that the phrase "on layoff from work" is not well understood. Addressing the problem then requires the determination of a substitute expression that captures the intended meaning, but is more globally comprehended. One might argue that a thorough evaluation should assess the value of cognitive interviewing in not simply detecting (*finding*) such problems, but additionally in leading to new questions that represent improvements (*fixing*). Thus, this view asserts that the ultimate result of cognitive interviewing should be a questionnaire draft that can be demonstrated to be in some manner superior to its initial form.

It is, however, difficult to provide such a demonstration. To the extent that *finding* and *fixing* problems are logically separate activities (possibly carried out by different individuals), it is difficult to conclude that a changed questionnaire is necessarily a product of the conduct of cognitive interviewing (i.e., that a change can be attributed to a specific practice such as the use of question paraphrasing). Rather, it could be that increased attention to the questions by qualified questionnaire designers, as part of the testing and development process, may be responsible for any measurable improvements in quality (Blair and Presser, 1993). The observation that an improved questionnaire exists, subsequent to cognitive interviewing and redrafting, is therefore not sufficient for demonstrating that cognitive interviewing was itself instrumental in bringing about this improvement.

9.3.1   Recommendations for Evaluation

First, based on extensive experience with cognitive techniques, we believe that both verification and discovery are important products of cognitive interviewing, and that each is worthy of systematic study as basic outcomes. However, when conducting an evaluation, one must determine at the outset which of these constitutes the focus. To date, most evaluations of cognitive interviewing have focused on discovery, as opposed to verification (Bolton, 1993; Fowler and Roman, 1992; Presser and Blair, 1994; Willis and Schechter, 1997). We propose that this may be too limiting a viewpoint, given that a substantial contribution of cognitive interviewing techniques may be the process of verifying (or, just as important, failing to verify) problems that are anticipated based on other means, such as a prior "expert" review of the questionnaire.

Likewise, we argue that evaluators must decide whether their focus will be on finding or fixing problems. Authors of cognitive interviewing studies often presume that questions that are flawed are to be fixed. However, *finding* rather than fixing problems may be emphasized for the more general purpose of simply contributing to our understanding of the survey response process. In particular, if one conducts cognitive interviewing in order to *measure* potential response error rather than as an attempt to *reduce* this error, as distinguished by Groves (1989) and by Esposito and Rothgeb (1997), then the issue of whether questions are fixed is somewhat irrelevant. In fact, several cognitive studies have been devoted to the understanding of sources of measurement error associated with particular survey questions, in order to inform the analysis of these questions (Beatty et al., 1996; Esposito and Rothgeb, 1997). From this viewpoint, the fact that a questionnaire that has been cognitively tested is *not* shown to have been "improved" does not necessarily indicate that cognitive interviewing is devoid of value.

9.4 ARE SURVEY QUESTIONS REALLY IMPROVED? MEASUREMENT OF QUESTIONNAIRE DATA QUALITY

Ultimately, in order to assess the value of cognitive interviewing (or any pretesting method) as a method for either finding or fixing problems in survey questions, the researcher must be able to adequately measure the key dependent variable of survey question quality. In other words, we must have some independent means for determining whether problems exist in these questions. Even if some of the limitations discussed above are addressed, and the products of cognitive interviewing are well-defined, well-documented, and quantifiable, it is not necessarily clear that the "problems" identified necessarily represent defects within the relevant domain (the field environment). In order to conclude this, we must also know whether questions which are identified as problematic through cognitive testing actually produce poor quality data in the field (Groves, 1996). However, the single most serious obstacle to the evaluation of cognitive interviewing techniques appears to be the scarcity of independent measures of survey question quality.

9.4.1 Recommendations for Evaluation

Evaluation of the validity of cognitive interviewing must ultimately rest on the active development of criterion measures of survey question quality (or conversely, of question problems) other than those typically obtained by simply carrying out the interviewing process and then recording the apparent results. The challenge of validating a method under conditions in which the critical criterion measure is itself difficult to assess is by no means unique to survey methodology. We believe it may be useful for survey methodologists to take cues from validation studies in the psychological literature, especially in the area of mental testing. Specifically, we consider means for evaluating cognitive interviewing that are based on three forms of evaluation used to assess psychological test instruments: (a) content validation, (b) criterion validation, and (c) construct validation (Anastasi, 1976).

Content Validation   The establishment of content validity relies on a model or theory concerning exactly what the to-be-evaluated test or method intends to measure, and determines whether the method in fact covers the appropriate domains. For example, a candidate test of mathematical ability can be examined to determine if the test items adequately represent the relevant components that are believed to comprise "mathematic aptitude." As applied to the evaluation of cognitive techniques, we can similarly examine the characteristic features of the methods we use, to determine whether these map onto the cognitive processes believed to be important to the survey response process.

Forsyth and Lessler (1991) proposed an approach that implicitly makes use of content validation. They conducted an in-depth analysis of both the specific features of cognitive pretesting methods and the features of survey questions that involve cognitive processing, to determine whether the correspondence between these is sufficient for concluding that cognitive methods are likely efficacious. Their overall conclusion was that different cognitive methods, such as use of question paraphrasing, response latency measurement, and rating tasks, appropriately tap different aspects of the survey response process. Further, their analysis gives rise to several hypotheses about the types of survey questions that will be most amenable to different forms of cognitive pretesting. We propose that the role of cognitive theory is critical in developing such hypotheses. However, few survey researchers who have evaluated cognitive techniques have adopted such a theoretical orientation, beyond simply making reference to the "four-stage" cognitive model proposed by Tourangeau (1984). Further reliance on a theory that specifically relates cognitive techniques to survey questions will give rise to testable hypotheses of the type suggested by the Forsyth and Lessler model.

Criterion Validation   A second basic methodological approach to validation is to obtain a direct criterion measure by which to assess the utility of a test or method. In this vein, survey question quality can be assessed through record check studies in which records are used as a "gold standard," and compared to respondent reports related to the same phenomena. Therefore, it might be worthwhile to conduct validation studies by developing questionnaires through the use of cognitive interviewing techniques, and then assessing the quality of the final questions by computing both measures of bias and variability.

Admittedly, record check studies tend to be expensive, are often impractical, and do not generally apply to knowledge and attitude questions (as these cannot normally be verified through records). In addition, records can themselves contain error, or may simply not exist for many categories of autobiographical questions (Davis and DeMaio, 1993; Martin and Polivka, 1995; Turner and Martin, 1984). For this reason, few studies have been done which assess the efficacy of cognitive interviewing in reducing either net or gross measurement error, using records as a basis.

However, acceptable substitutes to records may sometimes exist. Tucker (1992) and Silberstein (1989, 1991) have developed indirect measures of survey response error in consumer expenditure surveys. Further, Smith (1991) effectively created record data during the course of a study on dietary recall, using diary-keeping procedures in which subjects maintained behavioral logs related to food intake. These dietary diaries were collected prior to questionnaire administration, and were used to develop criterion measures in assessing the quality of survey questions requiring recall of dietary behavior. More attention to such synthetic indexes of data quality would represent an enhancement of our evaluation designs.

An alternative to the use of records to measure absolute quality of survey questions involves the demonstration that these questions are reliable; that is, when administered repeatedly to the same respondents, they tend to produce identical data. Relying on reliability as a criterion measure for evaluating cognitive techniques, one might attempt to demonstrate that questions developed through cognitive methods display adequate reliability, or even increased reliability, relative to a definable reference standard. Friedenreich, Courneya, and Bryant (1997) used the cognitive methods of Willis (1994) to develop a physical exercise questionnaire, and then pretested that instrument through reinterview (presentation of the questions to the same respondent at two points in time), in order to assess measurement reliability. They did obtain relatively high agreement between these two points, although without use of a control group or comparison pretesting method. Similarly, Hess and Singer (1996) utilized measures of reliability as assessed through reinterview to determine the effectiveness of behavior coding techniques in identifying problematic questions. We note that such a reinterview paradigm could be applied more generally to evaluate cognitive interviewing outcomes as well.

Construct Validation   Consistent with a tradition deriving from the area of instrument development in psychological testing, some researchers have not focused on an absolute criterion measure of data quality, but instead have determined whether findings from the cognitive laboratory are consistent with other measures purported to be related to the critical construct of "question quality." One approach is to study subject comprehension of critical terms or overall question intent. For example, in studies of labor force concepts, Campanelli, Martin, and Creighton (1989), and Martin and Polivka (1995) developed vignettes describing hypothetical individuals and work situations to study the degree of overlap between respondent interpretation and official survey definitions. Although not incorporating a direct measure of data quality, the level of correspondence between respondent interpretation and question intent is viewed as a measure likely to be related to data quality. A useful design, then, might be to determine whether a measure of misunderstanding of key terms is found to be reduced through the application of cognitive interviewing techniques.

A second, more recent development that may also be effective is the use of measures such as respondent reaction time as a critical indicator of "bad" questions (Bassili and Scott, 1996). For example, if cognitive interviewing is found to be useful for "flagging" questions that have already been identified as problematic, based on the finding that significant time tends to pass between question presentation and the subject's response to the question, one may have increased confidence in the use of these methods.

A final type of construct validation, applied by Lessler, Tourangeau, and Salter (1989), Davis and DeMaio (1993), and Willis and Schechter (1997) involves a determination of whether subject behavior related to particular survey questions, in the context of cognitive testing, parallels behavior of survey respondents in a field environment. Advocates of this approach assume that if subject behavior in the cognitive laboratory parallels behavior in the field, then one can conclude that the cognitive processes used by subjects in the lab must bear some resemblance to those used in the field environment.

For example, a preliminary study by Davis and DeMaio (1993) determined that dietary recall behavior was similar under laboratory and simulated field conditions, with similar numbers of food items reported as consumed in each. A recent study by Willis and Schechter (1997) further examined the degree of overlap between laboratory and field environments. Based on laboratory testing, they made several hypotheses concerning the effects that variation in question wording would have in a field environment, and found, through the use of split-ballot experiments, that respondents behaved as predicted. In particular, field respondents were as likely as laboratory subjects to produce guesses for an "impossible" knowledge question, to ignore critical parts of survey questions, and to give behavioral reports that varied greatly with modification of subtle biasing characteristics in question wording.

Similarly, Schechter, Beatty, and Willis (1997) found that the relatively high level of difficulty exhibited by elderly lab subjects when asked quantitative behavior frequency items was reflected in inflated missing data rates for elderly respondents in a field survey. Further experiments of this type which assess carry-over between lab and field situations are vital. In general, validation may depend largely on designs that make use of correlates of data quality, as opposed to the direct measurement of response error involving absolute criterion measures.

9.5 COMPARISON OF COGNITIVE INTERVIEWING WITH OTHER PRETESTING METHODS

Researchers who conduct questionnaire pretesting have developed or adapted several pretesting methods, such as expert review and behavior coding, as alternatives to cognitive interviewing (DeMaio, 1983; DeMaio and Rothgeb, 1996; Forsyth and Lessler, 1991). From a practical standpoint, the value of cognitive interviewing must ultimately be assessed with respect to these alternate techniques. Even if it can be shown that cognitive interviewing produces an unequivocal increase in question quality, it is not clear that this approach is superior.

Experiments emphasizing comparisons between pretesting methods, and in particular, expert review, behavior coding, and cognitive interviewing, have been conducted by Presser and Blair (1994), Fowler and Roman (1992), and Esposito and Rothgeb (1997). Such investigations illustrate a thicket of methodological complications. A critical issue in comparing techniques is the problem of developing an index of question quality that can be used as a common measure by which to assess the relative value of each method. For example, Presser and Blair (1994) relied mainly on the number of problems with survey questions identified by each pretesting technique as a critical dependent variable, and detected fewer problems with cognitive interviewing than with several other methods.

However, as discussed previously, it is difficult to determine whether, in the absence of independent measures of question quality, problems detected through the use of any pretesting method are "real." The use of number of problems found is of clear interest, but, in a comparison of methods, provides little basis for determining which method is superior. That is, this measure emphasizes sensitivity but not specificity; there is no control for false positive outcomes, in which problems that are identified through pretesting simply do not exist in the field environment (Willis and Schechter, 1997). Further, even nominally objective techniques such as behavior coding do not produce a single value representing the number of problems identified, as this value depends on (often subjective and arbitrary) variables such as the numerical criterion set for identification of a problem, and the level of aggregation of problem types used (Esposito et al., 1994).

Beyond developing a meaningful dependent variable, a second basic complication in comparing pretesting methods is that such comparisons must equalize the total amount of resources devoted to each, so that obvious confounds related to "level of effort" are avoided. However, even if one attempts to control (or at least measure) the resources applied, it is still unclear whether a comparison between methods constitutes a "fair test" that both satisfies the requirements of experimental science and simultaneously represents actual pretesting practice. Clearly, each technique must be given a chance to provide its "best shot" in the competition.

For example, in their comparative study of pretesting techniques, Presser and Blair (1994) used cognitive interviewers who were relatively inexperienced, and the results were therefore not necessarily representative of cognitive interviewing in general. As the authors point out, any comparative study may have very limited generality, as the procedural parameters represented in the study may not adequately reflect the levels that are employed in everyday usage.

Additionally, to constitute an equitable test, comparisons between methods must be careful to take into account such variables as the nature of the sample selected, and the way this influences the effective sample size for purposes of pretesting. Many questionnaires focus on specific subpopulations (e.g., people who use assistive devices such as canes, wheelchairs, and walkers), and critical questions may be asked of only a small number of respondents. As such, pretesting methods may vary greatly in the degree to which they effectively target the appropriate population, and so effectiveness may depend on this factor, rather than on other inherent features of the method. It may even be that the efficacy of any pretesting procedure is not absolute, but depends critically on the type (and specificity) of the population targeted.

Another issue related to "fairness" of method comparisons can be illustrated by referring to the practice of iteration. One cited advantage of cognitive interviewing techniques is that they can be conducted iteratively, so that cycles of testing and modification can be done relatively quickly, based on small samples (DeMaio et al., 1993; Royston, 1989; Willis, 1994). If one were to conduct a comparative study which restricted cognitive interviewing to a single round of testing, in order to simplify the design and to render the procedure comparable to another pretesting technique, a proponent of the iterative approach could rightly argue that the experimental design in effect cripples the procedure by introducing unrealistic constraints. On the other hand, allowing an iterative cognitive interviewing procedure to be compared with only one round of behavior coding or expert review also constitutes an inappropriate design, because of the inherent design confound.

The general point demonstrated by the above examples is that key beneficial features regularly viewed as associated with cognitive interviewing (or with any other pretesting procedure) may not actually constitute core features of that technique alone. For example, Converse and Presser (1986) have advocated the general use of small, iterative rounds of pretesting, independent of any mention of cognitive interviewing, and the use of iterative cycling for methods such as expert review has been reported (Schechter, Beatty, and Block, 1994). It may be then, that a good deal of the efficacy of cognitive interviewing derives not from properties that are inherent in that procedure, but from features that simply represent *good pretesting design in general*. Teasing out these factors involves an analysis of those features of cognitive interviewing that are unique to this technique, and are not applicable as well to other pretesting techniques. To date, CASM researchers have not sufficiently considered these complex, multivariate issues in contrasting pretesting methods.

A final view concerning the comparison of pretesting methods challenges the assumption that these should be "stacked up" against one another in the first place. Such comparisons may even be viewed as pointless, to the extent that questionnaire designers are interested in making use of multiple methods, rather than choosing between them (for example, see the discussion of the "relative confidence model" by Esposito and Rothgeb, 1997). In particular, a consideration of the distinction made earlier between discovery and verification suggests that cognitive interviewing may serve as a supplement to, rather than a competitor of, methods such as expert evaluation, and that (a) these methods may illustrate different strengths, and (b) they may be best applied at different points in the questionnaire development process (DeMaio et al., 1993; Willis, Trunzo, and Stussman, 1992).

Campanelli (1997) summarizes a number of potential strengths and weaknesses of various pretesting methods. For example, behavior coding may be more reliable than cognitive interviewing, but less broad in the types of problems it uncovers. Further, Campanelli notes that an emphasis on cost suggests that expert analysis is particularly efficient, followed by cognitive interviewing, and then behavior coding.

Recommendations for Evaluation   Following Bassili and Scott (1996), we propose that it may be possible to obtain measures of question quality useful for purposes of pretesting methods comparison. In particular, one might effectively "load the dice," and assess the relative usefulness of pretesting techniques in uncovering problematic questions that have been purposely "implanted" in tested questionnaires. A clear challenge here is to select target survey questions that are known to exhibit problems, but that are not so obvious as to render the exercise hopelessly artificial in nature.

In general, we advocate studies that compare the relative effectiveness of various pretesting techniques, as well as those that view them as mutually supportive. However, researchers must be clear about which set of underlying assumptions they are making about the relationships between different techniques, given that this drives both evaluation practice and the interpretation of results. A particularly vital issue is again the relative importance of finding versus fixing problems. It is obviously simpler to assess the effectiveness of competing pretesting techniques in identifying problems than to assess their contributions toward solving those problems, and therefore in improving data quality. However, the ultimate usefulness of any pretesting technique derives not only from its function in identifying (finding) problems, but also in providing a strong base for question modification (fixing).

To illustrate this point, one oft-noted limitation of behavior coding is that, although it is useful for identifying problems, it may be of limited value for the purpose of determining *why* a problem occurred, and therefore knowing how to fix it (Oksenberg, Cannell, and Kalton, 1991). On the other hand, a basic premise of cognitive interviewing is that the understanding of the cognitive basis for the problem detected may point to specific modifications that will serve to alleviate that problem (DeMaio et al., 1993; Sirken and Herrmann, 1996). As such, evaluations that compare pretesting methods should take into account not only the relative efficiencies of different techniques in detecting problems, but as well the richness of information obtained concerning the nature of the problems detected, and the way this information in turn contributes to improvements in the tested questions.

9.6   CONCLUSIONS

We have examined several key issues related to the evaluation of cognitive interviewing techniques, and have made suggestions concerning a number of promising approaches. Clearly, given the challenges to evaluation that we have identified, and the fundamentally divergent viewpoints that reflect different researchers' philosophies concerning the nature of evaluation, a program to assess the validity of cognitive interviewing will not derive from a single experiment, or even one experimental approach. Rather, it is likely that a useful picture will be obtained only as a result of a number of investigations that tackle different aspects of the problem.

Development of General Cognitive Principles

To conclude, we once again consider the overall purpose of cognitive testing. Referring to a distinction made at the beginning of the chapter between *pretesting* and the general study of cognitive *process*, we propose that practitioners of cognitive techniques broaden their views of the purposes of their investigations. Even if cognitive interviewing techniques are found to be valid and indispensable tools for purposes of pretesting questions, are we really limiting ourselves to the function of attacking each new questionnaire from scratch, from a purely empirical viewpoint? Or, as was hoped at the time that CASM was conceived, is it possible to combine the findings of a number of applied studies, in order to generate general hypotheses that apply over a range of survey questions, and that contribute to the development of useful cognitive principles of questionnaire design?

We suggest that such a contribution is possible, and strongly advocate that, based on almost fifteen years of experience with cognitive techniques, practitioners take time to step back and consider several broader questions: In particular—what have we learned in *general* about questionnaire design, based on the thousands of cognitive interviews that have been conducted, that can be used to inform the crafting of survey questions? Further—have we learned anything about the nature of cognition, as this relates to more generally applicable theories of comprehension, memory, reasoning, and judgment? Ultimately, we must work to diminish the chasm that has developed between the "assembly-line" process of pretesting questionnaires, and the more general search for understanding of the cognitive processes invoked by respondents who are presented with the rather daunting task of providing meaningful answers to common survey questions.

ACKNOWLEDGMENTS

At the time this chapter was written Gordon Willis was at the National Center for Health Statistics, Centers for Disease Control and Prevention (CDC), and Brian Harris-Kojetin was at the U.S. Bureau of Labor Statistics (BLS). The views expressed are those of the authors and do not represent official policy of CDC, BLS, or the Census Bureau.

REFERENCES

Anastasi, A. (1976). *Psychological Testing*. New York: Macmillan.

Bassili, J. N., and Scott, B. S. (1996). Response latency as a signal to question problems in survey research. *Public Opinion Quarterly, 60,* 390–399.

Beatty, P., Schechter, S., and Whitaker, K. (1996). Evaluating subjective health questions: Cognitive and methodological investigations. *Proceedings of the Section on Survey Research Methods, American Statistical Association*, 956–961.

Beatty, P., Willis, G. B., and Schechter, S. (1997). Evaluating the generalizability of cognitive interview findings. In *Office of Management and Budget Seminar on Statistical Methodology in the Public Service, Statistical Policy Working Paper 26*, pp. 353–362. Washington, DC: Statistical Policy Office.

Belson, W. A. (1981). *The Design and Understanding of Survey Questions*. Aldershot, England: Gower.

Bercini, D. H. (1992). Pretesting questionnaires in the laboratory: An alternative approach. *Journal of Exposure Analysis and Environmental Epidemiology, 2*, 241–248.

Bickart, B., and Felcher, E. M. (1996). Expanding and enhancing the use of verbal protocols in survey research. In N. Schwarz and S. Sudman (Eds.), *Answering Questions: Methodology for Determining Cognitive and Communicative Processes in Survey Research*, pp. 115–142. San Francisco: Jossey-Bass.

Biehal, G., and Chakravarti, D. (1989). The effects of concurrent verbalization on choice processing. *Journal of Marketing Research, 26*, 84–96.

Blair, J., and Presser, S. (1993). Survey procedures for conducting cognitive interviews to pretest questionnaires: A review of theory and practice. *Proceedings of the Section on Survey Research Methods, American Statistical Association*, 370–375.

Blixt, S., and Dykema, J. (1993). Before the pretest: Question development strategies. *Paper presented at the annual meeting of the American Association for Public Opinion Research*, St. Charles, IL.

Bolton, R. N. (1993). Pretesting questionnaires: Content analyses of respondents' concurrent verbal protocols. *Marketing Science, 12*, 280–303.

Bolton, R. N., and Bronkhorst, T. M. (1996). Questionnaire pretesting: Computer-assisted coding of concurrent protocols. In N. Schwarz and S. Sudman (Eds.), *Answering Questions: Methodology for Determining Cognitive and Communicative Processes in Survey Research*, pp. 37–64. San Francisco: Jossey-Bass.

Campanelli, P. (1997). Testing survey questions: New directions in cognitive interviewing. *Bulletin de Methodologie Sociologique, 55*, 5–17.

Campanelli, P., Martin, E., and Creighton, K. (1989). Respondents' understanding of labor force concepts: Insights from debriefing studies. *Proceedings of the Fifth Annual Census Bureau Research Conference*, 361–374. Washington, DC: U.S. Bureau of the Census.

Cannell, C. F., Fowler, F. J., and Marquis, K. (1968). The influence of interviewer and respondent psychological and behavioral variables on the reporting in household interviews. *Vital and Health Statistics*, Series 2, No. 26. Washington, DC: U.S. Government Printing Office.

Cantril, H. (1944). *Gauging Public Opinion*. Princeton, NJ: Princeton University Press.

Conrad, F., and Blair, J. (1996). From impressions to data: Increasing the objectivity of cognitive interviews. *Proceedings of the Section on Survey Research Methods, American Statistical Association*, 1–9.

Converse, J. M., and Presser, S. (1986). *Survey Questions: Handcrafting the Standard Questionnaire*. Beverly Hills: Sage.

Davis, W. L., and DeMaio, T. J. (1993). Comparing the think-aloud interviewing technique with standard interviewing in the redesign of a dietary recall questionnaire. *Proceedings of the Section on Survey Research Methods, American Statistical Association*, 565–570.

DeMaio, T. J. (Ed.). (1983). *Approaches to developing questionnaires*. Statistical Policy Working Paper No. 10. Washington, DC: Office of Management and Budget.

DeMaio, T., Mathiowetz, N., Rothgeb, J., Beach, M. E., and Durant, S. (1993). *Protocol for pretesting demographic surveys at the Census Bureau*. Unpublished manuscript, Center for Survey Methods Research, U.S. Bureau of the Census.

DeMaio, T. J., and Rothgeb, J. M. (1996). Cognitive interviewing techniques: In the lab and in the field. In N. Schwarz and S. Sudman (Eds.), *Answering Questions: Methodology for Determining Cognitive and Communicative Processes in Survey Research*, pp. 177–195. San Francisco: Jossey-Bass.

Ericsson, K. A., and Simon, H. A. (1980). Verbal reports as data. *Psychological Review, 87*, 215–251.

Ericsson, K. A., and Simon, H. A. (1984). *Protocol analysis: Verbal reports as data*. Cambridge, MA: MIT Press.

Esposito, J. L., and Rothgeb, J. M. (1997). Evaluating survey data: Making the transition from pretesting to quality assessment. In L. Lyberg, P. Biemer, M. Collins, E. de Leeuw, C. Dippo, N. Schwarz, and D. Trewin, *Survey Measurement and Process Quality*, pp. 541–571. New York: Wiley.

Esposito, J. L., Rothgeb, J. M., and Campanelli, P. C. (1994). *The utility and flexibility of behavior coding as a method for evaluating questionnaires*. Paper presented at the meeting of the American Association for Public Opinion Research, Danvers, MA.

Fisher, R. P., and Geiselman, R. E. (1992). *Memory-Enhancing Techniques for Investigative Interviewing: The Cognitive Interview*. Springfield, IL: Thomas.

Foddy, W. (1996). *An Empirical Evaluation of Probes Used for the In-Depth Testing of Survey Questions*. Unpublished manuscript, Monash University, Australia.

Forsyth, B. H., and Lessler, J. T. (1991). Cognitive laboratory methods: A taxonomy. In P. P. Biemer, R. M. Groves, L. E. Lyberg, N. A. Mathiowetz, and S. Sudman (Eds.), *Measurement Errors in Surveys*, pp. 393–418. New York: Wiley.

Fowler, F. J., and Cannell, C. F. (1996). Using behavioral coding to identify problems with survey questions. In N. Schwarz and S. Sudman (Eds.), *Answering Questions: Methodology for Determining Cognitive and Communicative Processes in Survey Research*, pp. 15–36. San Francisco: Jossey-Bass.

Fowler, F. J., and Roman, A. M. (1992). *A Study of Approaches to Survey Question Evaluation*. Unpublished manuscript, Center for Survey Research, University of Massachusetts, Boston.

Friedenreich, C. M., Courneya, K. S., and Bryant, H. E. (1997). *The Lifetime Total Physical Activity Questionnaire: Development and reliability*. Unpublished manuscript, Division of Epidemiology, Alberta Cancer Board, Canada.

Gerber, E. R., and Wellens, T. R. (1997). Perspectives on pretesting: "Cognition" in the cognitive interview? *Bulletin de Methodologie Sociologique, 55*, 18–39.

Groves, R. M. (1989). *Survey Errors and Survey Costs.* New York: Wiley.

Groves, R. M. (1996). How do we know what we think they think is really what they think? In N. Schwarz and S. Sudman (Eds.), *Answering Questions: Methodology for Determining Cognitive and Communicative Processes in Survey Research,* pp. 389–402. San Francisco: Jossey-Bass.

Hess, J. C., and Singer, E. (1996). Predicting test-retest reliability from behavior coding. *Proceedings of the Section on Survey Research Methods, American Statistical Association,* 1004–1009.

Jobe, J. B., and Herrmann, D. J. (1996). Implications of models of survey cognition for memory theory. In D. J. Herrmann, C. McEvoy, C. Herzog, P. Hertel, and M. K. Johnson (Eds.), *Basic and Applied Memory Research: Practical Applications: Vol. 2,* pp. 193–205. Mahwah, NJ: Erlbaum.

Jobe, J. B., Keller, D. M., and Smith, A. F. (1996). Cognitive techniques in interviewing older people. In N. Schwarz and S. Sudman (Eds.), *Answering Questions: Methodology for Determining Cognitive and Communicative Processes in Survey Research,* pp. 197–219. San Francisco: Jossey-Bass.

Jobe, J. B., and Mingay, D. J. (1991). Cognition and survey measurement: History and overview. *Applied Cognitive Psychology, 5,* 175–192.

Lessler, J., Tourangeau, R., and Salter, W. (1989). Questionnaire design in the cognitive research laboratory. *Vital and Health Statistics,* Series 6, No. 1 (DHHS Publication No. PHS 89-1076). Washington, DC: U.S. Government Printing Office.

Loftus, E. (1984). Protocol analysis of responses to survey recall questions. In T. Jabine, M. Straf, J. M. Tanur, and R. Tourangeau (Eds.), *Cognitive Aspects of Survey Methodology: Building a Bridge Between Disciplines,* pp. 61–64. Washington, DC: National Academy Press.

Martin, E., and Polivka, A. E. (1995). Diagnostics for redesigning survey questionnaires: Measuring work in the Current Population Survey. *Public Opinion Quarterly, 59,* 547–567.

Oksenberg, L., Cannell, C., and Kalton, G. (1991). New strategies for pretesting survey questions. *Journal of Official Statistics, 7,* 349–365.

Presser, S. and Blair, J. (1994). Survey pretesting: Do different methods produce different results? In P. V. Marsden (Ed.), *Sociological Methodology: Vol. 24,* pp. 73–104. Washington, DC: American Sociological Association.

Royston, P. N. (1989). Using intensive interviews to evaluate questions. In F. J. Fowler, Jr. (Ed.), *Health Survey Research Methods,* pp. 3–7 (DHHS Publication No. PHS 89-3447). Washington, DC: U.S. Government Printing Office.

Schechter, B., Beatty, P., and Block, A. (1994). Cognitive issues and methodological implications in the development and testing of a traffic safety questionnaire. *Proceedings of the Section on Survey Research Methods, American Statistical Association,* 1215–1219.

Schechter, S., Beatty, P., and Willis, G. (1997). *Recall and judgment issues in subjective health reports.* Paper presented at the Conference on Cognition, Aging, and Survey Measurement, Ann Arbor, MI.

Schechter, S., Blair, J., and Vande Hey, J. (1996). Conducting cognitive interviews to test self-administered and telephone surveys: Which methods should we use? *Proceedings of the Section on Survey Research Methods, American Statistical Association*, 10–17.

Schechter, S., and Herrmann, D. (1997). The proper use of self-report questions in effective measurement of health outcomes. *Evaluation and the Health Professions, 20*, 28–46.

Schwarz, N., and Bienias, J. (1990). What mediates the impact of response alternatives on frequency reports of mundane behaviors? *Applied Cognitive Psychology, 4*, 61–72.

Schwarz, N., Hippler, H. J., Deutsch, B., and Strack, F. (1985). Response scales: Effects of category range on reported behavior and comparative judgments. *Public Opinion Quarterly, 49*, 388–395.

Selltiz, C., Jahoda, M., Deutsch, M., and Cook, S. W. (1959). *Research Methods in Social Relations*. New York: Holt.

Silberstein, A. R. (1989). Recall effects in the U.S. Consumer Expenditure Interview Survey. *Journal of Official Statistics, 5*, 125–142.

Silberstein, A. R. (1991). Response performance in the Consumer Expenditure Diary Survey. *Proceedings of the Section on Survey Research Methods, American Statistical Association*, 338–343.

Sirken, M., and Herrmann, D. (1996). Relationships between cognitive psychology and survey research. *Proceedings of the Section on Survey Research Methods, American Statistical Association*, 245–249.

Smith, A. F. (1991). Cognitive processes in long-term dietary recall. *Vital and Health Statistics*, Series 6, No. 4 (DHHS Publication No. PHS 92-1079). Washington, DC: U.S. Government Printing Office.

Sudman, S., Bradburn, N. M., and Schwarz, N. (1996). *Thinking About Answers: The Application of Cognitive Processes to Survey Methodology*. San Francisco: Jossey-Bass.

Tanur, J. M., and Fienberg, S. E. (1992). Cognitive aspects of surveys: Yesterday, today, and tomorrow. *Journal of Official Statistics, 8*, 5–17.

Tourangeau, R. (1984). Cognitive sciences and survey methods. In T. Jabine, M. Straf, J. Tanur, and R. Tourangeau (Eds.), *Cognitive Aspects of Survey Methodology: Building a Bridge Between Disciplines*, pp. 73–100. Washington, DC: National Academy Press.

Tucker, C. (1992). The estimation of instrument effects on data quality in the Consumer Expenditure Diary Survey. *Journal of Official Statistics, 8*, 41–61.

Tucker, C. (1997). Measurement issues surrounding the use of cognitive methods in survey research. *Bulletin de Methodologie Sociologique, 55*, 67–92.

Turner, C. F., and Martin, E. (1984). *Surveying Subjective Phenomena*. New York: Russell Sage Foundation.

Von Thurn, D. R., and Moore, J. C. (1994). Results from a cognitive exploration of the 1993 American Housing Survey. *Proceedings of the Section on Survey Research Methods, American Statistical Association*, 1210–1214.

Willis, G. B. (1994). *Cognitive interviewing and questionnaire design: A training manual.* Working Paper No 7. National Center for Health Statistics. Hyattsville, MD.

Willis, G. B., Royston, P., and Bercini, D. (1989). Problems with survey questions revealed by cognitively-based interviewing techniques. *Proceedings of the Fifth Annual Census Bureau Research Conference*, 345–360. Washington, DC: U.S. Bureau of the Census.

Willis, G. B., Royston, P., and Bercini, D. (1991). The use of verbal report methods in the development and testing of survey questionnaires. *Applied Cognitive Psychology, 5*, 251–267.

Willis, G. B., and Schechter, S. (1997). Evaluation of cognitive interviewing techniques: Do the results generalize to the field? *Bulletin de Methodologie Sociologique, 55*, 40–66.

Willis, G. B., Trunzo, D. B., and Stussman, B. J. (1992). The use of novel pretesting techniques in the development of survey questionnaires. *Proceedings of the Section on Survey Research Methods, American Statistical Association*, 824–828.

CHAPTER 10

Income Reporting in Surveys: Cognitive Issues and Measurement Error

Jeffrey C. Moore
U.S. Bureau of the Census

Linda L. Stinson
U.S. Bureau of Labor Statistics

Edward J. Welniak, Jr.
U.S. Bureau of the Census

10.1 INTRODUCTION AND OVERVIEW

Because income data are germane to a wide array of important policy issues, income questions are almost a constant in government-sponsored surveys. This chapter briefly reviews research on cognitive factors affecting income reporting and research concerning the actual quality of survey reports of income. Its intent is to link knowledge about cognitive factors specific to the reporting of income with knowledge regarding the magnitude and nature of the errors made by respondents in response to income questions.

We begin with a review of the cognitive investigations, and find many possible cognitive contributors to inaccurate income reporting, including definitional issues, recall and salience problems, confusion, and sensitivity. These factors are potentially the source of substantial errors in income reports, although it is not always obvious that they would produce errors in predominantly one or the other direction.

Cognition and Survey Research, Edited by Monroe G. Sirken, Douglas J. Herrmann, Susan Schechter, Norbert Schwarz, Judith M. Tanur, and Roger Tourangeau.
ISBN 0-471-24138-5   © 1999 John Wiley & Sons, Inc.

Next, we turn to the quality of survey income data, beginning with comparisons of survey estimates of income to independent "benchmarks." This research shows consistent shortfalls in the survey estimates across a wide variety of income types, although the nature of this research does not justify the conclusion that the under*estimates* are due to survey respondents' under*reports*. We also look at income nonresponse in this section, a clear indicator of reporting problems.

Last, we present a summary of research, primarily record check studies, which offer direct evidence concerning the quality of respondents' reports, and which confirms a general tendency toward income amount underreporting. This research also suggests, however, that the underreporting of income sources, and random error in income amount reports, may be more distinguishing features of the quality of income reports than the underreporting of income amounts.

10.2 COGNITIVE FACTORS IN INCOME REPORTING

In keeping with common practice (e.g., Tourangeau, 1984), we distinguish several stages in the cognitive "work" necessary to answer a survey question, which we label understanding, retrieval, and response production.

10.2.1 Understanding Income Concepts and Terms

Income survey design presents many challenges, not the least of which is the most basic step of the process: defining income constructs for survey respondents in clear and easily understood language. Certainly, analysts' technical definitions are unlikely to be well understood by survey respondents. For example, Dippo and Norwood (1992) show a wide range of interpretations of "nonwage cash payments," from the inclusion of noncash in-kind payments to the exclusion of even cashable checks. Another example of a technical definition is the fact that "income" is intended to include both gains and losses. However, owners of real estate and other income-generating property, for whom losses (e.g., failure to rent property, stock fluctuations, etc.) are commonplace, often fail to include such losses in their property income reports (Cantor, Brandt, and Green, 1991).

Other research shows, however, that even commonplace terms elicit variable understandings. Stinson (1997) asked respondents what they included in their "total family income." Some did not include their own or a spouse's part-time wages, or interest income, because the amounts were not significant to the family's income. Others excluded income not available to the immediate family, such as wages kept for individual use or handed over to extended family.

Bogen (1995) reports that idiosyncratic definitions of "income" cause respondents to omit sporadic self-employment, third or fourth jobs, and "odd jobs" such as occasional baby sitting. (Marquis, Duan, Marquis, and Polich (1981) also conclude that such jobs are underreported, based on their observation of a modest underreporting bias for annual earnings from all sources, in contrast to an absence of bias in reports of earnings from specific employers.) Moore, Marquis, and Bogen (1996) find that respondents' restricted (and perhaps technically correct) interpretations appear to have prevented many Food Stamps recipients from reporting their Food Stamps in response to an "income, pay, and other money" cue. Other research has demonstrated confusion over the distinction between income "earned" and "received" during a specified period (Marquis, 1990; Cantor et al., 1991; Bogen and Robinson, 1995).

10.2.2 Retrieval Issues

Assuming that income questions effectively convey their intended meanings to respondents, respondents next must retrieve the appropriate income information from memory. This process can introduce error into survey reports in a variety of ways.

Lack of Knowledge Respondents often simply do not have the necessary information stored in memory, resulting in "don't know" response and error-prone guessing. Cantor et al. (1991) note that those who do not use records, particularly when reporting asset income, often appear to be guessing almost blindly about asset types and amounts because the information is simply absent from memory.

The primary survey design solution to respondents' lack of knowledge is to bypass memory altogether through the use of records. Maynes (1968) reports an experiment involving known savings account holders, some of whom were instructed to consult their records to report account balances, while others were instructed *not* to consult records. Both treatments produced essentially unbiased reports, but there were striking differences in the extent of random error—85 percent of record users reported their account balances within ±1 percent of the true value, versus only 49 percent among those instructed not to use records.

Grondin and Michaud (1994) find major reductions in income amount reporting errors among respondents who used a premailed notebook or referred to tax forms, compared to those who answered the interview questions unaided. "Assisted" income reports were from two to seven times more likely to be accurate (within ±5 percent of the true value) than unassisted reports, and the memory aids reduced "don't know" responses to nearly zero. Moore and colleagues' (1996) experimental interview procedures using the Survey of Income and Program Participation (SIPP) achieved high levels of record use, and significantly improved income amount reports, although perhaps at the cost of higher survey nonresponse and higher interviewing costs.

An important potential drawback of record use is interviewers' lack of enthusiasm for such procedures. Interviewers have been observed to overtly discourage record use (Marquis, 1990), although Moore et al. (1996) report that interviewers implemented procedures emphasizing record use quite well. Nor are records always helpful. Marquis (1990) reports problems such as the frequent disorder of respondents' records and undecipherable record information. However, most of Marquis' respondents reported a willingness to use records, had interviewers requested them. Reasons for not using records included (a) difficulty finding them, (b) the ease of educated guessing about income amounts, and (c) the unavailability of records for some income types.

Low Salience of Income Information   Even when knowledge about income is present in memory, Moore et al. (1996) demonstrate that some income types might be overlooked in a cursory recall effort. Their experimental "free recall" interview procedures, without specific probes for specific types of income, were sufficient for eliciting almost all job, pension, and major transfer income source reports, but not asset income sources. In fact, only about one-quarter to one-third of all asset income sources were reported in response to free recall procedures. Most assets required specific "recognition" probes before respondents reported them.

Error-Prone Reconstruction Strategies   When survey questions require information not directly retrievable from memory, respondents must "reconstruct" that information. Such reconstruction strategies are variably successful. Cantor et al. (1991) find that very few respondents who are asked to report monthly earnings attempt direct retrieval. Most use one of three reconstruction strategies: (a) recalling paycheck amounts and calculating the number of paychecks received (the most common strategy), (b) estimating hours worked and multiplying by the hourly wage, or (c) dividing annual salary by twelve. Marquis (1990) describes the same strategies and several others, including: (a) identifying "average" pay and multiplying that by "typical" hours worked, (b) recalling the paycheck amount, rounding to hundreds, and translating into "monthly" estimates, (c) identifying an "average" pay amount and then adjusting up or down for each month, (d) recalling exact amounts for recent months and factors that would raise or lower those amounts for more distant months, and (e) recalling weekly amounts and multiplying by four. Bogen (1995) provides additional examples of income amount reporting strategies that result in inexact estimates, including: (a) reporting weekly pay, but failing to include extra paydays, (b) severe rounding of reported dollar amounts, and (c) reporting "average" amounts.

Confusion   Confusion among income category labels is common. This can be especially troubling for a detailed accounting of asset income, and in a number of different studies, Vaughan and colleagues (Goodreau, Oberheu, and Vaughan, 1984; Klein and Vaughan, 1980; Vaughan, 1978; Vaughan, Lininger, and Klein, 1983) document the tendency of respondents to confuse transfer program names. Marquis and Moore (1990) find that misclassification of Aid to Families with Dependent Children (AFDC) as "general welfare" causes a significant proportion of AFDC response errors. Other investigations have found that respondents have great difficulty recognizing and distinguishing official program names such as Medicaid, Supplemental Security Income (SSI), and Worker's Compensation (Bogen, 1994; Cantor et al., 1991).

Other Recall Issues One demonstrated correlate of transfer income reporting problems is brief "spells" of receipt. Livingston's (1969) investigation of reports of "public assistance" receipt finds that underreporting is associated with both a small total benefit amount and fewer months of receipt. Hu's (1971) examination of "welfare family" respondents finds that failure to report is associated with younger household heads and higher household income, and, therefore, less time receiving welfare. Goodreau et al. (1984) also find that "part-period" AFDC recipients are three times more likely to underreport their AFDC participation than those who receive benefits throughout the entire survey reference period.

Some evidence suggests that recall period length contributes to faulty recall. Withey (1954) compares "current income" reports in one year with "last year's income" collected one year later and finds only 61 percent agreement between the report pairs. On the other hand, Marquis and Moore (1990) compare reports about "last month" with reports about "four months ago" and find almost no evidence of increased failure to report transfer program participation with the passage of time and no consistent increases in error rates for the more distant reference month.

Proposed solutions to income report recall problems include the suggestion to use "last calendar year" as the specified time period, and to collect the data at tax return time (Sudman and Bradburn, 1982). Another strategy tries to avoid income fluctuations caused by holidays, vacations, illness, overtime, etc., by suggesting that respondents be asked only to report the most recent pay period (Rodgers, Brown, and Duncan, 1993). A third approach would lessen the emphasis upon precise recall and shift the focus toward reporting "usual" pay amounts. However, results of research comparing income estimates obtained from these types of reports with company records call into question the proposed solutions. Rodgers et al. (1993) report a substantially higher correlation of company records and respondent reports of yearly earnings (+0.79) than the correlation for reports of income received in the last pay period (+0.49) or the respondent's "usual" pay (+0.46).

10.2.3 Response Production—Sensitivity

While both misunderstanding and memory fallibility play important roles in the inaccurate reporting of income, these two factors do not completely explain reporting problems. Respondents' sensitivity about discussing their income may also lead to "motivated mis-remembering" and other reporting failures. This should not be surprising, since taboos against speaking about money are very strong in the United States. To even raise the topic is considered inappropriate; Gallagher (1992), a therapist specializing in the "psychology of money," reports that 90 percent of her clients say they never discuss money in their homes.

Evidence of these sensitivities are not difficult to find in the survey realm. A recent survey of U.S. sexual practices developed self-administration procedures for some sections of an exceedingly personal interview in order to reduce sensitivity. Family income was included in the private, self-administered section of the interview—wisely, since many respondents considered the income question the most personal and sensitive question asked (Laumann, Ganon, Michael, and Michaels, 1994). Van Melis-Wright and Stone (1993, p. 1129) studied attitudes toward money and reported that the two most frequently endorsed statements were, "*I think it is impolite to ask others about their financial situation.*" and "*Surveys asking about my finances should be completely anonymous.*"

10.2.4   Summary of Cognitive Factors in Income Reporting

Underreporting is perhaps the most common concern about income reports. However, the various cognitive factors described above do not, for the most part, strongly support a prediction that income sources and income amounts will necessarily be underreported, only that errors in general (and nonresponse) are likely. Definitional problems, for example, seem likely to result in both inclusion and exclusion errors. Retrieval problems, such as those stemming from lack of knowledge, imperfect reconstruction strategies, or confusion, would also seem likely to increase nonresponse and add general (rather than directional) "noise" to the reporting process. Of all the cognitive factors, only the low salience of some income sources and the sensitivity of income information clearly suggest that errors will tend to be underreports, although here, too, nonresponse may be as likely an outcome as response error.

10.3   THE QUALITY OF SURVEY ESTIMATES AND NONRESPONSE

In this section we review research on the quality of the data produced by income surveys. First we examine "benchmark" comparisons, which are consistent with (but certainly do not prove) the common assumption that income errors are predominantly of the underreporting variety. Next we review nonresponse results which offer a clear indication of income reporting problems.

10.3.1   Comparison of Survey Estimates with Independent Benchmarks

The Census Bureau regularly examines the quality of its income estimates through comparisons to independent benchmark estimates. These comparisons are rarely straightforward. Data from independent sources are never perfectly comparable to survey data, and the required adjustments are often inadequate. The flawed adjustments, and the errors in the independent estimates themselves, add substantial uncertainty to the comparisons. Nevertheless, these evaluations are often interpreted as indicators, at least to some extent, of the quality of the survey data.

Coder and Scoon-Rogers (1996) compare Current Population Survey (CPS) and SIPP estimates to various benchmarks, the primary results of which are summarized in Table 10.1. Although Table 10.1 reveals some variations in the comparability of survey and benchmark estimates, the more obvious story that emerges from these results is the consistency with which the survey estimates fall short of the benchmarks. The almost inescapable conclusion is that the consistent under*estimates* reflect an underlying tendency for survey respondents to under*report*.

We urge caution in drawing this conclusion. As Marquis et al. (1981) and others have pointed out, there are multiple competing explanations for the failure of survey estimates to match benchmark totals, many of which have nothing to do with errors of response to survey questions about income amounts. Surely the benchmark data are indicative of a problem, but without more information the exact nature of that problem remains uncertain.

10.3.2 Income Nonresponse

To the extent that nonresponse is both frequent and nonrandom, survey estimates are more prone to inaccuracy, even if respondents' reports are highly accurate. In this section we examine the extent of nonresponse to income questions and find it to be sufficiently severe to wreak considerable mischief with the alignment of survey and benchmark estimates.

Federal government surveys typically achieve higher levels of cooperation than nongovernment surveys (e.g., Bradburn and Sudman, 1989; Goyder, 1987; Heberlein and Baumgartner, 1978). Even with this built-in advantage, however, income questions in U.S. government surveys are prone to high rates of item nonresponse. Table 10.2 shows item nonresponse results in the form of imputation rates for several income categories from the 1996 CPS March Income Supplement. (An item's value is imputed during processing to fill in missing data resulting primarily from the failure of the interview to produce a valid entry. The vast majority of imputations are a result of blank, do not know, or refused entries.) These rates include both survey respondents who failed to provide answers to individual income items, and March Supplement nonresponding individuals in otherwise CPS-interviewed households (typically, about 10–12 percent of all eligible persons in otherwise-interviewed households fail to respond to the income supplement). The only nonresponse component not represented in Table 10.2 is CPS's initial "household noninterview" rate of around 7–8 percent.

Table 10.2 indicates that, even in the best case, about one in five survey "reports" is produced not by the intended response process but rather by the process used to fill the holes left by nonresponse; typically, the rate is closer to (or exceeds) one in four. As noted earlier, high rates of nonresponse are a likely

Table 10.1 Ratio of SIPP and CPS March Income Supplement Aggregate Income Estimates to Independent Aggregate Income Estimates for 1984 and 1990

Source of Income	1984 Independent Aggregate Estimate (billions)	SIPP (%)	CPS (%)	1990 Independent Aggregate Estimate (billions)	SIPP (%)	CPS (%)
EMPLOYMENT INCOME:						
Wages and salaries	$1,820.1	91.4	97.3	$2,695.6	91.8	97.0
Self-employment	192.6	103.1	70.2	341.4	78.4	66.8
ASSET INCOME:						
Interest	244.8	48.3	56.7	282.8	53.3	61.1
Dividends	59.3	65.9	51.8	126.3	46.1	31.3
Rents and royalties	19.4	211.3	95.4	44.1	102.9	87.8
GOVERNMENT TRANSFER INCOME:						
Social Security	160.5	96.2	91.9	225.5	98.3	93.0
Railroad Retirement	5.6	96.4	71.4	6.9	95.7	66.7
Supplemental Security Income	9.9	88.9	84.8	13.6	94.9	89.0
Aid to Families with Dependent Children	13.9	83.5	78.4	19.7	70.1	71.6
Other cash welfare	2.0	135.0	120.0	2.9	86.2	86.2
Unemployment Insurance	16.3	76.1	74.8	17.7	84.2	80.2
Workers' Compensation	14.1	56.7	48.2	14.6	86.3	94.5
Veterans' Pensions and Compensation	13.9	82.0	59.7	13.8	84.1	77.5
RETIREMENT INCOME:						
Private pensions	65.2	63.8	57.2	70.2	107.1	110.8
Federal employee pensions	20.3	98.0	84.7	30.4	73.4	82.6
Military retirement	15.6	105.1	98.1	20.4	92.2	89.2
State and local employee pensions	21.9	88.1	71.7	36.1	75.1	80.1
MISCELLANEOUS INCOME:						
Alimony	2.7	100.0	81.5	2.5	116.0	124.0

Source: Adapted from Coder and Scoon-Rogers, 1996

Table 10.2 Imputation Rates and Amounts for Selected Sources of Income from the 1996 CPS March Income Supplement

Source of Income	Total Number of Persons with Income (000's)	Number of Persons with Income Amount Reported	Number of Persons with Imputed Income Amount	Percent of Persons with Imputed Amounts
EMPLOYMENT INCOME:				
Wage and salary income	134,135	98,930	35,205	26.2
Nonfarm self-employment income	11,618	8,622	2,996	25.8
ASSET INCOME:				
Interest	107,871	60,354	47,518	44.1
Dividends	29,697	15,253	14,444	48.6
GOVERNMENT TRANSFER INCOME:				
Social Security	37,530	26,490	11,039	29.4
Supplemental Security Income	4,808	3,855	943	19.6
Public assistance	4,943	3,855	1,088	22.0
Workers' Compensation	2,064	1,498	566	27.4
Veterans' Pensions and Compensation	2,549	1,875	674	26.4
RETIREMENT INCOME:				
Pension income	14,350	10,013	4,337	30.2
MISCELLANEOUS INCOME:				
Alimony	463	329	134	28.9
Child Support	5,190	4,180	1,010	19.5

Source: Unpublished data from the U.S. Bureau of the Census.

outcome to survey questions on topics for which respondents lack knowledge or are reluctant to report, both of which characterize income survey tasks.

10.4 RESPONSE ERROR IN INCOME REPORTS

In this section we turn to direct evidence concerning the quality of income reports, primarily in the form of individual-level record check studies. This evidence also suggests a general tendency toward income amount underreporting, although income source underreporting and random error in amount reports may be even more important.

10.4.1 Response Quality for Income Source Reports

Income reporting in surveys is generally a two-stage process involving first the reporting of income sources, and then the reporting of amounts received from those sources. Here we examine research concerning response error in income source reports. We find a general tendency toward underreporting which seems to affect wage and salary income source reports only modestly (if at all), reports of transfer income sources somewhat more so, and asset income sources most of all. Random error propensities seem to follow this same trend, although data are more scant.

Wage/Salary Income Sources Most research on wage/salary income source reporting assesses the quality of reports of the receipt of *any* wage/salary income. Miller and Paley (1958) compare census income reports against matched tax forms, and find a 2–6 percent shortfall in the census-derived rate of receipt of wage/salary income. Kaluzny (1978) finds high levels of underreporting of wage/salary earnings among participants in an income maintenance experiment, although major limitations in the study design and analysis render the conclusions of this work somewhat questionable. Other studies using tax records (e.g., Coder, 1992; Grondin and Michaud, 1994; U.S. Bureau of the Census, 1970) suggest that bias in survey reports of any wage/salary income is typically much closer to zero, and that errors of any kind are quite rare.

Moore et al. (1996) directly assess wage/salary income reports about a specific employer by interviewing a sample of known employees. Virtually all job holders reported the presence of their job with the relevant employer, but respondents failed to report the receipt of job income from that employer in 4–11 percent (depending on the experimental treatment) of the months in which they were known to have received such income. (The study's design does not permit assessment of overreportings.)

Transfer Program Income Sources Transfer program income sources seem more prone to underreporting than wage/salary sources, although the "partial design" record checks which comprise much of the work in this area make net bias difficult to assess. Marquis et al. (1981) review several studies which suggest generally modest levels of underreporting error, ranging from about −2 percent for General Assistance (Weiss, 1969; David, 1962, reports a −7 percent rate among a similar population), to −13 percent for SSI recipients (Vaughan, 1978). Other studies of known program participants suggest substantially higher levels of underreporting among true program participants. For example, Livingston (1969) finds that 22 percent of known public assistance recipients failed to report any public assistance income in a local census. Similarly, Hu (1971) finds 27 percent underreporting in the survey reports of families known to have received cash assistance or medical assistance for children. Other researchers show underreporting error levels among known recipients of AFDC of −14 percent (Klein and Vaughan, 1980) and −9 percent (Goodreau et al., 1984).

These partial-design studies offer clear evidence that survey reporting of transfer program participation is far from perfect, but do not support conclusions concerning the overall frequency of response errors or the magnitude and direction of bias. Marquis et al. (1981, p. 22) review two complete design record check studies which "challenge the conventional wisdom that transfer participation is underreported in sample surveys." Of the four comparisons generated by the two studies, only one suggests a very small net underreporting bias—Oberheu and Ono (1975) find that survey reports of AFDC participation "last month" fall short of record data by about 2 percentage points. The other three comparisons actually indicate modest net overreporting of the receipt of General Assistance (by about 5 percentage points—Bancroft, 1940), yearly AFDC (8 percentage points—Oberheu and Ono, 1975), and Food Stamps (6 percentage points—Oberheu and Ono, 1975). As Marquis et al. point out, these small net biases mask substantial amounts of underlying error, with most studies showing low correlations between survey and record values. Grondin and Michaud (1994) also find substantial error levels in a match of Canadian tax return information to unemployment benefits reports in two surveys. Grondin and Michaud find that 12–19 percent of the survey reports are in error, with underreports exceeding overreports such that the survey-reported rates of participation fall well below the rate derived from the tax returns. (The report does not present sufficient detail to permit the assessment of underreporting and overreporting rates or the relative net bias in the survey estimate.)

Marquis and Moore (1990) report the results of a complete design record check study which finds low (2 percent or less) overall error rates for all programs examined. Nevertheless, high rates of underreporting among true program participants produce consistent net underreporting biases which are trivial for some programs (e.g., –3 percent for veterans' pensions and compensation), modest for others (e.g., –12 percent for SSI; –13 percent for Food Stamps), and quite high for still others (e.g., –18 percent for workers' compensation; –20 percent for Unemployment Insurance (UI); –39 percent for AFDC). In subsequent research, Moore et al. (1996) find comparably high rates of underreporting among a Milwaukee, Wisconsin sample of known recipients of AFDC, Food Stamps, UI, and SSI, but many false positive reports among true nonparticipants as well.

Asset Income Sources With but a single exception, all of the research concerning asset ownership is of the partial design variety, involving the survey reports of known asset holders. The near-unanimous conclusion of this research is that those who own such assets very often fail to report them. As noted before, however, this research cannot address the general tendencies of respondents to underreport or overreport asset ownership. However, the one example of a complete design study does suggest both high levels of random error and a substantial net underreporting bias.

In perhaps the only complete design research on the accuracy of asset ownership reporting, Grondin and Michaud (1994) report two comparisons of annual income survey reports with tax return data. They find high levels of error in asset ownership reports, and an overwhelming predominance of underreports—in one instance the survey estimate falls short of the tax record estimate by 39 percentage points; in the other the net difference is "only" 15 percentage points.

Partial design record check studies, on the other hand, are fairly common. Lansing, Ginsburg, and Braaten (1961) investigate response errors among known owners of savings accounts. In two separate studies, they find that about one-quarter of respondents fail to report the existence of a savings account. Ferber and colleagues have carried out a number of similar studies which also find extensive underreporting among true asset owners—between 19 percent and 46 percent in various savings account studies (Ferber, 1966; Ferber, Forsythe, Guthrie, and Maynes, 1969a); and 30 percent in a study of stock ownership (Ferber, Forsythe, Guthrie, and Maynes, 1969b). Maynes (1965), however, reports on a similar study whose results seem something of an outlier. Among approximately 3,300 known savings account owners in The Netherlands, only about 5 percent failed to report the sample account in a financial survey.

10.4.2   Response Quality for Income Amount Reports

This section examines income amount response errors. We find that wage and salary income response bias estimates are generally small and without consistent sign, with low random error indicators as well. Bias estimates for transfer income amount reporting vary in magnitude but are generally negative (indicating underreporting), and random error is also an important problem. Random error is also quite marked in asset income reports, although indicators of consistent bias are less clear.

Wage and Salary Income (and Total Family Income) Amounts   Marquis et al. (1981) review a substantial body of research on response error in reports of wage/salary and total income. They find evidence suggesting that small and irregular earnings sources might be underreported, but their primary conclusion is that wage/salary and total income reports are subject to very little net bias and very little random measurement error.

Welniak (1986) finds some evidence of random error in the frequent failure of summed CPS-reported income component amounts to fall into the same income interval as the same interview's reported total income—the global and summed reports match in less than half of the cases. Other research, however, tends to support Marquis et al.'s conclusion of minimal bias and minimal random error. Miller and Paley (1958) compare median census income reports for both individuals and families to matched tax returns and find a small net bias which is slightly negative (about −3 percent) for matched families, but slightly positive (about +4 percent) for matched unrelated individuals. A more recent tax return study by Coder (1992) also finds a modest net bias in wage/salary income reports of about −4 percent, and very high reliability, as indicated by a correlation of +0.83 between survey and tax amounts.

Studies conducted outside the United States also generally find low levels of bias and random error in wage/salary reports. Andersen and Christoffersen (1982) compare interview data to Danish tax records and find a slight positive net bias of about +2.5 percent in respondents' reports. Körmendi (1988) compares telephone and personal visit interview reports with Danish tax information and also finds very small net biases in mean amounts reported and very high correlations—from +0.8 to over +0.9—between the tax and survey reports. Grondin and Michaud's (1994) results seem to differ from the general findings noted above, although their report offers few details which would permit a careful comparison with other research. They find higher error frequencies in two comparisons of annual income data with tax information, with about one-third of interview reports differing from their matched tax reports by more than ±5 percent. (The report offers no evidence regarding the sign or magnitude of the resulting net bias in the survey reports.)

At odds with the tax comparison results, Kaluzny (1978) finds substantial negative bias in the wage/salary reports of income maintenance experiment sample cases matched to state unemployment agency data, ranging from a minimum of −13 percent (for steadily-working female heads of household) to −79 percent (for nonsteadily-working nonheads). Halsey's (1978, pp. IV–46) findings in similar income maintenance experiments differ markedly from Kaluzny's, showing very small negative biases on the order of 2–4 percent—statistically significant but "not large enough to be economically important." Halsey also finds very high correlations (approximately +0.90) between record and report values, indicating high reliability. Bound and Krueger (1991) summarize a study matching earnings reports from the CPS to Social Security Administration (SSA) earnings records, which finds essentially zero net bias in CPS income reports (for those whose incomes did not exceed the SSA's earnings maximum cutoff).

Employers' records provide perhaps the best means of detecting errors in respondents' wage/salary survey reports. Hoaglin (1978) reports on two such comparisons involving annual and monthly income reports. The net bias in respondents' annual income reports is very close to zero; errors in monthly reports seem slightly larger and more consistently negative than the annual reports. Carstensen and Woltman (1979) use employers' records to assess CPS income reports and find a nonsignificant tendency to overreport both annual job income (by about 5 percent) and monthly income (by about 10 percent), but to nonsignificantly *under*report weekly and hourly pay rates (by about 4–6 percent). Duncan and Hill (1985), in a validation study among employees of a single manufacturing company, find essentially zero bias (−0.2 percent in one instance, −1 percent in another) in respondents' annual earnings reports.

Rodgers, Brown, and Duncan (1993) also find little evidence of important net bias in employees' survey reports of annual income, the mean of which differed from the employer record mean by less than 1 percent (no indication of the sign of the bias is provided). Rodgers et al. find somewhat higher levels of random error than prior investigators, as indicated by relatively low survey-record correlations for earnings last pay period (+0.60), and for usual pay period earnings (+0.46), although the estimate for annual earnings remains quite high (+0.79). Moore et al. (1996) also provide some evidence in support of the notion that survey response errors for wage/salary income amount reports tend to be modest. They find substantial agreement between survey reports of monthly income and an employer's payroll records, with two-thirds to three-quarters of respondents' monthly reports accurate to within ±5 percent of the record value.

Transfer Program Income Amounts The evidence cited is scant, but Marquis et al.'s (1981) review presents a somewhat inconsistent picture of response error in survey reports of income from government transfer programs. Two studies of social security income reporting (Haber, 1966; Vaughan and Yuskavage, 1976) suggest modest levels of net bias (approximately 5 or 6 percent)—but the signs are inconsistent. David (1962) finds a net bias of about −18 percent for "general assistance" amounts; Oberheu and Ono (1975) find much more severe under-reporting (about −30 percent) of AFDC amounts.

More recent work shows fairly consistent underreporting response biases for transfer program income, but with substantial variation in magnitude across programs. Indicators of unreliability, however, are uniformly high, indicating the presence of much random error, even for income amount types that show little bias. Grondin and Michaud (1994) present data from two different comparisons of survey UI benefit reports and tax records which show evidence of fairly substantial error. In each survey, only slightly more than half of the survey reports agree with the tax record amount within ±5 percent. (No evidence is presented regarding the direction of errors for disagreement cases.) Dibbs, Hale, Love-rock, and Michaud (1995) offer more evidence on net bias in UI reports, showing a modest general tendency for respondents to underreport annual benefit amounts by about 5 percent.

The records of transfer program agencies have also been used to establish the nature and level of response error in transfer income amount reports; these studies suggest a general tendency for transfer program income to be underreported. Livingston (1969), for example, compares survey reports of the receipt of "public assistance" income with administrative records, and finds a net bias in the survey reports of −27 percent (in a comparison of medians). Livingston also considers AFDC and "old age assistance" (OAA) separately and finds both to be marked by rather severe underreporting—OAA recipients on average reported only 80 percent of their OAA income (i.e., the net bias is about −20 percent); for AFDC recipients the level of bias is −30 percent.

Halsey's (1978) summary of income reporting in two income maintenance experiments also suggests a substantial underreporting bias for transfer income amounts, as well as substantial random error. Halsey analyzes amount reporting among those for whom either the report or the record was nonzero and finds that reports of transfer program income are negatively biased—by approximately −25 percent for AFDC, and −50 percent for UI benefits. The reliability of the survey reports are also quite low, with correlations between record and report values only in the 0.40–0.60 range.

Hoaglin (1978) presents a very different picture, suggesting that net bias in annual and monthly transfer income reports is essentially zero. Hoaglin's presentation does not permit the derivation of precise estimates, but his results clearly indicate negligible net bias for both annual and monthly Social Security (about −3 percent in a comparison of medians); for annual and monthly SSI and "welfare," and monthly UI (no annual data are available), the median response error is $0.

Goodreau et al. (1984) compare administrative record data with the survey reports of known AFDC recipients. In keeping with Hoaglin's results, they find high accuracy overall—about two-thirds of all AFDC reports were accurate within ±2 percent. There is a modest underreporting tendency overall, with the total amount of survey-reported assistance reaching 96 percent of actually received amounts. Moore et al. (1996) also suggest that program participants who correctly report their participation, report most benefit amounts with considerable accuracy. In their study, about 70–80 percent of AFDC, Food Stamps, and SSI amount reports were accurate within ±5 percent of truth; UI amount reports were substantially less accurate, with accuracy rates predominantly in the 20–30 percent range.

Asset Income Amounts　Because asset income studies are so rare—this review uncovered only one—we must focus primarily on response quality for the *value* of reported assets rather than the amounts of income they produce. In general, these studies offer little evidence of consistent bias—the bias estimates are occasionally substantial, but about as likely to be positive in sign as negative.

Grondin and Michaud's (1994) comparison of survey reports with tax records is the only known study of response error in the amounts of income from asset holdings—in this case, interest and dividend income. Unfortunately, they only report the frequency with which the survey reports are in agreement (within ±5 percent) with the tax data—37 percent in one survey, 50 percent in another. This suggests that such reports carry a substantial random error component, but they offer no clues as to the magnitude or direction of any response bias.

Lansing, Ginsburg, and Braaten (1961) summarize two studies of response error in the survey reports of known owners of savings accounts. One finds a −14 percent net bias in reported savings account balances, which masks an even higher average absolute discrepancy approaching ±50 percent. Lansing et al.'s second study finds a much more moderate (and in this case positive) net bias of +2 percent. In several studies, Ferber (1966) also finds widely diverging bias estimates for reported asset values, ranging from substantially negative (−20 percent), to essentially zero (+0.3 percent), to the moderately positive (+8 percent).

Maynes (1965) finds a moderate negative bias of about −5 percent in the reported balances of known savings account owners in The Netherlands. Maynes also finds more error (but approximately equivalent net bias) for reports about the previous account balance 10 months ago as compared to current balance reports. Ferber et al. (1969a) also find very little net bias in the reports of savings account balances—less than +0.1 percent—but substantial random error. Only about 40 percent of reports fall within ±10 percent of the true amount. The same researchers (Ferber et al., 1969b) also find a trivial net bias (+0.2 percent) in the reports of stock holdings, as well as very little random error (approximately 80 percent of reports were reported with no error).

10.5 CONCLUSIONS

Our review of investigations of the cognitive aspects of providing income data in surveys suggests, first of all, that the field is far from having final and definitive information on how respondents understand, think about, and form answers to income questions. There is clearly much more work to be done to fully understand these processes, although existing work does highlight many possible cognitive contributors to inaccurate reporting, including lack of knowledge, misunderstanding and other definitional issues, recall problems and confusion, and sensitivity.

Studies of income survey data quality demonstrate widespread problems with the measurement of income in surveys, some of which are visible to the naked eye, and others whose existence can only be confirmed under the microscope. In the former category there are the consistent and often large shortfalls in nationally-weighted survey estimates as compared to independent benchmarks, and the very high levels of nonresponse. We suspect that many of the important cognitive problems faced by respondents when presented with income questions—incomplete knowledge, to state an obvious example—contribute to nonresponse outcomes.

In the category of less immediately obvious problems are the errors in individual respondents' survey reports of both income sources and income amounts that, collectively, signal the presence of both bias and random error effects. Research shows that the propensities for such problems vary substantially across different types of income. For wage and salary income (by far the largest component of personal income in the United States) indicators of both response bias and random error are quite low. For other types of income, however, the situation is notably worse. In particular, we find that both bias and random error are severe for reports of transfer program income; asset income has yet to be carefully examined, although research to date suggests that substantial random error afflicts these reports.

Taken all together, even a cursory review makes it apparent that asking respondents to report their income is taxing in many ways, although no single cognitive issue seems predominant. The positive side of this situation is that many avenues are available for making inroads on income measurement error problems. The more daunting specter that it raises, however, is that a large number of different problems must be solved in order to significantly improve measurement quality. Furthermore, some of those problems—for example, respondents' lack of knowledge of their own income situation, and their strongly-held sensitivity concerns—seem largely immune to "cognitive" solutions.

ACKNOWLEDGMENTS

The authors gratefully acknowledge the useful comments and suggestions of Martin David, Kent Marquis, and Betsy Martin. This chapter is a highly condensed version of the background paper prepared for use at the CASM II Seminar; contact Jeffrey C. Moore for a copy of the full background paper. The opinions expressed herein are the authors' and do not necessarily represent the official views or positions of either the Bureau of the Census or the Bureau of Labor Statistics.

REFERENCES

Andersen, B., and Christoffersen, M. (1982). Om sporgeskemaer. *Socialforsknings-instituttets studie 46*, Copenhagen, Denmark. [In Danish; cited in Körmendi (1988).]

Bancroft, G. (1940). Consistency of information from records and interviews. *Journal of the American Statistical Association, 35*, 377–381.

Bogen, K. (1994). *Documentation of SIPP CAPI cognitive interviews.* Unpublished U.S. Bureau of the Census memorandum, June 13, 1994.

Bogen, K. (1995). *Results of the third round of SIPP CAPI cognitive interviews.* Unpublished U.S. Bureau of the Census memorandum, April 28, 1995.

Bogen, K., and Robinson, J. (1995). *Results from the fourth round of SIPP redesign cognitive interviews.* Unpublished U.S. Bureau of the Census memorandum, October 17, 1995.

Bound, J., and Krueger, A. (1991). The extent of measurement error in longitudinal earnings data: Do two wrongs make a right? *Journal of Labor Economics*, 9, 1–24.

Bradburn, N., and Sudman, S. (1989). *Polls and Surveys.* San Francisco: Jossey-Bass.

Cantor, D., Brandt, S., and Green, J. (1991). *Results of first wave of SIPP interviews.* Unpublished Westat report to the U.S. Bureau of the Census (memorandum to Chet Bowie), February 21, 1991.

Carstensen, L., and Woltman, H. (1979). Comparing earnings data from the CPS and employers records. *Proceedings of the Section on Social Statistics, American Statistical Association*, 168–173.

Coder, J. (1992). Using administrative record information to evaluate the quality of the income data collected in the Survey of Income and Program Participation. *Proceedings of Statistics Canada Symposium 92: Design and Analysis of Longitudinal Surveys*, pp. 295–306. Ottawa: Statistics Canada.

Coder, J., and Scoon-Rogers, L. (1996). *Evaluating the quality of income data collected in the annual supplement to the March Current Population Survey and the Survey of Income and Program Participation.* U.S. Bureau of the Census: SIPP Working Paper #215.

David, M. (1962). The validity of income reported by a sample of families who received welfare assistance during 1959. *Journal of the American Statistical Association, 57*, 680–685.

Dibbs, R., Hale, A., Loverock, R., and Michaud, S. (1995). *Some effects of computer-assisted interviewing on the data quality of the Survey of Labour and Income Dynamics*. Statistics Canada: SLID Research Paper Series No. 95-07.

Dippo, C., and Norwood, J. (1992). A review of research at the Bureau of Labor Statistics. In J. M. Tanur (Ed.), *Questions About Questions: Inquiries into the Cognitive Bases of Surveys*, pp. 271–290. New York: Russell Sage Foundation.

Duncan, G., and Hill, D. (1985). An investigation of the extent and consequences of measurement error in labor-economic survey data. *Journal of Labor Economics, 3*, 508–532.

Ferber, R. (1966). *The reliability of consumer reports of financial assets and debts*. Urbana, IL: Bureau of Economic and Business Research, University of Illinois.

Ferber, R., Forsythe, J., Guthrie, H., and Maynes, E. (1969a). Validation of a national survey of financial characteristics: Savings accounts. *The Review of Economics and Statistics, LI*, 436–444.

Ferber, R., Forsythe, J., Guthrie, H., and Maynes, E. (1969b). Validation of consumer financial characteristics: Common stock. *Journal of the American Statistical Association, 64*, 415–432.

Gallagher, N. (1992). Feeling the squeeze. *Networker, 16*, 16–23, 26–29.

Goodreau, K., Oberheu, H., and Vaughan, D. (1984). An assessment of the quality of survey reports of income from the Aid to Families with Dependent Children (AFDC) program. *Journal of Business and Economic Statistics, 2*, 179–186.

Goyder, J. (1987). *The Silent Minority: Nonrespondents on Sample Surveys*. Boulder, CO: Westview.

Grondin, C., and Michaud, S. (1994). Data quality of income data using computer-assisted interview: The experience of the Canadian Survey of Labour and Income Dynamics. *Proceedings of the Section on Survey Research Methods, American Statistical Association*, 830–835.

Haber, L. (1966). Evaluating response error in the reporting of the income of the aged: Benefit income. *Proceedings of the Section on Social Statistics, American Statistical Association*, 412–419.

Halsey, H. (1978). Validating income data: Lessons from the Seattle and Denver income maintenance experiment. *Proceedings of the Survey of Income and Program Participation Workshop—Survey Research Issues in Income Measurement: Field Techniques, Questionnaire Design and Income Validation*, pp. (IV)21–(IV)51. Washington, DC: U.S. Department of Health, Education and Welfare.

Heberlein, T., and Baumgartner, R. (1978). Factors affecting response rates to mailed questionnaires: A quantitative analysis of the published literature. *American Sociological Review, 43*, 447–462.

Hoaglin, D. (1978). Household income and income reporting error in the housing allowance demand experiment. *Proceedings of the Survey of Income and Program Participation Workshop—Survey Research Issues in Income Measurement: Field Techniques, Questionnaire Design and Income Validation*, pp. (IV)52–(IV)95. Washington, DC: U.S. Department of Health, Education and Welfare.

Hu, T. (1971). The validity of income and welfare information reported by a sample of welfare families. *Proceedings of the Section on Social Statistics, American Statistical Association*, 311–313.

Kaluzny, R. (1978). Validation of respondent earnings data in the Gary income maintenance experiment. *Proceedings of the Survey of Income and Program Participation Workshop—Survey Research Issues in Income Measurement: Field Techniques, Questionnaire Design and Income Validation*, pp. (IV)1–(IV)20. Washington, DC: U.S. Department of Health, Education and Welfare.

Klein, B., and Vaughan, D. (1980). Validity of AFDC reporting among list frame recipients. In J. Olson (Ed.), *Reports from the site research test*, Chapter 11. Washington, DC: U.S. Department of Health and Human Services/ASPE.

Körmendi, E. (1988). The quality of income information in telephone and face to face surveys. In R. M. Groves, P. P. Biemer, L. E. Lyberg, J. T. Massey, W. L. Nicholls II, and J. Waksberg (Eds.), *Telephone Survey Methodology*, pp. 341–356. New York: Wiley.

Lansing, J., Ginsburg, G., and Braaten, K. (1961). *An investigation of response error*. Urbana, IL: Bureau of Economic and Business Research, University of Illinois.

Laumann, E., Ganon, J., Michael, R., and Michaels, S. (1994). *The Social Organization of Sexuality: Sexual Practices in the United States*. Chicago: The University of Chicago Press.

Livingston, R. (1969). Evaluation of the reporting of public assistance income in the special census of Dane County, Wisconsin: May 15, 1968. *Proceedings of the Ninth Workshop on Public Welfare Research and Statistics*, 59–72.

Marquis, K. (1990). *Report of the SIPP cognitive interviewing project*. Unpublished U.S. Bureau of the Census report, August 1990.

Marquis, K., Duan, N., Marquis, M., and Polich, J. (1981). *Response Errors in Sensitive Topic Surveys: Estimates, Effects, and Correction Options*. Santa Monica, CA: Rand Corporation.

Marquis, K., and Moore, J. (1990). Measurement errors in SIPP program reports. *Proceedings of the Annual Research Conference*, pp. 721–745. Washington, DC: U.S. Bureau of the Census.

Maynes, E. (1965). The anatomy of response errors: Consumer saving. *Journal of Marketing Research, II*, 378–387.

Maynes, E. (1968). Minimizing response errors in financial data: The possibilities. *Journal of the American Statistical Association, 63*, 214–227.

Miller, H., and Paley, L. (1958). Income reported in the 1950 census and on income tax returns. *An Appraisal of the 1950 Census Income Data*, pp. 179–201. Princeton, NJ: Princeton University Press.

Moore, J., Marquis, K., and Bogen, K. (1996). *The SIPP cognitive research evaluation experiment: Basic results and documentation*. Unpublished U.S. Bureau of the Census report, January 11, 1996.

Oberheu, H., and Ono, M. (1975). Findings from a pilot study of current and potential public assistance recipients included in the Current Population Survey. *Proceedings of the Section on Social Statistics, American Statistical Association*, 576–579.

Rodgers, W., Brown, C., and Duncan, G. (1993). Errors in survey reports of earnings, hours worked, and hourly wages. *Journal of the American Statistical Association,* *88,* 1208–1218.

Stinson, L. (1997). *The subjective assessment of income and expenses: Cognitive test results.* Unpublished U.S. Bureau of Labor Statistics report, January 1997.

Sudman, S., and Bradburn, N. (1982). *Asking Questions.* San Francisco: Jossey-Bass.

Tourangeau, R. (1984). Cognitive sciences and survey methods. In T. Jabine, M. Straf, J. Tanur, and R. Tourangeau (Eds.), *Cognitive Aspects of Survey Methodology: Building a Bridge Between Disciplines,* pp. 73–100. Washington, DC: National Academy Press.

U.S. Bureau of the Census (1970). *Evaluation and research program of the U.S. census of population and housing, 1960: Record check study of accuracy of income reporting.* Series ER60, Vol. 8, Washington, DC.

van Melis-Wright, M., and Stone, D. (1993). Psychological variables associated with respondents' sensitivity to income questions—A preliminary analysis. *Proceedings of the Section on Survey Research Methods, American Statistical Association,* 1124–1129.

Vaughan, D. (1978). Errors in reporting Supplemental Security Income in a pilot household survey. *Proceedings of the Section on Survey Research Methods, American Statistical Association,* 288–293.

Vaughan, D., Lininger, C., and Klein, R. (1983). Differentiating veterans' pensions and compensation in the 1979 ISDP panel. *Proceedings of the Section on Survey Research Methods, American Statistical Association,* 191–196.

Vaughan, D., and Yuskavage, R. (1976). Investigating discrepancies between Social Security Administration and Current Population Survey for 1972. *Proceedings of the Section on Social Statistics, American Statistical Association,* 824–829.

Weiss, C. (1969). Validity of welfare mothers' interview responses. *Public Opinion Quarterly, 32,* 622–633.

Welniak, E. (1986). Unpublished U.S. Bureau of the Census research, May 1986.

Withey, S. (1954). Reliability of recall of income. *Public Opinion Quarterly, summer,* 197–204.

SECTION C

CHAPTER 11

Casting a Wider Net: Contributions from New Disciplines

Roger Tourangeau
The Gallup Organization

At the 1983 CASM Seminar, Norman Bradburn and I were commissioned to write papers that described the relevance of the cognitive sciences to survey methods problems (Bradburn and Danis, 1984; Tourangeau, 1984). Afterwards, Andrew Ortony remarked that neither paper nor the first CASM Conference as a whole had drawn on the full range of cognitive sciences; instead, both the papers and the conference focused almost exclusively on the potential contributions from cognitive psychology. Ortony's observation contained more truth than I cared to admit. And, in fact, he could easily have gone even further in his critique. In some ways, the first CASM Conference drew not from cognitive psychology in general, but rather from a couple of specific areas, from which it drew rather heavily—namely, the psychology of memory and judgment.

The chapters in this section of the book represent an attempt to address Ortony's complaint, by broadening the base of disciplines from which the CASM movement draws. Four of the chapters represent viewpoints that were either omitted entirely from the first seminar or were relegated to bit parts. The disciplines that were shortchanged include linguistics, whose relevance Charles Fillmore's chapter discusses; artificial intelligence, for which Arthur Graesser's chapter makes the case; connectionism, to which Eliot Smith's chapter serves as an introduction; and ethnography, which is represented by Eleanor Gerber's chapter. The final chapter in this section, by Robert Groves, has a somewhat different purpose; it attempts to integrate the insights from CASM-style methodological research into statistical estimation procedures.

Cognition and Survey Research, Edited by Monroe G. Sirken, Douglas J. Herrmann, Susan Schechter, Norbert Schwarz, Judith M. Tanur, and Roger Tourangeau.
ISBN 0-471-24138-5 © 1999 John Wiley & Sons, Inc.

As is often the case, it is possible to characterize the accomplishments of the CASM movement in apparently contradictory terms. On the one hand, one can plausibly argue that a paradigm shift has taken place during the period following the first CASM Conference—a fundamental change in the way we view survey errors. If a scientific paradigm determines the questions researchers focus on, the methods they use to address those questions, and the concepts and theories in which they frame their answers, then something like a paradigm shift does seem to have occurred within survey methodology. Clearly, researchers have adopted a new vocabulary to describe measurement errors in surveys, developed new tools to investigate them, and brought new theoretical ideas to bear in understanding and reducing the different sources of measurement error.

There is, on the other hand, a widespread recognition that the acceptance of the new ideas has been quite superficial in many respects. Although the movement began to be institutionalized more than a decade ago (as one friend of mine has put it, "cognitoriums" began springing up all over Washington, DC at the federal statistical agencies), it was not always clear that the research conducted at the new cognitive laboratories was all that cognitive. Researchers often claimed to use "cognitive" methods in pretesting questionnaires, but in many cases they were referring to methods like focus groups that have no particular basis within any of the cognitive sciences (see Beatty, Willis, and Schechter, 1997; Conrad and Blair, 1996, for especially thoughtful discussions of "cognitive interviewing" as a method of pretesting). Clearly, there has been a great shift in terminology—it seems virtually mandatory now to use the word "cognitive" in a methodological paper, but the change in the labels often masks an absence of real change in survey practice.

There are some serious obstacles to a deeper acceptance of the new paradigm—and to more profound accomplishments from the movement. One of these obstacles to change is the existence of an earlier paradigm, one that is, arguably, more successful than the new cognitive approach at informing survey practice. Let me refer to the existing paradigm as the Hansen-Hurwitz-Bershad approach, after one of the key papers articulating that viewpoint (Hansen, Hurwitz, and Bershad, 1961). That model and its descendants are still the current reigning models of survey error which guide survey practice. Colm O'Muircheartaigh effectively lampooned this approach, noting that every time a new methodological problem is discovered, a new variance component is duly named and added to the model (see Chapter 4 of this book). In some ways, this reflects the flexibility and comprehensiveness of the old paradigm. More than thirty years later, the paper by Hansen and his colleagues still stands as one of the major milestones in the history of survey research, representing the development of a rigorous statistical conception of error and the adoption of this conception in the design of surveys and the estimation of survey results. A key difference between the old and the new paradigms is that the earlier paradigm is woven into survey practice—in estimation procedures, in reinterview programs, in examinations of interviewer variance; by contrast, the cognitive paradigm often seems to have been grafted on only superficially.

In this context of competing paradigms, it will be extremely useful to find ways to integrate the new conception of survey error with the existing conception and to reconcile or combine the statistical and psychological conceptions of error wherever it is possible to do so. The chapter by Robert Groves in this section represents an important first step in that direction.

A second barrier to the acceptance of the new paradigm and to additional practical accomplishments by the CASM movement is the lack of explicitness of many of the new ideas. Of course, with any new theoretical approach, it may take some time for the implications to be fully worked out. In the case of the CASM movement, it is still often not very clear as to what is being predicted in a given situation or what is being prescribed for survey practice. Once again, O'Muircheartaigh has anticipated this point; he notes in Chapter 4 that survey researchers using cognitive pretesting methods will talk with great confidence about what their findings mean, but be completely unwilling to predict what will happen in the next survey. One way to address this lack of explicitness is to try to develop models that are sufficiently constrained that they yield concrete predictions.

Work in artificial intelligence often takes this form. A theory about how some mental task is carried out by people (or how it could be done by computers) is expressed in a detailed and rigorous computational model—often embodied in a program—which is, by its very nature, forced to yield concrete predictions. One such model especially relevant to the survey enterprise is the QUEST model developed by Arthur Graesser and his colleagues and discussed in Chapter 13. QUEST is probably the most detailed model of how people answer questions, including the type of questions that appear in surveys. (For an earlier attempt to model the process of answering questions, see Lehnert, 1978.)

Aside from the existence of a prior paradigm for understanding survey errors and the lack of explicitness of the new one, a third obstacle to the acceptance of the new ideas has been the incompatibility of the cognitive approach with certain key survey realities. In general, cognitive psychology (and many of the other cognitive sciences) have tended to downplay individual differences. But survey designers are constantly confronted with such differences, even if they cannot always cope with them very well. One of the key survey realities that cognitive psychology is not especially well-suited to deal with is cultural diversity. Survey researchers acknowledge that the questions in surveys often presuppose a conceptual framework that the respondents simply do not share. For instance, questions about who "usually lives here" simply may not apply very well in certain types of households with loose and extended ties among the members (Gerber, 1993). But recognizing the problem is not the same as knowing the solution. And, unfortunately, it is not clear that cognitive psychology provides much in the way of useful tools for identifying or dealing with the conceptual incompatibilities that cultural diversity can produce. What is required instead (or in addition) are tools drawn from anthropology, such as the ethnographic interview. Chapter 14, by Eleanor Gerber, a cognitive anthropologist who works at the Census Bureau, describes the role of ethnography in survey design. She notes that cognitive interviewing is sometimes used as a substitute for ethnography in the planning stages, but that the goals of cognitive and ethnographic interviewing are not just different but often incompatible.

Another survey reality—closely related to cultural diversity—is linguistic barriers. Once again, the tools popularized by the new movement are not necessarily the right ones for dealing with the linguistic diversity and the barriers it creates for surveys. Much of the research inspired by CASM involves memory and judgment, but many of the problems in surveys, including Belson's results on the misunderstanding by survey respondents of everyday terms (Belson, 1981), seem to involve the comprehension of the questions. And although there has been a great deal of useful psycholinguistic research applied to surveys, there is an additional discipline—linguistics—that also provides useful tools for understanding comprehension problems. In Chapter 12, Charles Fillmore gives a sample of the insights into comprehension problems produced when an eminent linguist looks at some typical survey questions.

Even within psychology—the discipline which the CASM movement has drawn from most heavily—a relatively narrow set of ideas has been applied to survey problems; most of the work done under the CASM banner has relied on concepts taken from the study of memory and judgment, with some borrowings from psycholinguistics. In the period since the original CASM Seminar, a new approach has won wide acceptance within cognitive psychology. This new approach—connectionism—has been, at best, poorly represented within the CASM movement. Connectionism starts from radically different conceptions of memory structure and retrieval processes from those that prevailed at the time of the first CASM Conference. The connectionist models see memories as distributed across multiple "units" and they represent learning as a change in the strength of the connections among these units. The new models also postulate a different architecture—with different memory systems—from that assumed in the traditional models. Connectionism has had tremendous impact on cognitive psychology but limited impact on the CASM movement. Chapter 16, by Eliot Smith, a leading proponent of connectionism within social psychology, gives us an introduction to the connectionist viewpoint and discusses its relevance to survey research.

Because survey research has never emerged as a separate academic discipline, it has always drawn both new practitioners and new ideas from a range of other disciplines, including statistics, psychology, economics, political science, market research, and sociology. As the chapters here demonstrate, new disciplines have their contributions to make as well.

REFERENCES

Beatty, P., Willis, G. B., and Schechter, S. (1997). Evaluating the generalizability of cognitive interview findings. In *Office of Management and Budget Seminar on Statistical Methodology in the Public Service, Statistical Policy Working Paper 26*, pp. 353–362. Washington, DC: Statistical Policy Office.

Belson, W. (1981). *The Design and Understanding of Survey Questions*. Aldershot, England: Gower.

Bradburn, N., and Danis, C. (1984). Potential contributions of cognitive research to survey questionnaire design. In T. Jabine, M. Straf, J. Tanur, and R. Tourangeau (Eds.), *Cognitive Aspects of Survey Methodology: Building a Bridge Between Disciplines*, pp. 101–129. Washington, DC: National Academy Press.

Conrad, F., and Blair, J. (1996). From impressions to data: Increasing the objectivity of cognitive interviews. *Proceedings of the Section on Survey Research Methods, American Statistical Association*, 1–9.

Gerber, E. R. (1993). Understanding residence terms: The meaning of census terms to respondents. *Proceedings of the Section on Survey Research Methods, American Statistical Association*, 150–155.

Hansen, M. H., Hurwitz, W. N., and Bershad, M. A. (1961). Measurement errors in censuses and in surveys. *Bulletin of the International Statistical Institute*, *38*, 359–374.

Lehnert, W. G. (1978). *The Process of Question Answering: A Computer Simulation of Cognition*. Hillsdale, NJ: Erlbaum.

Tourangeau, R. (1984). Cognitive science and survey methods. In T. Jabine, M. Straf, J. Tanur, and R. Tourangeau (Eds.), *Cognitive Aspects of Survey Methodology: Building a Bridge Between Disciplines*, pp. 73–100. Washington, DC: National Academy Press.

CHAPTER 12

A Linguistic Look at Survey Research

Charles J. Fillmore
University of California at Berkeley

12.1 PREAMBLE

In my room at the Boar's Head Inn, the fine hotel that housed the CASM II Conference, I found a brief questionnaire with the familiar tell-us-how-you-like-us function. So, in the spirit of cooperating with survey designers, I resolved to fill it out.

The first two questions were easy: I merely needed to state that this was my first visit, and that the experience of staying there was, overall, Excellent rather than Good, Fair or Poor. But with the third and fourth questions shown in Figures 12.1 and 12.2, I immediately started to experience what I had learned to call "respondent burden." I was amused by the implied contrast in the first two choices shown in Figure 12.1, and I wondered what I would have said if I had been vacationing there with children whose presence I enjoyed. Those choices presuppose a contrast which I do not take for granted, but they didn't apply to me. I was there for a conference; I was there on business. But the third and fourth choices required me to decide *between* Conference and Business. The authors apparently had in mind some relevant distinction between conference attending and business, and if I wanted my response to serve their purposes, I needed to figure out what that distinction was. Since CASM II was clearly a conference, and since I was clearly not there for any money-making purpose, I assumed my choice had to be Conference.

But then question #4 shown in Figure 12.2 asked me how I spent my time.

Cognition and Survey Research, Edited by Monroe G. Sirken, Douglas J. Herrmann, Susan Schechter, Norbert Schwarz, Judith M. Tanur, and Roger Tourangeau.
ISBN 0-471-24138-5   © 1999 John Wiley & Sons, Inc.

3.   What's the primary purpose of your visit to The Boar's Head Inn?

    __Pleasure/Vacation

    __Vacation (with children)

    __Conference

    __Business

Figure 12.1   Third question from customer survey at the Boar's Head Inn.

4.   How did your spend your time in Charlottesville?

    __Business

    __Siteseeing

    __Just passing through the city

    __Using The Inn's amenities and recreational facilities

    __Other (please specify)

Figure 12.2   Fourth question from customer survey at the Boar's Head Inn.

Since a special semantic intention in one part of a document is not likely to be sub-
verted or contradicted in the very next part, I had to assume that if Business was
the wrong answer for question #3 (Figure 12.1), it could not be the correct answer
for this question. But why did my situation, which appeared as one of the listed
choices in question #3, get relegated to "Other" in question #4? Maybe I needed
to reconsider my reasoning about question #3: Perhaps the Conference they were
thinking of was the kind devoted to, say, spirituality or New Age healing. Atten-
dance at such a conference would probably not be classified as Business.

    The choices we are given in language, either by conventions of the lan-
guage's grammar and vocabulary, or, as in this case, by someone's use of language, make up what semanticists refer to as *contrast sets* (see Lyons, 1977,
Vol. I, chapter 9). Generally contrast sets are taken as exhaustively partitioning
a set of possibilities into mutually exclusive categories, in the way that (for the
most part) the categories Male and Female exhaustively and exclusively sort
human beings. The first two choices in question #3 presuppose a division of
possibilities that some hotel guests might not accept, since they imply that vacationing with one's children is not pleasurable; and the juxtaposition of questions
#3 and #4 causes the confusion we have just considered by including category
names that in ordinary talk are overlapping rather than mutually exclusive. In
the end, of course, I could not complete the questionnaire.

In the pages that follow, I will be raising issues about the reliability of survey results that will turn on such matters as whether we can be sure that respondents understand what they were being asked, and if not, what might have gone wrong.

12.2 THE SKEPTICAL LINGUIST

As someone professionally concerned with the process by which people use language to communicate their beliefs and feelings, I have become skeptical of that derivative process by which agencies try to learn about what people believe and feel through the methods of survey research. My skepticism, it appears, even exceeds that of the late astronomer and teacher Carl Sagan. In *The Demon-Haunted World* (1996), Sagan urges his readers to arm themselves against hoaxes and pseudoscience[1] by being skeptical about everything they are told and by consulting a checklist in what he calls "The Fine Art of Baloney Detection." In deploring the present state of public awareness of science, Sagan dishes out some discouraging statistics: that, for example, more than half of Americans claim to believe in the Devil, that 10 percent of them claim to communicate with him (p. 123), that more than a quarter of Americans believe in astrology, and a third of *them* believe that Sun-sign astrology is scientific (p. 303).

Now all of these claims about what people believe might be true, and I should not expect references to details of survey methods in a popular book, but I have the general impression from the frequency of such statistics in his book that Sagan takes these claims as beyond question, and I wondered why he did not aim his skepticism at such reports. If I read that more Americans believe in angels than in evolution, I would worry not simply about whether it is true, but also about what it could mean. Is a belief in angels a distinct mental state from believing in ghosts or fairies? Biblical angels tend to be men who have been in the angel business forever, greeting-card angels tend to be women, cartoonists' angels are men or women who were once among us but now sit on clouds and play harps ("Grandpa won't be with us any more because he's an angel now"). How much of all this belongs in believing in angels? In short, I have no idea, from reading such news, about the actual content of the beliefs that can be attributed to people who claim to believe in angels. I would like such reports to be accompanied by exact copies of the questions respondents were given, I would like to see the contrast sets from which they drew their choices, and I would like the chance to check the wording of the surrounding questions.

My own, usually unwilling, participation in interview questionnaires has almost always been abandoned midstream. I typically find that with the interesting questions, none of the answers allow me to give a meaningful answer, and choices like "Not Applicable" and "I don't know" are often not available.

---

[1] I am not suggesting that these terms apply to survey research.

12.3 LINGUISTICS AND SURVEY RESEARCH

A great deal of cognitive work goes into participating in a questionnaire-based interview. This includes trying to imagine the motives and interests of the agency or sponsor that commissioned the survey, trying to decide what "voice" one wishes to have represented when the survey results get made public (the Gallup Organization's slogan is "Helping people be heard"), and trying to figure out what exactly is being asked. All of these issues touch on the work, interests, and experiences of linguists.

A vastly oversimplified review of the field of linguistics might distinguish what can be called *microlinguistics*, which includes knowledge and practices internalized in each individual language-user, from what might be called *macrolinguistics*, intended to include both the effects of context on producing or interpreting language, and a view of language from the outside.[2] Microlinguistics, then, covers the production and perception of the sounds of speech as they are organized in the speaker's language (*phonology*), the repertory of meaningful minima in the speaker's language system (the *lexicon*—roughly the words in the speaker's vocabulary and the elements that go into the formation of words), and the rules or constructions by which the speaker produces complex words (*morphology*) or phrases and sentences (*syntax*). Knowledge of the lexicon includes knowledge of the meanings that lexical items encode and the system of contrasts that such meanings exhibit (*lexical semantics*), and knowledge of syntax is accompanied by knowledge of the principles by which the meanings of sentences are built up out of the meanings of their constituent parts (*compositional semantics*).

What I have called macrolinguistics includes knowledge and processes that involve speakers in the act of communicating, especially where the context conveys information which supplements or subverts the apparent meaning of the spoken or written message (*linguistic pragmatics, sociolinguistics, psycholinguistics*) as well as externally viewed features of speakers and their language (*dialectology, language typology, linguistic history*, etc.).

Understanding the language of survey questions involves the lexicon and lexical semantics ("Do I know the words in this question and do they have the meanings I think they do?"), syntax and compositional semantics ("Am I taking this sentence apart the way the speaker or writer intended?"), pragmatics ("Do I understand the point of the question; do I know what is going to be made of the things that I say?"), and sociolinguistics ("Will I make a bad impression if I admit that I don't understand the question?"). The offered parenthesized soliloquies might suggest that I think much of this cognitive work is conscious, but

---

[2]A general overview of linguistics as a whole may be found in such standard undergraduate textbooks as Akmajian, Demers, Farmer, and Harnish (1995) and O'Grady, Dobrovolsky, and Aronoff (1997). Specialist surveys of linguistic semantics are Allan (1986) and Lyons (1977); careful studies of concepts and distinctions in lexical semantics are Cruse (1986) and Lehrer (1974). Linguistic pragmatics is treated in Green (1989) and Levinson (1983), and certain philosophical underpinnings can be seen in Grice (1975), and Searle (1969). Psychological aspects of linguistic meaning are dealt with, very readably, in Aitchison (1987).

in most cases conversation participants have no articulatable "metacognitive awareness" of the process by which they draw conclusions from a communicative transaction. And where dialect differences and historical change are responsible for any of the differences between the interviewer's and the interviewee's language, there could be miscommunication without either side being aware that anything is amiss.

In preparing for this conference, I examined several packages of survey research questions in order to see what a linguist might say about them. The material that will form the basis of my discussion here includes several modules of the General Social Survey of the National Opinion Research Center (GSS), and documents from a draft form of the National Health Interview Survey (NHIS) Redesign, dated from January, 1996. (The numbering of the question examples are for ease of reference only.)

The observations I present in this chapter will be experience-based judgments about the language and context of some of the questions I have reviewed, and guesses about how they might be interpreted. This is obviously a far cry from careful and controlled "cognitive interviewing" to be sure. But cognitive interviewing is expensive, and if my judgments are correct, there are sure to be at least some people who, by giving reasonable but unintended interpretations of the language of the questions, perhaps by listening or reading *more* carefully than they are intended to, will generate response patterns that cannot be trusted. Even in cases where it could be shown that the vast majority of respondents "correctly" interpret a passage that I show to be unclear, contradictory, or ambiguous, I would still recommend replacing it with something that reduces the possibility of confusion. I believe, therefore, that it might be worthwhile for writers of survey questions to pay attention to certain linguistic issues before turning to cognitive interviewing techniques.

Some Observations

An important distinction in linguistic pragmatics is that between *sentence meaning*, based on the sentence's semantics, from *speaker meaning*, having to do with assumptions about why the sentence was uttered (Grice, 1957; Levinson 1983, p. 17f). In a question in GSS, respondents are asked to identify their degree of agreement or disagreement with respect to the statement given in question #5.

> **5.** Irish, Italian, Jewish, and many other minorities overcame prejudice and worked their way up. Blacks should do the same without special favors. (GSS)

It is possible that someone might agree with the actual language of question #5—blacks *should* overcome prejudice and "work their way up"—but disagree with the sentiment they impute to the utterer of such a pronouncement. Agreeing

6.   If you had to choose, which thing on this list would you pick as the most important for a

child to learn to prepare him or her for life?

> 1.   To obey.
>
> 2.   To be well-liked or popular.
>
> 3.   To think for himself or herself.
>
> 4.   To work hard.
>
> 5.   To help others when they need help.

Figure 12.3   Question from General Social Survey.

with a statement and agreeing with a person who utters a statement are not the same thing.

Guessing about the intentions behind a question will also come into play as we look at question #6 in Figure 12.3. How literally or how seriously is the reader likely to take the qualification "the most important for a child to learn to prepare him or her for life"? If this question is interpreted as a way of discovering people's life values, I suspect that "to help others when they need help" is a praiseworthy answer; but that is surely not something anyone would think of when considering things a child can "learn" to "prepare for life." "To be well-liked or popular" would be a helpful trait in someone "preparing for life" but it is hardly something that can be "learned" in the usual way—and it is probably not a highly valued response. In short, this question requires the respondent to cope with a tension between wanting his or her ranking of the moral and social values of these traits to be represented in the survey, and worrying about how and whether children can learn them.

When members of a contrast set are presented as values on a scale, relative position in a list helps to shape the interpretation given to the terms; the informing pragmatic context is the ordered list. In introductory semantics classes this point is traditionally illustrated with conventions for evaluating school performance. The judgment Good has quite a positive meaning when it is positioned in the three-step system Good—Fair—Poor; but not so if it is positioned third in a five-step scale Excellent—Outstanding—Good—Fair—Poor. And in such a five-step-scale, the order of the first two could be either Excellent–Outstanding or Outstanding–Excellent; it is the position in the list, and not simple knowledge of the words' meanings, that will communicate which of these words represents the highest point on the scale.

There are places in our sample interviews where respondents need to depend, not on the language itself, but on the sequence in which response choices are listed. One clear case involves the intended distinction between "pretty well satisfied" and "more or less satisfied." One GSS question asks for the degree of satisfaction with the respondent's "financial situation." It is only by noticing that the answer choices are listed in the order "pretty well satisfied," "more or less satisfied," and "not satisfied at all" that one can know that the first of these is more positive than the second. Given my understanding of how these questions are administered, I doubt that this could ever be a problem, but we can at least imagine a situation in which a respondent happens to say "pretty well satisfied" before the interviewer reads the choices out loud and is quickly checked off as having given that response, without the respondent ever finding out that "pretty well satisfied" was the highest value on the scale. In general, for terms that are not independently interpretable (i.e., where their position in a list is not identified), a report of the results of a survey might not be appropriately informative: if I am told that a certain percentage of some group of citizens are "pretty well satisfied" with their financial situations, I still do not know enough.

A structurally similar situation occurs with an NHIS question about the respondent's employment status in the previous week. For this question, there are four responses from which to choose:

(1) Working at a job or business.
(2) With a job or business but not at work.
(3) Looking for work.
(4) Not working at a job or business.

The links to the follow-up questions make it clear that response (2) is supposed to cover things like vacations or sick leave and response (4) is supposed to cover a situation where the respondent is unemployed. I wonder, of course, whether the potential confusion in interpreting response (2) is worth it. First, is it really important to know the difference between (1) and (2)? If so, would not "employed but not at work" for (2) and "unemployed" for (4) be clearer? If the difference between (1) and (2) is not important, would it not be sufficient to offer the following three choices: (i) Employed, (ii) Unemployed but looking for work, and (iii) Unemployed and not looking for work?

The pragmatic notion of *presupposition* concerns expressions which, from their form or their context, suggest that certain facts or beliefs are taken for granted. Questions with false presuppositions can be quite harmless if they can be quickly rejected. If I ask you if you always get along with your spouse and you tell me you are not married, the matter gets cleared up instantly, but if I insist on a simple yes or no answer, you will feel burdened.

Another question (#7) in NHIS has such a presupposition, repairable in context:

7. As a result of this injury, how much work did you miss?

This question technically presupposes that some work was missed, but fortunately, one of the options is "Not employed at the time of the injury." In troublesome cases, a question, or one or more of the offered answers, could convey presuppositions which the respondent wishes to reject without being given the means to signal that rejection.

A GSS question asks respondents to give reasons for being proud of America, and one of the options is that life in America is characterized by the "fair and equal treatment of all groups in society." In this case, the rejection of the answer may be because of a presupposition. One respondent's failure to express pride in regard to this item might be due to a belief that what the phrase presupposes is false (suppose the respondent is an American Indian); another respondent's reason might be a resentment of the democratization of American society, a belief that such a state of affairs is achieved only by damaging the status and well-being of the privileged group to which the respondent so deservingly belongs.

Let us turn to another GSS example (question #8) which is about images of God:

8. We'd like to know the kinds of images you are likely to associate with God.

This question takes for granted that the respondent believes in God and has some image about what God is like. That may not be true.

There can be a troubling asymmetry in how to make sense of the answers to questions with presuppositions. We can illustrate this by considering the difference between a "yes" and a "no" answer to question #9.

9. If your party nominated a woman for President, would you vote for her if she were qualified for the job? (GSS)

The presuppositions are in the phrase "your party" and in the assumption that voting for one's own party's candidate is the default. A "no" answer to this question is clear: it would mean: "Even if a female candidate is proposed by my own party, I still would not vote for her, no matter how qualified she was." But people who do not feel that way are going to be divided between "yes" and "I don't know," and we can not know for sure how to interpret their answers. On hearing the question, I might believe that some other candidate, male or female, might strike me as more appealing or more qualified, and in that case the most honest and straightforward answer would necessarily be "I don't know." But if I think that the purpose of the question is to find out if I would reject female politicians out of hand, I might end up choosing "yes" just to distinguish myself sharply from the position associated with the "no" answer. In other words, respondents might choose "yes" not because that literally represents their views, but because it is the only clear contrast with the "no" answer with which they wish to be disassociated. If there is a reason to care about who would vote against a female candidate no matter what, there might be an equal reason to care about who would vote for their party's candidate no matter what. The question should be rewritten so that each answer, including the "I don't know," has a clear interpretation.

There are many passages in the surveys where respondents might not know what is included in the semantic categories posed by the question. Certain categories have "fuzzy" boundaries. If someone asks you if members of your family are limited by difficulty remembering, you are likely to say "well, yes, sometimes, depending on what you mean by limited." But if the word "limited" comes with an intensifier, like "in any way," the standards for interpretation suddenly become much stricter. Consider question #10 from the NHIS:

10. Is anyone in the family LIMITED IN ANY WAY by difficulty remembering or by experiencing periods of confusion?

Here we notice not only the presence of the intensifying phrase just mentioned, but that it is in UPPER-CASE LETTERS, reinforcing the need for strict interpretation. Given these tighter standards, you might think of the fact that you are never sure about how to spell "accommodate" or "asymmetry" and you might honestly feel that this problem with your memory limits your chances of making it as a proofreader. But if such an interpretation fits the intentions of the question, the question is pointless. The language of the question seems to fight against reasonable assumptions about its purpose, precisely because of a qualifying expression that appears very deliberate but which cannot be taken seriously.

Words can have different meanings in different regions of the speech community; the description of such facts is part of the work of dialectologists. The existence of such words in survey questions could in some cases invite doubts about the reliability of responses based on questions using those words. A possibly trivial example from the questions I have seen is with the terms "brother-in-law" and "sister-in-law." These words are found in the list of kin terms offered in NHIS for identifying members of the household unit. While it is very unlikely that any confusion about them would be considered relevant to the main purposes of the household survey, it happens that these terms include slightly different sets of kinfolk in different parts of the United States. In particular, dialects differ according to whether the terms include the "spouse of a sibling of one's spouse" interpretation. To everybody, a brother-in-law can be the brother of one's spouse or the husband of one's sister, and a sister-in-law can be the sister of one's spouse or the wife of one's brother; but in some dialects of American English, a brother-in-law can be the husband of the sister of one's spouse, and a sister-in-law can be the wife of the brother of one's spouse. Thus, if the husband of the sister of the wife of the "head of household" lives in the same household, some respondents would count him as family and others would exclude him, because the term includes different kin in different dialects.

Some terms are as close as we can get to a concept, but in their primary meaning they do not fit the concept we are trying to express. I remember being in a situation where people were confused by Mao Tse-tung's name. On learning that "Mao" was his family name, it was difficult for the conversants to find a way of describing "Tse-tung," since neither "first name" nor "Christian name" seemed to fit. Even in an American context, the term "first name" is a possible source of confusion. It can be used in contrast with "middle name"; or it can apparently refer to a personal name as opposed to a family name. In an NHIS question, the interviewer asks if a family member might have "another first name." My guess is that in this question, "first name" is to be understood as contrasted with "family name": when Walter J. is called "Walter" by his colleagues and "Johnny" by his family, he could probably be said to have two first names. This is only technically a potential problem: This is not in a part of the questionnaire where interviewers would be prevented from explaining what the question means, but I did not understand it at first. (And I am not sure that my guess of what it means is correct.)

The word "race" and the classification of people into races is a really great problem. One of the questions in the NHIS draft is the following (question #11).

> **11.** Which one of these groups . . . would you say BEST represents your race?

The question is accompanied by a list of familiar names of "races." Many people have mixed-race backgrounds, yet this question perpetuates the practice by which each individual has to select the one which "BEST represents" his or her ancestry. There is a huge and troubling presupposition here. I understand that there are valid official statistical interests in finding out how people choose to group themselves among the approved "races"; but this leaves no satisfying choice for people who wish to reject such classifications. There has been a recent decision issued by the Office of Management and Budget (Federal Register Notice, 1997) that instructs all federal agencies to accept the reporting of more than one race. But the core problems of how one interprets this sort of question and how one selects the appropriate category(s) remain.

Distinct from the vagueness or uncertainties involved in words or phrases with fuzzy boundaries, we also find that some expressions are *ambiguous*, in that they can be interpreted in two or more distinct ways and the context does not dictate the intended sense. I think this is true of the phrase "the work you do" in question #12.

> **12.** On the whole, how satisfied are you with the work you do. . .? (GSS)

Out of context I think I would interpret the question as referring to the quality of the work I produce, but I realize that in the context of the questionnaire it probably should be interpreted as referring to the job I have. Since I could be satisfied with the job I have, but not satisfied with the work I do on that job, I could imagine giving different answers to this question depending on which interpretation happened to strike me when I read or heard the question.

In some cases a difference between two senses exists because the word belongs to both a general and a specialist vocabulary. This happens frequently in those areas of technical discourse that impinge on everyday life, where the same term comes to have both a strict and a lax interpretation. Different interpretations of the words "stroke" and "surgery" might interfere with the accuracy of answers in the NHIS medical history records. Should a "transient ischemic attack" be counted as a "stroke?" This event is sometimes described as a light stroke, or as a mini-stroke, but it is also sometimes described as a precursor of a stroke. One NHIS question concerns the number of times a respondent has had surgery in the past twelve months. There are certain catheter procedures in heart patients that are referred to in some contexts as noninvasive surgery and in others as alternatives to surgery. I suspect that with these questions it is the stereotypical stroke and surgery that are intended, but respondents with particular medical histories might not know that.

Often the ambiguity of an expression is clarified by its juxtaposition with other members of the contrast set. In a department store context, the question of whether you are paying by "cash or check" requires the interpretation of cash as involving coins and banknotes, but if the question is "cash or credit" the word "cash" does not rule out payment by check. There are times when our experience with a word is so strong that a disambiguating context does not get noticed. For example, there are many contexts in which we are asked whether we own or rent the place where we live, and given that pair of choices, the word "own" is taken as including situations in which the house is mortgaged. There are two acceptable ways of talking about the home mortgage situation: one is that we own the house but are paying off a loan to the bank; but we can also say that the bank owns the house (and can act on that fact if we fall behind in our payments) but we are in the (long) process of buying it. In question #13 from the NHIS draft, we find the word "own" as a member of a contrast set which could suggest the stricter interpretation:

13. Is this house/apartment (i) owned, (ii) being bought, (iii) rented or, (iv) occupied by some other arrangement by you (or someone in the household)?[3]

Here the respondent's decision is apparently supposed to be influenced by the fact that the second element in the list is the phrase "being bought." That phrase should probably be understood as referring to ongoing mortgage payments. Its presence as the second response option would likely induce someone who did not immediately grab the "owning" option to choose "being bought" if the house was mortgaged. But some mortgage holders might out of habit choose "own" before hearing, or without paying attention to, the full list of choices, and might not suspect the need to adjust their interpretation in the context of the larger field of alternatives.

A self-contradiction is something which, because of its meaning, cannot be true. Some survey questions in GSS that identify polar opposite positions on a scale oddly associate the middle value with the words "I AGREE WITH BOTH

---

[3]Itemizing with i–iv added by the author for illustration purpose.

ANSWERS." Probably most people would not even notice the words that label the intermediate value, and will choose that intermediate value if they cannot go with either end of the scale. But there might be some fussy people, who have had experiences in school with trick questions, who read the text carefully, realize that they would be contradicting themselves by choosing the middle value, and are troubled by the fact that the escape choice this time is "I haven't made up my mind"!

One of these examples is presented here in Figure 12.4. The choice at one end is that the government is doing too much; the opposing choice is that the government is not doing enough. The interviewer reads the text and asks the respondent to look at a card with five positions identified on it. The trick, from the point of view of the suspicious interviewee, is that the language of the text that is read aloud is different from the language on the card he or she is supposed to look at. People who choose the middle value based on what the interviewer reads to them are doing something reasonable, but they are not following the instructions to "Read card 23." Those who choose the middle value while reading the card either do not understand what is happening or are willing to contradict themselves.

While we are all aware of the occasional dangers of agreeing to the terms of contracts that we have not carefully read, we also know that in most of the friendly communicative transactions we take part in, we have no need to rely on a close reading of the written material we encounter. If a friendly interviewer

14. [read aloud by interviewer] "Now look at Card 23. Some people think that the government in Washington is trying to do too many things that should be left to individuals and private businesses. Others disagree and think that the government should do even more to solve our country's problems. Still others have opinions somewhere in between. Where would you place yourself on this scale, or haven't you made up your mind on this?"

Card 23:

I STRONGLY AGREE THAT THE GOVERNMENT SHOULD DO MORE.	I AGREE WITH BOTH ANSWERS.	I STRONGLY AGREE THAT THE GOVERNMENT IS DOING TOO MUCH.

Figure 12.4 Question from General Social Survey.

tells us that the middle choice in a scale means "somewhere in between" the two extremes, and if that same interviewer tells us to "look at Card 23," we are likely to assume that we will see what we have been told is there and so we do not need to look carefully. It may be that nobody in the history of this survey has ever noticed what is written on "Card 23," but it could not hurt to make the text of the list of choices congruent with the purpose of the question.

Some problems of language interpretation have to do with putting the pieces of a sentence together. What exactly is the function of the first phrase in question #14?

> **14.** Because of a health problem, does anyone in the family have difficulty walking without using any special equipment? (NHIS draft)

It would seem that if someone needs help to get around, that in itself is evidence of "a health problem." I simply do not know the function of that initial qualifying phrase in this question.

In a related example, how should one interpret the "if" clause in question #15?

> **15.** Do you strongly agree [. . . etc.][4] that methods of birth control should be available to teenagers between the ages of 14 and 16 if their parents do not approve? (GSS)

It would be more natural to have said "EVEN if their parents do not approve." It is known that people generally hear conditional sentences as expressing two-way implications, and if that applies in this case, the question might raise in people's minds the bizarre possibility of providing children with birth control technology if their parents do not approve, and of denying it to them if their parents do approve. I am sure that people can guess the intention, in spite of the awkward phrasing, but they should not have to guess.

The "because" clause in question #16 introduces a presupposition that might be confusing to some.

> **16.** Family life often suffers because men concentrate too much on their work. (GSS)

If we are merely being asked to comment on the truth of this statement, we know that it presupposes that men concentrate too much on their work and we are asked to decide whether this is a frequent cause of suffering in family life. The question could also be asking us to decide whether we should agree or disagree with someone who makes that statement, and in such a case we are free to reject the presupposition. To avoid problems with the presupposition, the

---

[4]The "[. . . etc.]" here stands for a listing of the whole range of choices from "strongly agree" to "strongly disagree."

question really has to be broken into two: "Do men concentrate too much on their work?" and if the answer to that is positive, "Does that fact cause family life to suffer?"

In a number of cases, questions in the surveys exhibit some sort of semantic bizarreness. If this has the effect of adding to other sources of confusion, it should be avoided. Consider question #17.

> **17.** Given the world situation, the government protects too many documents by classifying them as SECRET and TOP SECRET. Do you strongly agree [. . . etc.]? (GSS)

What part of the sentence is qualified by the phrase "Given the world situation?" Is it that the current world situation (on the nature of which the sentence tells us nothing) motivates the government to be overly protective? Or is it that the present world situation motivates the author's judgment that the government's protection of documents is excessive? The question would be less confusing, and would probably yield more reliable answers, if the phrase "Given the world situation" were omitted.

Words that belong in the same category can be thought of as *coordinate*. Examples of sets of coordinate terms are color names like blue, green, red, or kin terms like father, mother, uncle, sister, and so on. A certain amount of cognitive dissonance appears when one expects but does not find coordinate terms. Most of the proffered responses to question #18 are the names of countries.

> **18.** From what countries or part of the world did your ancestors come? (GSS)

But one of the response categories listed is "American Indian." Of course there is no way to ask the question to have it include American Indians in a natural way. Maybe the choice "American Indian" is to be thought of as a way of rejecting the question's presupposition.

In some cases we find that a sentence is simply grammatically awkward, as in question #19.

> **19.** I often do not tell my friends something that I think will upset them. (GSS)

The intended meaning must be "I seldom say things to my friends that I think might upset them." What is strange about this sentence is that the word "often" precedes the negation of a verbal expression that does not designate a prototypically periodic activity. Eating spinach is, for most of us, not something which occurs at regular intervals, and it is therefore odd to say "I often don't eat spinach," whereas "I often don't eat breakfast" sounds quite natural; occasions of "not eating breakfast" can, in a sense, be counted. Such periodicity should not exist in the act of giving bad news to friends, so in question #19 the structure is odd.

Another example of grammatical aberrance is seen in the slightly strange use of the simple past tense form in questions #20 and #21.

20. Did you have a head cold or chest cold that started during the past two weeks?

21. Did you have a stomach or intestinal illness with vomiting or diarrhea that started during the past two weeks?

The simple past tense ("Did you have") is out of place unless there is a definite temporal reference period that does not include the present moment. The questions should read "Have you had a head cold [etc.] in the past two weeks which started during this period?"

In the cases of semantic bizarreness or grammatical aberrance I have just examined, it is unlikely that the actual wording would make it impossible for the respondent to understand what is being asked, or to know how to answer it. Such uses of language reinforce the idea that the way language is used in these encounters is so different from ordinary English that special interpretation strategies are called for; unfortunately, we cannot always find out what interpretation strategies individual respondents create for themselves.

12.4 CONCLUSIONS

A striking fact about the examples I have commented on is their variety. Difficulties have been found in (a) the use of words that are vague or ambiguous and grammatical constructions whose semantic functions are unclear, (b) the presentation of contexts which invite tension between the ordinary meanings of words and assumptions about the purposes of the survey encounter, (c) conflicts between ordinary meanings and the intentional or unintentional contrasts evoked by the surrounding text, and (d) contextually given or linguistically signaled presuppositions which respondents cannot always escape. It might seem that a complete list of the places where things can go wrong could be prepared, and from that, one could write up a checklist of suggestions and warnings on chart paper to be tacked up on the wall behind each item writer's computer, and that would be the end of it. But in too many cases, the problems have emerged from unanticipated interactions of various aspects of linguistic structure and communicative context. I would prefer to think that, in the survey research industry, a strong interest could be created in making sure that item writers and evaluators become sensitized to all aspects of the workings of language, and that, maybe, roles for linguistic professionals might be defined to fit the needs and objectives of survey research.[5]

[5]There are dangers in being an outsider. In an earlier version of this chapter, I commented on some of the questions in a manner that revealed my ignorance of the ways in which the questionnaires were administered. I have no reason to be confident that I have not done the same this time around.

REFERENCES

Aitchison, J. (1987). *Words in the Mind.* Oxford, England: Blackwell.

Akmajian, A., Demers, R., Farmer, A., and Harnish, R. (1995). *Linguistics: An Introduction to Language and Communication,* 4th ed. Cambridge, MA: MIT Press.

Allan, K. (1986). *Linguistic Meaning* (2 volumes). London, England: Routledge and Kegan Paul.

Cruse, D. A. (1986). *Lexical Semantics.* Cambridge, England: Cambridge University Press.

Federal Register Notice (Oct. 30, 1997). Revisions to the standards for the classification of federal data on race and ethnicity. *Notices, 62(210),* 58781–58790.

Green, G. (1989). *Pragmatics and Natural Language Understanding.* Hillsdale, NJ: Erlbaum.

Grice, H. P. (1957). Meaning. *Philosophical Review, 66,* 377–388.

Grice, H. P. (1975). Logic and conversation. In P. Cole and J. L. Morgan (Eds.), *Syntax and Semantics: 3. Speech Acts,* pp. 41–58. New York: Academic Press.

Lehrer, A. (1974). *Semantic Fields and Lexical Structure.* Amsterdam: North-Holland.

Levinson, S. (1983). *Pragmatics.* Cambridge, England: Cambridge University Press.

Lyons, J. (1977). *Semantics* (2 volumes). Cambridge, England: Cambridge University Press.

O'Grady, W., Dobrovolsky, M. and Aronoff, M. (1997). *Contemporary linguistics: An Introduction,* 3rd ed. New York, NY: St. Martin's Press.

Sagan, C. (1996). *The Demon-Haunted World: Science as a Candle in the Dark.* New York, NY: Ballantine.

Searle, J. (1969). *Speech Acts.* Cambridge, England: Cambridge University Press.

CHAPTER 13

The Use of Computational Cognitive Models to Improve Questions on Surveys and Questionnaires

Arthur C. Graesser, Tina Kennedy, and
Peter Wiemer-Hastings
The University of Memphis

Victor Ottati
Purdue University

No one would dispute the value of a computer aid to assist the designer of surveys and questionnaires. The computer aid would have particular modules that critique each question on the various levels of language, discourse, and world knowledge. For example, the critique would identify cumbersome syntax, words that are unfamiliar to most respondents, terms that are ambiguous in the discourse context, questions that overload working memory, and questions that seem unmotivated (out of the blue). In an ideal world, the computer aid would offer suggestions on how to correct problematic questions. Ideally, the tool would be so sophisticated, reliable, and accurate that it would print out improved questions automatically.

The bad news is that it is extremely difficult to develop a computer model that can comprehend questions at all levels: language, discourse, and world knowledge. During the last 10 years, the Department of Defense has evaluated the best computer models of information extraction in the fields of artificial intelligence, computational linguistics, and cognitive science (DARPA, 1995; Jacobs, 1992; Lehnert, 1997). There has been noticeable progress in automating components of language analysis that lie within the span of a sentence and

Cognition and Survey Research, Edited by Monroe G. Sirken, Douglas J. Herrmann, Susan Schechter, Norbert Schwarz, Judith M. Tanur, and Roger Tourangeau.
ISBN 0-471-24138-5   © 1999 John Wiley & Sons, Inc.

short discourse segments, such as identifying the correct senses of words with multiple senses, parsing sentence syntax for sentences that are short or moderate in length, and extracting important information with respect to a narrow topic (e.g., financial news, news articles about terrorism). However, the systems are very limited in handling deep comprehension and lengthy stretches of discourse. Although the progress in the field is impressive from the standpoint of science and perhaps very specific practical applications, the field is not at the point where it can supply an impressive automated computer aid for designers of questionnaires and surveys.

Nevertheless, there is some good news on the horizon for those who can appreciate incremental gains. The development of a cognitive computational model does not need to be perfect in order to be useful. It need not solve all of the problems that confront the designers of questionnaires. The aid could offer advice about those components in which it can deliver reliable feedback. Some of these components are so complex, technical, or subtle that they are invisible to the unassisted human eye, even the eye of an expert in question-naire design or the eye of an accomplished computational linguist. For exam-ple, it would be impossible for these experts to catch all of the problems in sentence syntax and working memory load. Very few experts would have the time and patience to dissect each question at such a fine grain. A computer aid would be useful even if it produced occasional errors in diagnosis, such as identifying a misleading presupposition that really does not pose a problem to the respondent. Such faulty diagnoses would be eliminated when the human experts scrutinize the computer output. We envision a computer aid that is used collaboratively with a human expert on questionnaire design, so the human can always supersede and make the final decision about each suggestion offered by the computer. The computer aid would be analogous to a "spellcheck" facility in most word processing packages; the computer suggests incorrect spellings, but it is the human writer who ultimately decides the proper spelling of each word.

In this chapter, we hope to accomplish two goals. First, we want to defend the objective of building a computer aid to benefit designers of questionnaires. The computer aid could either be used as an automated tool (if it ends up being accu-rate, reliable, and fully automated), or questionnaire designers could be trained to apply the theories that lurk behind the tool. Second, we will describe some contemporary computer systems that have the potential to correct some of the problems with bad questions on surveys, forms, and questionnaires. The com-puter models that we have in mind are grounded in cognitive science, a field that integrates contributions from artificial intelligence, computational linguis-tics, and psychology. In the past, efforts to apply cognitive science principles to questionnaire design have not been very systematic. These principles from cognitive science have the potential of theoretically fortifying the practice of questionnaire design in addition to providing practical tools.

13.1 THE VALUE OF COGNITIVE COMPUTATIONAL MODELS FOR QUESTIONNAIRE DESIGN

Cognitive scientists have developed a number of computer models of question asking and question answering during the last 20 years (Graesser and Franklin, 1990; Graesser, Gordon, and Brainerd, 1992; Graesser and Person, 1994; Kass, 1992; Lauer, Peacock, and Graesser, 1992; Lehnert, 1978; Robertson, 1994; Schank, 1986; Woods, 1977). Most of these computer models first "parse" the question by transforming the surface sequence of words into a logical form that can be handled by the computer language. Then the computer "interprets" the meaning of the logical form and matches this interpretation with the contents of world knowledge. The world knowledge is normally structured in the form of a large network of nodes (i.e., concepts, propositions) that are connected by relations (e.g., TYPE-OF, IS-A, CAUSES). Finally, a question–answering procedure searches through the structured database of world knowledge and fetches an answer to the question.

These computational models are potentially useful if they uncover representations and processes that are relevant to the design of questions in surveys and questionnaires. For many of the components (i.e., both representations and processes) of the computer model, there will be a corresponding component that is potentially relevant to the enterprise of designing questionnaires. Ironically, much can sometimes be learned from a malfunctioning or limited computer model. It is an interesting computational exercise to systematically break or remove the various components of a computer model and to observe what happens to the quality of the output. Analogously, a bad question on a questionnaire can be scrutinized from the standpoint of malfunctioning components in the computer model. This application of a computer model hopefully enhances the process of diagnosing faulty questions on questionnaires.

Graesser, Bommareddy, Swamer, and Golding (1996) recently reported how Graesser's QUEST model of human question answering can be used in the enterprise of questionnaire design. The details of this QUEST model are not discussed in the present chapter because they are discussed extensively in Graesser et al. (1996) and elsewhere (Graesser et al., 1992; Graesser and Franklin, 1990; Graesser, Lang, and Roberts, 1991; Graesser, McMahen, and Johnson, 1994). QUEST specifies the cognitive procedures and strategies that humans execute when they answer 19 different categories of questions. Some of these categories are open-class questions that permit a small number of legal response alternatives, such as *verification* questions (Is X true? "Are you a citizen of the United States?") and *disjunctive* questions (Is X, Y, or Z the case? "Are you male or female?"). Some question categories invite short answers, such as *concept completion* questions (Who? What? When? Where? "Who is your physician?") and *quantification* questions (How many? What is the value of a variable? "How many children do you have?"). Most of the question categories invite lengthy descriptions in the answers, such as *causal antecedent* questions (What caused event X to occur? "Why did you lose your job?"), *goal orientation* questions (What goals motivated action X? "Why did you move to Tennessee?"), and *comparison* questions (How is X similar to/different from Y? "What is the difference between a dividend and interest?"). A hybrid question is an amalgamation of two question categories. For example, the following question would be a hybrid between the goal orientation and disjunctive categories: "Why did you move to Tennessee? ____ for a job; ____ for family; ____ other."

The QUEST model has four major components which together generate the answers to questions:

1. *Question interpretation.* QUEST parses the question syntactically, identifies referents of nouns, segregates presuppositions, interprets predicates (i.e., verbs, adjectives), and isolates the focus of the question. The question category is also identified in this component.

2. *Access to relevant information sources.* QUEST activates the relevant generic knowledge structures (e.g., scripts, stereotypes, and other packages of world knowledge) and specific knowledge structures (i.e., episodic memories).

3. *Pragmatics.* QUEST identifies the common ground (shared knowledge) and the goals of the questioner and respondent.

4. *Convergence to relevant answers.* QUEST searches through the vast landscape of relevant knowledge structures and produces the very small subset of nodes that constitute the good answers to the question.

Although most of the components and subcomponents of QUEST have been successfully implemented on a computer, others have been too challenging to automate with much success. Some of these components are similar, but not strictly identical, to the models of the question response process in the questionnaire design literature (Cannell, Miller, and Oksenberg, 1981; Sudman, Bradburn, and Schwarz, 1996; Tourangeau, 1984).

Graesser et al. (1996) identified 12 potential problems with questions that would be anticipated by the QUEST model. These are listed and briefly described in Table 13.1. The brief descriptions of the problems are self-explanatory, but interested readers can consult Graesser et al. (1996) for more details and concrete examples. Most of these potential problems are quite familiar to experts in survey methodology who have devised checklists and other methods for diagnosing specific flaws with problematic questions (Bickart and Felcher, 1996; Fowler, 1993; Jobe and Mingay, 1991; Lessler and Forsyth, 1996). However, the fact that the 12 problems are grounded in a detailed computational model has some advantages that will be discussed shortly. It should be noted that our list of 12 problems with questions is probably not exhaustive, but it did handle 96 percent of the problems that we identified when examining dozens of forms and questionnaires. The list of problems will presumably grow somewhat as the science of questionnaire design evolves further. Although the 12

Table 13.1   Problems with Questions

1. Complex syntax	The grammatical composition is embedded, dense, or structurally ambiguous
2. Working memory over-load	Words, phrases, or clauses impose a high load on immediate memory
3. Vague or ambiguous noun-phrase	The referent of a noun-phrase, noun, or pro-noun is unclear or ambiguous
4. Unfamiliar technical term	There is a word or expression that very few respondents would know the meaning of
5. Vague or imprecise pred-icate or relative term	The values of a predicate (i.e., main verb, adjective or adverb) are not specified on an underlying continuum
6. Misleading or incorrect presupposition	The truth value of a presupposed proposition is false or inapplicable
7. Unclear question cate-gory	It is difficult to determine what class of question is being asked
8. Amalgamation of more than one question category	The question may be assigned to two or more different classes of questions
9. Mismatch between ques-tion category and answer option	The question invites one set of answer options that is different from the question options in the questionnaire
10. Difficult to access specific or generic knowledge	A typical respondent would have difficulty re-calling the information requested in the question
11. Respondent unlikely to know answer (no information source)	A typical respondent would not know the in-formation requested in the question
12. Unclear question purpose	The respondent would not know why the question is being asked

categories are conceptually distinct, they are sometimes interdependent and correlated. For example, a question might suffer from having an unclear purpose (category 12) if there is an unfamiliar technical term (category 4) or if the respondent is unlikely to know an answer (category 11). Any given question can suffer from multiple problems, as will be demonstrated below.

In order to illustrate some of the problems in Table 13.1, consider the following problematic question. This question is on a questionnaire that hundreds of women have completed in a women's health clinic in Memphis.

Did your mother, father, full-blooded sisters, full-blooded brothers, daughters, or sons ever have a heart attack or myocardial infarction?
( ) NO
( ) YES

We would argue that this questions suffers from problems 2, 3, 4, 6, 10, and 11 in Table 13.1. This question imposes *working memory overload* in at least two ways. The first noun-phrase is long and cumbersome; the respondent is forced to keep track of a long list of 6 or more family members. The respondent is asked whether each of these family members has had a heart attack or myocardial infarction so there is a 6×2 matrix of implicit embedded questions for those respondents who believe that a heart attack might be different from a myocardial infarction. A long list or matrix of questions is too much to keep track of in a working memory that has limited capacity (Baddeley, 1986; Just and Carpenter, 1992). The question potentially has an *ambiguous noun-phrase* for respondents with adoptive parents. This is especially the case for those who do not induce the purpose of the questionnaire, namely to assess whether there are particular medical problems in the respondent's biological history. The expression "myocardial infarction" is undoubtedly an *unfamiliar technical term* for the majority of the respondents. For most respondents who are childless and from small families, there would be *incorrect presuppositions*; they would not have any full-blooded sisters, full-blooded brothers, daughters, and/or sons. It might be difficult or impossible to know whether some family members have had a heart attack or an infarction, so the question potentially suffers from problems 10 and 11 in Table 13.1. This is especially true for respondents who were not raised by their biological parents.

Graesser et al. (1996) reported that approximately 1 out of 5 questions on everyday forms and questionnaires suffer from at least one of the problems in Table 13.1. They conducted a study in which expert judges (who were trained on the QUEST model and the 12 problems in Table 13.1) were asked to identify problematic questions and the specific problems with each problematic question. In one of the studies, there were five forms: the 1040 income tax form (75 questions), the 1990 census form (102 questions), an application for graduate admission to the University of Memphis (44 questions), a dentist intake form (74 questions), and an application for a job at Kinko's (42 questions). The likelihood of a question having a particular problem listed in Table 13.1 varied from 0.006 to 0.057. At least one question suffered from each of the 12 problems anticipated by the QUEST model. However, it is important to acknowledge that the problems identified by QUEST or the trained judges might not prove to be problems for the actual respondents. The respondents could potentially correct or adjust to the problems diagnosed by QUEST.

13.1.1 Novices Cannot Identify Most of the Problems That QUEST Identifies

The above results suggest that the QUEST model is a promising approach to identifying bad questions on questionnaires. However, there are much stronger tests that bolster the argument that a computational model is worth its weight in gold. For example, it might be the case that a typical literate adult could readily identify the 12 problems. If so, then the model would not be very useful. It would be better to simply interview a sample of respondents and have them critique the questions on the questionnaire. In fact, survey researchers have frequently advocated the collection of think-aloud protocols from a sample of respondents during pretesting (Jobe and Mingay, 1991; Lessler and Sirken, 1985; Willis, Royston, and Bercini, 1991). We are quite sympathetic to the collection of think-aloud protocols during pretesting, but we believe that a computational model takes us one step further in identifying bad questions.

We conducted a study that assessed how readily the 12 problems in Table 13.1 could be identified by college students who are novices about questionnaire design. The respondents were 24 undergraduates at the University of Memphis. Each respondent gave a written critique of 38 questions that were extracted from the census form, dentist intake form, application to graduate school, and Kinko's job application form. These were all questions that had at least one problem in Table 13.1, according to the trained judges in Graesser et al. (1996). The respondent was instructed to write down two problems with each question. We analyzed whether these novices identified the problems that would be identified by QUEST.

Table 13.2 reports the results from the study. The first column of num-

Table 13.2 Mean Proportion of Items from Sample Questionnaires Which Are Identified as Having Specific Problems, Segregated by Trained Versus Novice Judges

	Mean Proportion		Discrimination
	Trained	Novices	Score of Novices
1. Complex syntax	0.158	0.041	0.03
2. Working memory overload	0.447	0.013	0.03
3. Vague or ambiguous noun-phrase	0.368	0.198	0.35
4. Unfamiliar technical term	0.289	0.126	0.32
5. Vague or imprecise predicate or relative term	0.237	0.023	0.05
6. Misleading or incorrect presupposition	0.526	0.264	0.06
7. Unclear question category	0.105	0.049	0.15
8. Amalgamation or more than one question category	0.198	0.008	0.01
9. Mismatch between question category and answer option	0.236	0.027	0.02
10. Difficult to access specific or generic knowledge	0.211	0.093	0.01
11. Respondent unlikely to know answer (no information source)	0.289	0.087	0.07
12. Unclear question purpose	0.658	0.167	0.03

bers presents the likelihood that a question suffered from a particular problem, according to two expert judges (i.e., graduate students who were thoroughly trained on the QUEST model and the 12 problems in Table 13.1). For the present purposes, we will define these expert judgments as the output from the QUEST model. The two judges agreed on 80 percent or more of judgments as to whether a particular question had a particular category of problem. The data in the first column indicate that the most frequent problems were: unclear question purpose, misleading/incorrect presupposition, and working memory overload. Complex syntax and unclear question category were the two least frequent problems.

The second column in Table 13.2 presents the mean proportion of questions in which the novice college students articulated the 12 problems. These proportions were consistently lower than the proportions of the expert judges, as we would expect. The third column is a measure of how discriminating the novices were in identifying a problem. The discrimination score is perfect (1.0) if the novices always articulate that some problem A exists for those questions that QUEST (i.e., the two expert judges) regards as having problem A, and never articulate problem A for those questions that QUEST regards as unproblematic with respect to A. More specifically, suppose that the likelihood of *correct identification* equals the conditional probability [p(problem A identified by novice|QUEST identifies problem A)], and that the likelihood of inappropriate identification equals the conditional probability [p(problem A identified by novice|QUEST does *not* identify problem A)]. The discrimination score equals: [(correct − inappropriate)/(1 − inappropriate)]. A discrimination score of 0 indicates that the subject is totally nondiscriminating in identifying whether or not a question has a particular problem.

The novices' data in Table 13.2 supports the claim that untrained literate adults are very poor at identifying the problems that QUEST identified. The only problem that novices can readily identify are problems 3 (vague or ambiguous noun-phrase) and 4 (unfamiliar technical term). Their identification scores and discrimination scores were highest for these two problems and approached zero for the remaining 10 problem categories. The novices did have a modest proclivity to identify questions with incorrect presuppositions and with unclear purposes (the two most frequent problems identified by QUEST and the trained judges), but the novice adults were not discriminating in doing so. The novices virtually never commented that a question overloaded memory, the third most frequent problem identified by QUEST. To conclude, novices are very poor at identifying the problems that QUEST identifies. This empirical outcome demonstrates the utility of QUEST, over and above the use of think aloud protocols and other forms of pretesting from samples of respondents.

At this point we are uncertain how well the QUEST model would compare to other analytical schemes for identifying problematic questions, such as the system developed Lessler and Forsythe (1996). There are no data that compares QUEST and other analytical schemes with a sample of questionnaire designers. Empirical research needs to be conducted to contrast alternative systems for diagnosing problematic questions. We also are uncertain whether the problems identified by QUEST significantly threaten the validity and reliability of the questions in a bona fide sample of respondents. The problems identified by QUEST may or may not be problems for a sample of respondents; similarly, problems in a sample of respondents may or may not be identified by QUEST.

13.1.2 QUEST Provides a Guide for Revising Questions

If QUEST is a useful tool, it should help the designer revise questions in a fashion that provides more reliable and valid answers. We recently conducted two studies that confirm that QUEST-inspired revisions do produce more reliable answers. A question is defined as being reliable if it produces the same answer when the respondent answers the same questions on two occasions. A reliability score for question Q is defined as the proportion of subjects who do not change their answers when they answer Q on two occasions.

In one of the two studies, we had 43 college students answer the 38 problematic questions (discussed above) on two occasions. However, we orthogonally varied whether they received the original problematic question (version O) or a QUEST-inspired revised question (version R) on the two occasions. Therefore, there were four conditions: OO, RR, OR, and RO. The assignment of conditions to the questions was counterbalanced across respondents so that each subject received approximately an equal number of questions in each condition and each question was assigned to each condition for approximately an equal number of respondents. If the QUEST-inspired revisions are the most reliable questions, then the highest reliability scores should be in the RR condition; in this condition, the subjects receive the revised question on both occasions. This expected outcome was in fact found when we analyzed the data. The reliability scores significantly differed among the four conditions, with means of 0.95, 0.90, 0.79, and 0.79 in the RR, OO, OR, and RO conditions, respectively, $F(3,126) = 29.48$, $p < 0.05$. Statistical tests uncovered the following ordering of means: RR > OO > OR = RO.

A second study focused on problem 12 (unclear question purpose). As a reminder, this was the most frequent problem that QUEST-trained judges found to be the case for the questions in our sample (see Table 13.2). The present study investigated whether the reliability of answers would be affected if the purpose of the question was given in the questionnaire. Unlike our previous studies, the present study used questions about sensitive material, such as sexual habits, alcohol, and drugs. For example, two of the questions were "Did you have any unwanted sexual experiences over the past year?" and "Were you physically abused by being hit, slapped, pushed, shoved, punched, or threatened with a weapon by a family member or close friend?" We expected that the purpose of the study and question would be critical information in determining whether respondents would volunteer information about such sensitive information. For example, the purpose of asking the above two questions would be "we are interested in knowing how much control you have over most of your daily activities and sexual experiences." The respondents were 16 women who participated in the Women's Health Initiative in Memphis. The 30 sensitive questions consisted of a subset of the questions on the questionnaires that are normally administered in the clinic.

The results revealed that the reliability of the answers is strongly affected when respondents are presented the purpose of the question. When the respondents received no purpose on one occasion of answering a question and then were given a purpose on another occasion, the intertrial reliability showed a significant 26 percent decline. This was a dramatic decline compared to a 1 percent decline in a condition in which no purposes were generated on one occasion and the respondents generated perceived purposes of the question on another occasion. (Unfortunately, the respondents were not given another occasion in which they were either presented or not presented with the purpose of the question.) The major point we wish to convey is that the answers to the sensitive questions are radically affected by the announced purposes of asking the questions. The reliability (and no doubt validity) of the answers are dramatically affected if the purpose of the question is vague or if the induced purpose fluctuates at different stages of a questionnaire.

13.1.3   A Computational Model Provides Analyses of Questions That Are Too Subtle for Experts in Questionnaire Design

Our strong claim is that a computational model, such as QUEST, provides a deeper and more detailed analysis of questions than is supplied by an expert in questionnaire design. The weaker claim is that the insights provided by the computational model are different from the human experts, but nevertheless useful. In either case, colleagues would not quibble with the wisdom of having expert human designers collaborate with the computer aid during the course of designing and revising questions.

Our weak claim is substantiated to the extent that we can show how computer tools provide useful analyses that are too cumbersome, complex, or subtle for the expert designer. We hope to provide this defense in the next section. Regarding our strong claim, additional research is needed to hold the race between the computational models and the expert designers.

13.1.4   Computational Models That Facilitate the Design of Questions

This section describes some of the recent computational models and software tools that should do some useful work in a computer aid that diagnoses bad questions. These models stretch beyond our QUEST model and have not yet been implemented in the context of survey questions per se. However, we are convinced that an effort that augments QUEST with these components would be a very worthwhile project.

Lexicons   Dozens of lexicons are available commercially and in research settings. A lexicon is a large dictionary of words (or familiar phrases) along with meanings, alternative senses (i.e., a *bat* is both a sports object and a mammal), syntactic frames that surround the word, semantic features, synonyms, and other types of content. The world-wide web currently has many special purpose dictionaries, including those that cover computer terms, internet terms, acronyms, slang expressions, geographical regions and landmarks, a NASA thesaurus, and so on. Even phone books are on the internet, so it is possible to identify addresses, telephone numbers, and email addresses of individuals. Some of these lexicons extract characteristics of words from a large corpus of printed text, such as Wordnet (Miller, Beckwith, Fellbaum, Gross, and Miller, 1990). This allows us to identify what senses of words are typically associated with particular knowledge domains (e.g., sports versus zoos) and which words are most frequent in the language. Wordnet can identify the most likely sense of a word in a sentence context at well above 90 percent accuracy. The system by Brill (1996) is quite accurate in identifying the parts of speech of words in context. It is not necessary for these computer systems to be perfect in their classification of words because other components in a natural language processing system might correct any errors made by the lexicon.

These lexicons could be readily integrated with our computer aid. It would be a trivial exercise to have the computer flag low-frequency lexical items as "unfamiliar technical terms" and thereby help solve problem 4 in Table 13.1. Only a subset of these flagged lexical items might be truly technical and some compound expressions might be missed (e.g., "private parts"). However, once again, the tool need not be perfect in order to be useful. A computer aid that is 70 percent accurate would end up doing some useful work.

One direction for questionnaire designers would be to develop a lexicon that contains vague or imprecise relative terms (see problem 5 in Table 13.1). Respondents sometimes have trouble accurately interpreting predicates whose meanings are relative rather than absolute, as is the case with adjectives (wealthy, large), adverbs (occasionally, often, frequently), and certain classes of verbs (e.g., hurt, slept, healed). Wordnet and other computerized lexicons have the potential of being expanded to identify problem 5 because they already classify words in adjectives, adverbs, and verbs (and subcategories of these). Each lexical entry would have a feature +/− RELATIVE. Whenever the predicate is +RELATIVE, the computer aid would print out an expression, such as "when you say *often*, does the respondent know how *often* compares to other values?"

The computer aid could potentially identify noun-phrases that are vague or ambiguous (problem 3). The computer would develop an accumulating discourse log of the noun referents in the entire questionnaire and how recently each referent appeared in a question. When a pronoun is used and there are multiple referents in the accumulating log that could match the referent, then the pronoun would be flagged as potentially ambiguous; the computer would print out the message "When you say *it*, does the respondent know whether this pronoun refers to "diet" or "evening?" When a noun is introduced for the first time in the questionnaire, the computer could mark it as potentially vague and print out the message: "Does the respondent know what you mean by "diet" or what "diet" you are referring to?" Kieras (1989) has developed a computer aid for technical writing that could be expanded to implement this type of computer aid. A simple clue like a definite article ("the" diet) is supposed to mean that the diet has already been introduced in the questionnaire (and is in the log). An indefinite article ("a" diet) is supposed to mean that the referent has not yet been introduced. This accumulating log might also be useful in simulating some of the empirical findings that have shown how answers are affected by context and the order of questions in a survey (Ottati, Riggle, Wyer, Schwarz, and Kuklinski, 1989; Schwarz, 1994; Sudman et al., 1996; Tourangeau and Rasinski, 1988).

Syntactic Parsers   The development of syntactic parsers is a cottage industry in the field of computational linguistics. These parsers vary in sophistication and computational architectures. The FASTUS system (Hobbs, Appelt, Tyson, Bear, and Israel, 1992) relies on simple pattern matching operations that identify particular words and word sequences. However, most parsers translate a surface string of words into a structured meaning representation (such as a deep structure or propositional representation) by performing elaborate sequences of symbolic operations (Allen, 1987; Charniak, 1996; Winograd, 1983). A parser by Berg (Miikkulainen, 1996) adopts a neural network computational architecture rather than a symbolic architecture.

Most of these parsers can scale questions on syntactic complexity. Scales of syntactic complexity would be useful for our computer aid because it would help solve problem 1 in Table 13.1. A sentence would be flagged as having complex syntax if it exceeds some threshold of complexity. Alternatively, one or more syntactic complexity scores could be computed for each question on the questionnaire. The designer could then revise the question in a fashion that reduces the complexity. As it turns out, there are multiple scales of complexity. A deep structure description is complex if there is a high ratio of nonterminal nodes (i.e., intermediate constituents) per explicit word. The complexity of *processing* is high if there is a high ratio of processing operations per explicit word. For example, "garden path" sentences have high complexity because the comprehender initially constructs the wrong syntactic structure and ends up reinterpreting the sentence (e.g., "the horse raced past the barn fell.")

A syntactic parser would also be helpful for repairing two other categories of problems in Table 13.1: Misleading or incorrect presuppositions (problem 6) and unclear question category (problem 7). Some syntactic parsers have made substantial headway in segregating propositions that are asserted (or new) versus presupposed (given). For example, the question "When was your last physical exam?" has the presupposition that "the respondent has had a physical exam" whereas the focal question queries when this occurred. The computer aid would isolate each presupposition and print out the following sort of expression to the question designer: "Is it true for all respondents that the respondent has had a physical exam?" Regarding problem 7, the syntactic parser would be used to help identify question categories. Even a rudimentary pattern match parser (Hobbs et al., 1992) would identify verification questions as having word sequences such as "Are you," "Is there," "Do you," and so on. In contrast "How many" and "How much" signals a quantification question and "Why" signals either a causal antecedent, goal-orientation, or expectation category (see Graesser et al., 1996, for a definition of these question categories). Moreover, the syntactic parser would play a role in diagnosing an amalgamation of more than one question category (problem 8 in Table 13.1) and a mismatch between a question category and an answer option (problem 9).

Working Memory Management   It is widely acknowledged that the human mind is constrained by a working memory that is limited in capacity (Baddeley, 1986; Ericsson and Kintsch, 1995; Just and Carpenter, 1992). The capacity of working memory limits both the number of processing operations that can be executed during a time span and the number of units that can be preserved in a passive storage buffer. The implications of these working memory limitations on questionnaire design are perfectly obvious. Questions should be written in a fashion that minimizes the load on working memory. Unfortunately, most designers of questionnaires have not taken this limitation very seriously. They frequently write single questions that include a large number of clauses, qualifiers, and prepositional phrases.

Consider the following problematic question from the 1990 United States census: *Do you have a physical, mental, or other health condition that has lasted for six or more months and which limits the kind of work you do at the job?* This question would undoubtedly overload the working memory of most respondents. It would be easier to process if the qualifying information was unpacked in separate sentences; shorter sentences minimize the load on working memory (Haberlandt and Graesser, 1985; Just and Carpenter, 1992). Shorter sentences, with few new ideas per clause, are more like conversational language and are easier to understand, whereas the language of print is highly embedded, dense and turgid (Clark, 1996; Tannen, 1989). The following sentence would be less demanding on working memory: *Some people have a physical, mental, or other type of "health condition." Do you have a health condition that has lasted for six or more months, and that has limited the kind of work you can do at your job?*

As another example, sentences with "right branching" syntax are easy to process because they first present the main clause (e.g., assertion or question) and subsequently add on clauses and phrases that qualify the first clause. In contrast, sentences with a "left embedded" syntax are difficult because the main clause is never finished until the end of the sentence and working memory must maintain the unfinished information. The following revision has a left embedded construction and would therefore be very difficult to process: *Is there any physical, mental, or other health condition, which has lasted for six or more months and limits the kind of work done on a job, that you have?* Left embedded constructions were alarmingly frequent in the sample of questionnaires that we examined.

Researchers in psycholinguistics and discourse psychology have developed sophisticated models that measure the amount of working memory resources that are consumed by sentences and text during comprehension (Ericsson and

Kintsch, 1995; Goldman, Varma, and Cote, 1996; Just and Carpenter, 1992). For example, Just and Carpenter's CAPS/READER model measures the load on working memory as each word is processed in a sentence. When the load is too high, comprehension breaks down or comprehension time increases dramatically. These computational models of working memory have had impressive fits to human data. For example, the predicted working memory load for the words in sentences have had an extremely high correlation with self-paced word reading times and with gaze durations in eye tracking studies. Readers with a low working memory span are seriously affected by sentences that are demanding on working memory.

Designers of questionnaires can benefit from these computational models that simulate the load on working memory during sentence comprehension. After each question is received by the program, the computer would measure working memory load as each word is processed. The load has been measured in many ways: number of nodes held in working memory, number of new nodes constructed for word W, number of processing operations performed at word W, and number of processing cycles that scan the contents of working memory. The questions would be revised until they met some established standards of working memory load. A sentence would be flagged if any word in the sentence exceeded a standard threshold of working memory load.

Latent Semantic Analysis   So far we have avoided the large abyss of world knowledge and deep comprehension. Attention to both of them is ultimately necessary for complete solutions to all of the 12 problems in Table 13.1. Yet these are the difficult challenges to those who develop cognitive computational models of comprehension (Graesser, Singer, and Trabasso, 1994; Graesser, Millis, and Zwaan, 1997). There currently is not a single computer program that would impress anyone in its ability to understand natural language and integrate the appropriate packages of world knowledge.

Once again, however, a partial solution may be useful to those who design questionnaires. In this section, we report one partial solution that seems promising. The traditional approach to representing world knowledge in artificial intelligence has been structured representations, such as semantic networks and conceptual graphs (Lehmann, 1992). However, world knowledge is frequently open-ended, imprecise, vague, and incomplete, so these traditional approaches have not adequately handled the role of world knowledge in understanding language. "Latent Semantic Analysis" (LSA) has recently been proposed as a statistical representation of a large body of world knowledge (Foltz, Britt, and Perfetti, 1996; Landauer and Dumais, 1997). LSA capitalizes on the fact that particular words appear in particular texts; the co-occurrence of words in texts reflects the constraints that exist in world knowledge.

The input to LSA is a co-occurrence matrix that specifies the number of times that a word W_i occurs in a text T_j. These frequencies are adjusted with a logarithm transformation that also corrects for the base rates of words appearing across texts; a word is a distinctive index for a text to the extent that its occurrence in the text is above the base rate for that word across texts. Statistical data reduction methods compress the large $W \times T$ co-occurrence matrix to K dimensions (typically, 100–300 dimensions, depending on the scope of the knowledge domain). Each word, sentence, or text ends up being a weighted vector on the K dimensions. The *match* (i.e., similarity in meaning, conceptual relatedness) between two words, sentences, or texts is computed between the two vectors, with values ranging from 0 to 1. The match between two language strings can be high even though there are few if any words in common between the two strings. This is because LSA goes well beyond simple string matches; the meaning of a language string is partly determined by the company (other words) that each word keeps.

The empirical success of LSA has been remarkable even though it does not achieve a complete understanding of a topic. Landauer and Dumais (1997) created a LSA representation with 300 dimensions from 4.6 million words that appeared in 30,473 articles in Grolier's Academic American Encyclopedia. They submitted to the LSA representation the synonym portion of the TOEFL test, a test developed by the Educational Testing Service to assess how well nonnative English speakers have mastered the words in the English language. The test has a four-alternative, forced choice format, so there is a 25 percent chance of answering the questions correctly. The LSA model selected the alternative that had the highest match with a comparison word. The LSA model answered 64.4 percent of the questions correctly, which is essentially equivalent to the 64.5 percent performance for college students from non-English speaking countries. In another study, Foltz et al. (1996) created a LSA representation with 100 dimensions from 31 texts and encyclopedia articles on the Panama Canal. College students read a sample of the texts and then wrote an essay that summarized the readings on the Panama Canal. The quality of the essays were scored by LSA by computing an average similarity score between each sentence in the essay and the highest matching sentence in the corpus. The quality of the summaries scored by LSA were nearly as reliable and valid as the quality ratings of expert human raters.

LSA may be meaningfully applied to the design of questionnaires that focus on a particular knowledge domain. It is possible to evaluate the *truth* of propositions, presuppositions, questions, and answers to questions with respect to the LSA spaces. When a linguistic expression has a low match with LSA space, the computer aid would raise the question to the designer of whether the respondent will know the meaning of the question: "This question does not have much to do with the knowledge base. Will the respondent know what this question means?" The *relevance* of an expression can also be evaluated. Feedback to the designer can be given when there is a low match between (a) question Q and the previous set of questions on the questionnaire and (b) a question and an answer (or response option). LSA can also be used to score the quality of essays and lengthy verbal answers to a question. We believe that the LSA approach will prove to be quite successful wherever a questionnaire taps a restricted semantic domain.

13.2　CLOSING COMMENTS

There are many skeptics who doubt whether computational models will ever comprehend language. No doubt there will be skeptics among those who conduct research on surveys and questionnaires. Our prognosis is much more optimistic, but hardly naive. Each decade, computers inch closer to being able to comprehend language, but it is unlikely that we will see impressive products in our lifetimes. Nevertheless, an imperfect computer aid has the potential of doing some useful work for questionnaire designers.

REFERENCES

Allen, J. (1987). *Natural Language Understanding*. Menlo Park, CA: Benjamin/Cummings.

Baddeley, A. D. (1986). *Working Memory*. New York: Oxford University Press.

Bickart, B., and Felcher, E. M. (1996). Expanding and enhancing the use of verbal protocols in survey research. In N. Schwarz and S. Sudman (Eds.), *Answering Questions: Methodology for Determining Cognitive and Communicative Processes in Survey Research*, pp. 115–142. San Francisco: Jossey-Bass.

Brill, E. (1996). Some advances in transformation-based part of speech tagging. *Proceedings of the 13th National Congress on Artificial Intelligence*, 722–727. Cambridge, MA: MIT Press.

Cannell, C. F., Miller, P. V., and Oksenberg, L. (1981). Research on interviewing techniques. In S. Leinhardt (Ed.), *Sociological Methodology*, pp. 389–437. San Francisco: Jossey-Bass.

Charniak, E. (1996). Tree-bank grammars. *Proceedings of the 13th National Congress on Artificial Intelligence*, 1031–1036. Cambridge, MA: MIT Press.

Clark, H. H. (1996). *Using Language*. Cambridge, England: Cambridge University Press.

DARPA (1995). *Proceedings of the Sixth Message Understanding Conference* (MUC-6). San Francisco: Morgan Kaufman Publishers.

Ericsson, K. A., and Kintsch, W. A. (1995). Long-term working memory. *Psychological Review*, *102*, 211–245.

Foltz, P. W., Britt, M. A., and Perfetti, C. A. (1996). Reasoning from multiple texts: An automatic analysis of readers' situation models. *Proceedings of the 18th Annual Conference of the Cognitive Science Society*, pp. 110–115. Mahwah, NJ: Erlbaum.

Fowler, F. J. (1993). *Survey Research Methods*. Newbury Park, CA: Sage.

Goldman, S. R., Varma, S., and Cote, N. (1996). Extending capacity-constrained construction integration: Toward "smarter" and flexible models of text comprehension. In B. F. Britton and A. C. Graesser (Eds.), *Models of Understanding Text*, pp. 73–114. Hillsdale, NJ: Erlbaum.

Graesser, A. C., Bommareddy, S., Swamer, S., and Golding, J. M. (1996). In N. Schwarz and S. Sudman (Eds.), *Answering Questions: Methodology for Determining Cog-

nitive and Communicative Processes in Survey Research, pp. 143–174. San Francisco: Jossey-Bass.

Graesser, A. C., and Franklin, S. P. (1990). QUEST: A cognitive model of question answering. *Discourse Processes*, *13*, 279–303.

Graesser, A. C., Gordon, S. E., and Brainerd, L. E. (1992). QUEST: A model of question answering. *Computers and Mathematics with Applications*, *23*, 733–745.

Graesser, A. C., Lang, K. L., and Roberts, R. M. (1991). Question answering in the context of stories. *Journal of Experimental Psychology: General*, *120*, 254–277.

Graesser, A. C., McMahen, C. L., and Johnson, B. K. (1994). Question asking and answering. In M. Gernsbacher (Ed.), *Handbook of Psycholinguistics*, pp. 517–538. San Diego, CA: Academic Press.

Graesser, A. C., Millis, K. K., and Zwaan, R. A. (1997). Discourse comprehension. In J. T. Spence, J. M. Darley, and D. J. Foss (Eds.), *Annual Review of Psychology*, Vol. 48, pp. 163–189. Palo Alto, CA: Annual Reviews Inc.

Graesser, A. C., and Person, N. K. (1994). Question asking during tutoring. *American Educational Research Journal*, *31*, 104–137.

Graesser, A. C., Singer, M., and Trabasso, T. (1994). Constructing inferences during narrative text comprehension. *Psychological Review*, *101*, 371–395.

Haberlandt, K., and Graesser, A. C. (1985). Component processes in text comprehension and some of their interactions. *Journal of Experimental Psychology: General*, *114*, 357–374.

Hobbs, J., Appelt, D., Tyson, M., Bear, J., and Israel, D. (1992). Description of the FASTUS system used for MUC-4. *Proceedings of the 4th Message Understanding Conference*. San Mateo, CA: Morgan Kaufmann.

Jacobs, P. S. (Ed.). (1992). *Text-Based Intelligent Systems: Current Research and Practice in Information Extraction and Retrieval*. Hillsdale, NJ: Erlbaum.

Jobe, J. B., and Mingay, D. J. (1991). Cognition and survey measurement: History and overview. *Applied Cognitive Psychology*, *5*, 175–192.

Just, M., and Carpenter, P. (1992). A capacity theory of comprehension: Individual differences in working memory. *Psychological Review*, *99*, 122–149.

Kass, A. (1992). Question asking, artificial intelligence, and creativity. In T. Lauer, E. Peacock, E., and A. C. Graesser (Eds.), *Questions and Information Systems*, pp. 303–360. Hillsdale, NJ: Erlbaum.

Kieras, D. (1989). An advanced computerized aid for the writing of comprehensible technical documents. In B. Britton and S. Glynn (Eds.), *Computer Writing Environments*, pp. 143–168. Hillsdale, NJ: Erlbaum.

Landauer, T. K., and Dumais, S. T. (1997). A solution to Plato's problem: The latent semantic analysis theory of acquisition, induction, and representation of knowledge. *Psychological Review*, *105*, 211–240.

Lauer, T. W., Peacock, E., and Graesser, A. C. (Eds.). (1992). *Questions and Information Systems*. Hillsdale, NJ: Erlbaum.

Lehmann, F. (Ed.). (1992). *Semantic Networks in Artificial Intelligence*. New York: Pergamon.

Lehnert, W. G. (1978). *The Process of Question Answering: A Computer Simulation of Cognition.* Hillsdale, NJ: Erlbaum.

Lehnert, W. G. (1997). Information extraction: What have we learned? *Discourse Processes, 23,* 441–470.

Lessler, J. T., and Forsyth, B. H. (1996). A coding system for appraising questionnaires. In N. Schwarz and S. Sudman (Eds.), *Answering Questions: Methodology for Determining Cognitive and Communicative Processes in Survey Research,* pp. 259–291. San Francisco: Jossey-Bass.

Lessler, J. T., and Sirken, M. G. (1985). Laboratory-based research on the cognitive aspects of survey methodology: The goals and methods of the National Center for Health Statistics study. *Milbank Memorial Fund Quarterly/Health and Society, 63,* 565–581.

Miikkulainen, R. (1996). Subsymbolic case-role analysis of sentences with embedded clauses. *Cognitive Science, 20,* 47–74.

Miller, G. A, Beckwith, R., Fellbaum, C., Gross, D., and Miller, K. (1990). Five papers on WordNet. Cognitive Science Laboratory, Princeton University, No. 43.

Ottati, V. C., Riggle, E. J., Wyer, R. S., Schwarz, N., and Kuklinski, J. (1989). The cognitive and affective bases of opinion survey responses. *Journal of Personality and Social Psychology, 57,* 404–415.

Robertson, S. P. (1994). TSUNAMI: Simultaneous understanding, answering, and memory interaction for questions. *Cognitive Science, 18,* 51–86.

Schank, R. (1986). *Explanation Patterns: Understanding Mechanically and Creatively.* Hillsdale, NJ: Erlbaum.

Schwarz, N. (1994). Judgment in social context: Biases, shortcomings, and the logic of conversation. In M. Zanna (Ed.), *Advances in Experimental Social Psychology, Vol. 26,* pp. 123–162. San Diego, CA: Academic Press.

Sudman, S., Bradburn, N. M., and Schwarz, N. (1996). *Thinking About Answers: The Application of Cognitive Processes to Survey Methodology.* San Francisco: Jossey-Bass.

Tannen, D. (1989). *Talking Voices: Repetition, Dialogue, and Imagery in Conversational Discourse.* Cambridge, MA: Cambridge University Press.

Tourangeau, R. (1984). Cognitive sciences and survey methods. In T. Jabine, M. Straf, J. Tanur, and R. Tourangeau (Eds.), *Cognitive Aspects of Survey Methodology: Building a Bridge Between Disciplines,* pp. 73–100. Washington, DC: National Academy of Sciences.

Tourangeau, R., and Rasinski, K. A. (1988). Cognitive processes underlying context effects in attitude measurement. *Psychological Bulletin, 103,* 299–314.

Willis, G., Royston, P., and Bercini, D. (1991). The use of verbal report methods in the development and testing of survey questionnaires. *Applied Cognitive Psychology, 5,* 25–267.

Winograd, T. (1983). *Language as a Cognitive Process.* Reading, MA: Addison-Wesley.

Woods, W. A. (1977). Lunar rocks in natural English: Explorations in natural language question answering. In A. Zampoli (Ed.), *Linguistic Structures Processing,* pp. 201–222. New York: Elsevier.

CHAPTER 14

The View from Anthropology: Ethnography and the Cognitive Interview

Eleanor R. Gerber

U.S. Bureau of the Census

The aim of this chapter is to discuss the contributions which anthropology has made and might make to the field of survey research, and in particular to cognitive methods for testing questionnaires. Evaluating survey responses often requires an understanding of the context of beliefs and social knowledge within which survey responses are situated. Cognitive anthropology can provide systematic ways of describing this broader context. Some concepts are discussed in the chapter which assist in the description of sets of terms used naturally by respondents. However, the investigation of sets of terms is of limited usefulness in fully describing a conceptual domain. The concept of a "schema" provides a more complex and flexible model of the way that cultural domains are structured.

14.1 ETHNOGRAPHY

Anthropological methods are primarily qualitative in nature, and are designed to produce descriptions of culture. The written descriptions may themselves be called "ethnography," or the term may refer to the process of gathering the information on which anthropological analysis rests. The techniques associated with ethnography stress complete and highly specific knowledge of a

Cognition and Survey Research, Edited by Monroe G. Sirken, Douglas J. Herrmann, Susan Schechter, Norbert Schwarz, Judith M. Tanur, and Roger Tourangeau. ISBN 0-471-24138-5 © 1999 John Wiley & Sons, Inc.

group of people. Typically, the term refers to both observation and interviewing. In this paper, the term "ethnography" refers primarily to the latter technique: in-depth, unstructured, informal interviewing. Through such interviewing and other extended social contact, the anthropologist's understanding will become more complex, more grounded, and interconnections between events, people, and ideas will emerge.

The capacity of ethnography to produce detailed and intimate knowledge of a culture was termed "thick description" by the anthropologist Clifford Geertz, following the philosopher Gilbert Ryle. Geertz (1977, pp. 9–10) describes the ethnographic endeavor in these terms: ". . . ethnography is thick description. What the ethnographer is in fact faced with—except when he is pursuing the more automatized routines of data collection—is a multiplicity of complex conceptual structures, many of them superimposed upon or knotted into one another, which are at once strange, irregular, and inexplicit, and which he must contrive somehow first to grasp and then to render. And this is true at the most down-to-earth, jungle field work levels of his activity: interviewing informants, observing rituals, eliciting kin terms, tracing property lines, censusing households . . . writing his journal." Because ethnography is capable of producing insights into complexity that can only arise from an intimate knowledge of a group of people, it makes a natural complement to the more distant way of collecting data represented by surveys.

14.1.1 Combining Ethnography and Survey

Two models for combining ethnographic and survey techniques have been discussed in the literature. In the first model, the "ethnographic" part of the research is done first, and informs subsequent research, for example, locating a sample for further research (Stepick and Stepick, 1990) or as an initial step which aids in the design of a survey instrument (e.g., Bently et al., 1988). The ethnography is used to provide a quick social description of the phenomenon under study. Such studies may be considered "preliminary," and not be thought of as an integral part of the research process. The ethnographic studies done in this mode may be essentially conceived of as a separate piece of research, conceptually distinct from the later survey research project.

Another model of combined research seeks to fully integrate the ethnographic component into the research process. It has been called the "ethnosurvey" (Massey, 1987; Massey, Alarcon, Durand, and Gonzalez, 1987.) This survey instrument was designed to be ethnographically sensitive and was administered by individuals trained in fieldwork. Thus, the "questionnaire" that was developed did not contain standard question wordings, but consisted of a grid of topics which could be reworded and introduced in whatever order the ethnographer–interviewer deemed most appropriate. This technique seems somewhat similar to recommendations made a few years later by Suchman and Jordan (1990). However, this aspect of the ethnosurvey did not arise out of concern for the linguistic and conversational qualities of the interview situation. Rather, it was primarily intended as a way of reducing the sensitivity of questions about illegal immigration.

Other systematic ways of combining ethnography with survey results have been suggested. One interesting example is the attempt by Friedenberg, Mulvihill, and Caraballo (1993) to use vignettes as a way of assessing the cultural sensitivity of a questionnaire. They suggest that the aim of a culturally sensitive survey is to capture the "inside perspective" of the respondent (p. 151). In order to measure how well this aim has been achieved, the authors designed research in which ethnography, a vignette study, and a survey were carried out in parallel. The degree of congruence between the data collected in each method could then be assessed mathematically. Friedenberg and her associates were somewhat surprised to find areas where the ethnography was not as accurate as other data collection techniques. "Our main assumption was that the ethnography was essentially 'closer to the ground' and thus more accurately reflected the 'truth' of a culture" (p. 155).

I would suggest that this is the underlying assumption which most combinations of ethnography and survey techniques share. In these descriptions of the fusion of ethnography and survey research, an interesting dichotomy arises. The ethnography is seen as providing the "truth" or "groundedness" (see above) or "richness and detail." These are contrasted with "rigor and generalizability" (Massey et al., 1987, pp. 11–12) or being "convincing" to policy makers and other researchers (Friedenberg et al., 1993, p. 152). Behind this dichotomy, of course, lies the problem of representativeness of the group in which the ethnography is done. Ethnographies generally are written about bounded, interacting communities, and do not claim to be "representative" of wider populations from which the communities are drawn. What ethnography can provide is a sense of the interconnectedness of beliefs and events in the social world. It is the very fact that it can capture the complexity of social life that makes ethnography useful. Survey questions by necessity segment and isolate events: they become sequential and bounded, though the actions and ideas that they represent are part of an interconnected web of significances. Thus, the very function of ethnography is what Geertz is referring to when he speaks of "thick description." The usefulness of ethnography in the survey context arises from its ability to represent complexity.

14.1.2 Surveys and Ethnography: Current Practice

In order to examine how ethnography has been used recently in a survey context, we conducted a search of data bases to find abstracts of published articles which describe the use of both survey methodology and ethnography. Four data bases were searched: Social Science Abstracts, Psychological Abstracts, Dialog, and ERIC (an education-oriented data base). Journal articles were examined if the abstracts or key words contained any of the terms "ethnography," "anthropology," or "in-depth interviews" along with either of the terms "survey" or "questionnaire." If the journal represented a data collection in which both of those methods were used on (roughly) the same population during (pretty much) the same research project, the article was accepted as an instance of the combination of the two techniques. Unpublished work or ethnographies published as separate research would not have been identified in this search. It should be noted that some instances of both "surveys" and "ethnography" might not meet strict standards. However, in order to count as a "survey" in this research, some standardized questions had to be asked, and in order to count as "ethnography," some period of observation or unstructured interviewing had to have taken place. Using these methods, 126 examples of publications, mostly journal articles, describing the combination of surveys and ethnography were identified. The journal articles were then coded for three characteristics: the location of the research, the general subject matter examined, and, for those studies done in the United States, the group of people who were the primary subjects.

Location   Table 14.1 shows the country where research employing both methodologies was carried out. Forty-nine of the 126 papers reported on research carried out in other countries. Africa and Asia account for more than half of the locations outside of the United States.

Subject Matter   Table 14.1 also indicates the general subject matter of surveys found in these data bases. Probably the most striking aspect of this distribution is the heavy concentration of research in two topic areas, medicine and education. Of the 126 papers examined, 99 were in these two fields (51 in medicine and 48 in education). The use of ethnography in medical research is an established practice. "Medical anthropology" is a subdiscipline, with separate curricula in many anthropology programs. Within the United States, half of the medical papers using both ethnography and survey techniques (13 of 26) relate to alcohol and drug abuse. The use of ethnography to study drug users can be seen as a natural extension of the concerns of medical anthropology, since the "subculture" of drug users affects both the course of the disease and medical attempts at intervention. Ethnography has been used in education research primarily in the form of what is called "classroom ethnography." Classrooms have been studied by observation in part to better understand the implicit messages which are communicated to school children by teachers, curricula, and the classroom environment in general. It has been pointed out that student teachers traditionally spend many hours in "observation," and classroom ethnography has provided a way to structure and intensify this experience (Levin, 1997).

Groups   The racial or ethnic groups which were the target of combined ethnographic and survey research were also of interest. Racial and ethnic groups are seldom specified in research conducted outside the United States, so this analysis only examined the 77 studies done here. Whites were generally not specified

Table 14.1 Number of Studies Using both Ethnographic and Survey Techniques in United States and nonUnited States Locations, by Topic of the Study

				Topic			
	Medical	Education	Migration/ Demography	Organization	Attitudes	Other	Total
United States	26	33	4	3	6	5	77
NonUnited States	25	15	0	0	4	5	49
Total	51	48	4	3	10	10	126

as a studied group[1] but were perhaps referred to as "college students," "super-visors in a large plant," etc. Thus, we have compared papers mentioning minor-ity race/ethnic groups to papers where no race/ethnic identification is given. Thirty-four of 77 United States-based studies specify research on a minority group population. The composition of the "no racial/ethnic group mentioned" category is also of interest. The subjects of these studies were often identified by social characteristics. Almost half (20 of 43) of these studies were con-cerned with what we might consider to be "special populations" rather than mainstream roles. They included drug users, disabled persons, homeless indi-viduals, residents in poor communities, and immigrants. If the 20 studies of "special populations" are added to the 34 studies of minority racial and ethnic groups, they account for about two-thirds of the 77 United States-based studies combining ethnography and survey techniques.

In fact, authors who discuss the relation of surveys and ethnography often assume that blended ethnographic and survey techniques are intended for the study of "special populations" (Friedenberg et al., 1993; Hines, 1993; Massey et al., 1987). It appears that ethnography is primarily used where groups are thought to be either unreachable (like IV drug users) or different enough to present difficulties in understanding. In general then, the conclusion of this brief analysis is that formal ethnographic techniques are combined with surveys pri-marily in studies of minority and other special populations, and generally on topics related to medicine and education, where ethnography has an established history. This indicates that anthropology in general, and formal ethnographic techniques in particular, have not been widely absorbed into survey research.

Despite its formal use in only a few restricted areas, a form of ethnography has become part of the design and testing of survey questionnaires. It is not seen as a distinct research method, and is seldom labeled as ethnography or even understood to be involved in the study of culture. This is the blending of ethnography and the cognitive pretesting of questionnaires.

14.2   ETHNOGRAPHY AND THE COGNITIVE INTERVIEW

Cognitive interviews often take on the function of providing ethnographic infor-mation to survey researchers. From this point of view, one of the functions cur-rently served by cognitive interviewing is what was elsewhere called "back-up ethnography" (Gerber and Wellens, 1996). One of the reasons that cognitive interviewing is currently valued is that it reveals ethnographic detail affect-ing respondents' understanding of the questions. When a cognitive researcher reports that one term is more appropriate than another to the way respondents think and talk, the statement is essentially based on a small-scale description of culture. This function of cognitive interviews is recognizably ethnographic

---

[1] In a few instances where minority populations were compared with Whites, they have been coded under the minority population.

when the terms are associated with a special population. For example, when a survey uses the term "outfit" to mean "IV injection apparatus" or "running buddy" to mean "friends and associates" (Carlson, Seigal, and Wang, 1996), it is easy to see that the terms have been selected because of their ethnographic appropriateness. It is less easy to see the ethnography underlying terms which are not associated with special populations. The suggestion, for example, that the decennial census term "usual residence" is unfamiliar to respondents (Gerber, 1994) because they think in terms of "permanent address" or "home" may not seem to depend on a description of culture. Yet, the only real difference between the two examples is how exotic the terminology seems.

Cognitive interviews have acquired this ethnographic function in part because survey sponsors and authors do not see a need for any prior cultural description on most survey topics. (They may see ethnography as primarily appropriate in exotic small-scale societies.) When the terms and concepts in the original questionnaire instrument are inappropriate, the first place in the research process where this can be discovered is often the cognitive interview. Culturally inappropriate terms may not be discovered earlier in the design process because the questionnaire authors may not be aware of the difficulties that these terms cause for respondents.

The ethnographic functions of cognitive interviews are evident in other ways:

1. *Recruiting for cognitive interviews.* Cognitive interviews take on an implicit ethnographic role when recruiting is based on groups in the population, such as racial or ethnic groups. The assumption underlying this practice appears to be that shared, nonrandom, cultural characteristics in these groups might affect survey responses. The cognitive interview then becomes the vehicle for identifying and evaluating such cultural differences.

2. *Using cognitive interviews to collect information about other cultural variations.* The small number of cognitive interviews which we are able to carry out sometimes makes it unlikely that we will encounter all of the situations which may affect responses to questions. We have used vignettes to present respondents with a wide range of situations relevant to such topics as grandparents caring for grandchildren and living situations of mobile and tenuously attached persons. Vignettes are ethnographic in two different ways. First, if they are to be effective, they must be created with a knowledge of the situations which may actually influence responses to the question. Second, one purpose of using them in this context is to extend the ethnographic "range" of the research to events, roles, and circumstances that available respondents may not otherwise provide.

Thus, cognitive interviews frequently serve ethnographic functions, although most cognitive researchers never call them "ethnography." The question which remains is the following: how well do ethnography and the cognitive interview fit together? Does the cognitive interview provide enough ethnographic detail to evaluate questions fully? Does ethnographic questioning interfere with the central purposes of the cognitive interview in monitoring the response process? These questions are considered in the next section.

14.2.1 Blending Ethnography and the Cognitive Interview

It is our belief that the ad hoc blend of ethnography and cognitive interviewing may not be the most efficient way to capture the ethnography relevant to survey questions, and may also interfere with the primary purposes of the cognitive interview. Ethnography requires questioning strategies which elicit matters that go beyond the specific question context. However, such questioning may interfere with the main aim of the cognitive interview, because it will introduce other concepts and terms to which respondents might not otherwise attend. This could influence responses to subsequent questions. By the same token, the narrow view given by cognitive probing may not provide enough flexibility to elicit all the terms and concepts which have a bearing on the subject matter of the question. Thus, the ethnographic interview requires too broad a definition of context to serve as good cognitive research, and the cognitive interview requires too narrow a definition of context to serve as good ethnographic research. An example of each kind of interviewing will clarify the point.

Ethnographic interviews on the concept of residence revealed semantic differences between the words "live" and "stay," which had both been used, somewhat interchangeably, on Census Bureau questionnaires and surveys. These semantic meanings were revealed during in-depth interviews in which respondents discussed a variety of living situations to respondents (Gerber, 1994) and their own history of personal experiences with living situations (Gerber, 1990). A questioning strategy using vignettes and semistructured protocols elicited some of the relevant factors associated with the terms. For example, we discovered that where a person's belongings (such as furniture and clothing) are located is an important element in deciding the permanence of an attachment. Other factors like the quality of the individual's relationship with other household members and particular expectations about the future also play a role in deciding about residence. We introduced and manipulated hypothetical situations to discover concept boundaries and the salience of the criteria for respondents. Thus, our strategy included questions like "Suppose that person had only moved out half of his stuff, would he still be 'living' there?" and "What if you expected the person to be there temporarily, but now it's been a year, is that still 'staying' there?" In addition, we also elicited related terminology, and discovered related residence terms that were used by some of the respondents (such as "floaters" to describe persons with attachments to many different places).

Subsequent cognitive interviews we did on rosters containing the terms used a very different interview strategy. During the cognitive portion of the interview, we allowed respondents to complete the section of the roster containing the terms, and then probed only briefly about the meaning of the terms "live" and "stay" in that particular context (Gerber, Wellens, and Keeley, 1996).

It is difficult to imagine eliciting full ethnographic detail about residence in the same interview that also investigates question wording and format. If we were to interrupt the flow of the questionnaire to ask the respondent about ethnographic detail (such as the location of belongings and expectations about the length of a stay in the previous example), these discussions would very likely contaminate subsequent responses to cognitive interview probes related more narrowly to the questionnaire content. It is believed that both forms of questioning and the kinds of information they elicit are important to a full evaluation of a survey question. A good example of this can be found in the cognitive research we did on the decennial census race and ethnicity questions. In this research, an understanding of both the narrow question context and the wider ethnographic background are useful.

In our cognitive interviews, we probed the meaning of the concept of "race" ("Can you tell me in your own words what 'race' means to you in this question?"). A common strategy respondents used was to define race in terms of "ancestral geography" (Gerber, de la Puente, and Levin, 1997). A typical definition in this mode was illustrated by the comment "Race; that's where your family was originally from." Our respondents' use of geography to define the concept of race may have been conditioned by the presence of the large number of geographical terms in the decennial race question, which includes a long list of Asian countries of origin. Indeed, one respondent was aware of the effects of the question context on her definition of race: "Well, in *this* question it looks like you are talking about nationality." Geographical thinking also influenced respondents' other reactions to the questions, such as the suggestion that the term "White" be replaced by "European American," or writing responses such as "American" into any available write-in space.

However, the reactions discussed above do not seem to be entirely an artifact of the way the survey question was asked. In fact, when questions were asked that included a smaller number of geographically based terms (although there were always some), we still observed many definitions of race in this geographical mode. If a geographical element is a part of American cultural understandings about race, reactions such as wanting a "European American" category or writing "American" into an available write-in box are unlikely to disappear completely even if the question is reformatted or revised.

It is a standard aim of cognitive-oriented survey research to understand how respondents interpret relevant concepts within the specific context of the question. The point emphasized here is that comparing the specific question context and the broader ethnographic description may also be useful. This comparison permits us to judge how planned question changes may affect response. A more complete ethnography could inform the way future questions might be written about this subject matter.

Therefore, analysis of survey questions can make use of the comparison of two independent, yet related sources of data: ethnographic description and specific probes about the question in context. Both are important for a complete assessment of respondents' reactions to survey questions. The rest of the chapter discusses some questioning strategies and concepts which may be useful in examining the ethnographic context of surveys.

14.2.2 Cognitive Anthropology and Cognitive Interviews: Terminology

Cognitive anthropology arose out of the attempt to study culture by examining terminological systems. Sets of terms were assumed to codify important persons, relationships, objects, and events, so discovering the structure of terms was considered an important ethnographic step (Frake, 1969). D'Andrade (1995) has described the development of the study of terminological systems. Various techniques and modes of analysis were developed for understanding sets of terms. Much of the scientific effort was aimed at finding ways to uncover the systematic relationships between terms. When terminological systems are analyzed, the aim is often to discover the principles (called features and dimensions) underlying the classification system. In the current context, however, a formal analysis of the underlying principles of classification may not be particularly useful, but it is often useful to elicit additional terms in the domain that the survey question involves.

For example, the terms "live" and "stay" discussed earlier proved to be in contrast with the term "visit." Understanding how the term "visit" (which was never used in our questionnaires) contrasts with the term "stay" was useful in assessing how respondents used the latter term. Respondents often needed to discuss "visits" in order to define what they meant by "live" and "stay." In particular, the term "visit" provides a boundary for the term "stay;" that is, stays are temporary, but visits are *really* temporary. A sensitivity to contrast relations is often useful in analyzing interview protocols from cognitive interviews. When a respondent is asked about living and staying, and begins to describe "visiting" and "homes," this is not off the topic. Rather, the respondent is providing important information about the meaning of terms in which the research is primarily interested.

The various questioning strategies to elicit sets of terms and systematic ways of analyzing them have been described extensively elsewhere (D'Andrade, 1995; Frake, 1969; Hines, 1993; Spradley, 1979). The main point for this argument is that there may be advantages to knowing something about the terms which surround the particular terms used in survey questions.

14.2.3 Cognitive Anthropology—Schemas

Cognitive anthropology no longer focuses exclusively on sets of terms. This change resulted from the inability of this mode of analysis to describe complex areas of culture or complex cognitive patterning. In particular, sets of contrasting terms are not an efficient way to understand the relationships which exist between persons, objects and events in the social world (D'Andrade, 1995). One concept which has been adopted by some anthropologists to represent such complexity is the "schema."[2] A good definition for our purposes is the following: "A schema is a conceptual structure which makes possible the identification of objects and events . . . Basically, a schema is . . . a procedure by which objects or events can be identified on the basis of simplified pattern recognition . . . To say that something is a 'schema' is a shorthand way of saying that a distinct and strongly interconnected pattern of interpretive elements can be activated by minimal inputs. A schema is an interpretation which is frequent, well organized, memorable . . ." (D'Andrade, 1992, pp. 28–29).

The concept arose from cognitive psychology and linguistics (D'Andrade, 1992, 1995). There is no commonly accepted definition of a schema, which may result in methodological difficulties, especially in specifying the content of a particular schema. However, the psychological uses of the concept are not central to the current argument. What is of importance is that schemas necessarily contain a great deal of cultural content. The reasons for this are apparent. If schemas allow us to interpret the world, and the world and its significances are primarily culturally constituted, then schemas allow us to interpret culture. Schemas also codify unique, personal experience, but the central insight of anthropology is that we essentially make sense out of experience through cultural means, so culture influences perception and action at every level. The relationship between schemas and culture is so close that, as D'Andrade points out, psychologists studying schemas were forced to become ethnographers to a certain extent: "One interesting by-product of the work on schemas in psychology and artificial intelligence was the careful formulation of particular pieces of American culture. Cultural schemas for restaurants, doctor–patient office visits, chess gambits, social contracts, American–Russian nuclear confrontation, the evaporation of water, story grammars, and birthday parties were all used in psychology experiments and artificial intelligence simulations" (D'Andrade, 1995, p. 126).

Anthropologists have been interested in using the concept of schemas for a variety of reasons. It provides a way of coordinating goals and motivation into the description of cultural understandings, which may in turn help to integrate cultural understandings with behaviors (Strauss, 1992). It also provides a way of examining the salience of various cultural elements to particular individuals, the variation of schemas from one person to another, and the way in which cultural systems are learned (Holland, 1992). Strauss and Quinn (1994) have outlined some of the implications of the concept of schemas for our understanding of the concept of culture.

1. Schemas are learned through a combination of personal, idiosyncratic learning and standardized public means. Thus, we learn our concepts of parenting both from our own experience as children and from public representations.

2. Schemas are shared. Although each person learns schemas through personal experience, as the authors point out, cultures are organized to ensure that

---

[2]D'Andrade points out that other terms, such as frame, scene, scenario, and script have been used to express much the same concept (1995, p. 122).

people have many of the same experiences. However, there is a "continuum of sharedness": some schemas are shared by nearly everyone in the same culture, and others by smaller groups, like a "workplace culture."

3. Schemas are relatively durable. They are durable within individuals because they are "well-learned, automatic, and self-reinforcing." They are also somewhat stable historically, as they are passed down through the generations.

4. Schemas are "relatively thematic"; that is, certain schemas may be applied repeatedly in a wide variety of contexts. They will be evoked over many situations and enacted in a wide range of specific ways. An American example is the concept of "self-reliance." This "thematic" schema can be encountered in numerous social acts and events. Thus (to extend Strauss and Quinn's example), the idea of "self-reliance" may be enacted in the context of child rearing ("It's time you learned to do it for yourself!"); medicine ("I am NOT sick and needy!") and politics ("Put welfare recipients to work!"); and these are only a few of the possibilities.

Another important point made about cultural schemas is that, in order for events to make sense, they must first be referred to a relevant schema. There are often many schemas within a culture in which a single event might find a significance. It is necessary to choose, to locate the event within one of them, before acting on it mentally or socially. Hutchins' study of Trobriand reasoning (Hutchins, 1980) stresses this point. In his account, litigants in a local land dispute must first refer perceived events to one of several available schemas. Hutchins refers to this process as "instantiation," that is, determining which of the available schemas a particular event exemplifies.

Thus, when a Trobriander is thinking about land tenure, identifying an event as an instance of one of the schemas for owning or transferring land is critical. The same event, a gift of yams, may be regarded in one schema as return reciprocity with no consequences for land transfer, or in another, as the reactivation of powerful land rights. A less distant example might be found in an event like a politician staying in a villa in the Bahamas belonging to a long-time associate. To make sense of the event, it can either be referred to the "friendship" schema or the "bribery" schema.

Some of these specific concepts may have consequences for the way that survey questions are written and cognitively pretested, and the next section will discuss some of these. It is also worth pointing out that the "schema" concept provides a good model for ethnography. It is capable of representing the complex interconnections between cultural understandings. It bridges the gap between public and private significances, and between shared and unique patterning. A good description of a schema provides the richness and explanatory power that is the hallmark of ethnography.

14.2.4 Schemas and the Cognitive Interview

Thinking in terms of schemas may provide a way to understand how ethnographic background affects responses to survey questions. This section describes some schema-related problems, and gives examples of the way that they can be usefully analyzed using the schema model.

Instantiation Survey questions often elicit responses in terms of more than one schema. An example arises in our own work on residence rules for the decennial census. Residence rules impinge on many cultural schemas, and it is not always clear to respondents which one ought to apply. The choice may have different consequences for survey response. For example, a roomer or boarder might count as a resident, if the instantiation is in terms of the rent he pays. However, as a nonrelative, a roomer is likely to be excluded from a household list if the "family" schema is invoked (Gerber, 1994).

To determine how much of a problem alternate instantiations can cause for a question, we need to know the relative strength of associations the question creates with the alternate schemas. This does not depend on question context alone, but also on the salience and "durability" of the various cultural schemas. Another residence-based example may illustrate this point. Decennial residence rules require that live-in workers, such as housekeepers and nannies, should be counted in the places where they work, following the census logic that people should be counted where they stay most of the time. Similarly, "commuter workers" who spend the week away from home to be closer to their jobs are intended to be counted where they stay during the week. Time does affect residence judgments for respondents, but other factors seem to have more of an effect. In this instance, interpretations involving the work schema are powerful. Respondents are nearly unanimous that residences "just for work" do not count as real residences. This appears to be based on beliefs about the nature of the social relations in employment. For example, no matter how long they have persisted, relations of employment are considered temporary and are not to be trusted (Gerber, 1994). This schema is powerful and coherent, and much more likely to affect responses than alternate interpretations, such as the amount of time spent in the place.

It might be thought that the solution to the problem of multiple schemas is to provide instruction to the respondent about what to do in ambiguous cases. This solution may not always be effective. Research indicates that the likelihood of respondents following such instructions about residence varies by the content of the rule (Gerber et al., 1996). Residence situations where the "just for work" idea applied were the least likely of all the situations included in our study to be decided in line with the census residence rules (whether or not the respondents were given an instruction telling them what to do). However, when rules were given for family members living away from home in institutional settings like college dorms, the instructions made a statistically significant difference in their performance.

The consistency of the alternate instantiations which are available may explain why showing respondents the rules makes a difference only in certain cases. For the work-related residences discussed above, a second path of interpretation reinforces the "just for work" interpretation. If a housekeeper or commuter worker is described as returning periodically to family, respondents generally think of the "family" place as the residence. Thus, both of the available paths of interpreting work-related residences lead respondents to conclusions which are contrary to the census rule. However, in the case of the institutional residences, the available interpretations lead in different directions. For example, it seems probable that college students living in campus dormitories may be instantiated into schemas about "maturing and becoming independent" or perhaps even a schema for "moving out." Thus, even though many parents would naturally place college students' residence with family, there exists at least one schema which would favor the opposite interpretation. Perhaps the existence of these opposite paths renders the interpretation less certain, and the respondent therefore more likely to seek or follow instructions.

A fuller understanding of the available schemas that surround a question would allow us to view the question–answering process (at least in some cases) as a choice between alternate interpretations. This research might also have consequences for interpreting cognitive interview results: it may help to decide when to ignore an interpretation that comes up only once in a set of cognitive interviews, and when to take it seriously.

Networks of Association   Cultural schemas are broad networks of associations, and when survey questions activate one part, the associative network leads out to the entire schema. Therefore, associations they create cannot be limited to the intended question content. This can affect response patterns. Understanding response patterns may therefore require the investigator to understand the schema more fully than a narrow interpretation of question content would demand. Our work on race provides a good example of this phenomenon (Gerber et al., 1997). The term "race," while not entirely clear to respondents, was clear enough so that the general question content was not ambiguous. However, respondents are acutely conscious of aspects of this schema which go beyond the criteria for choosing a race category to describe someone. Respondents may react to the question as a part of an American dialogue about race, rather than simply as a request to put themselves into a racial category.

One phenomenon which this awareness produced was a minute examination of the offered categories to determine if one's own group is being treated "fairly." In our study, respondents paid close attention not only to the category which applied to their own group, but to the categories available for others as well. They wanted to make sure that the categories were evenhanded, and did not give preferential treatment to specific groups. For example, White respondents sometimes noted that theirs is the only category containing a single term, and on this basis recommended using "White or Caucasian" as a category label. Note that this is not the result of misunderstanding "White" or of thinking that

"Caucasian" means something different or is clearer; it is the result of wanting to be treated the same as other respondents. Similarly, the term "African American" was sometimes problematic for Black respondents because it stressed "Americanness" when, for other groups it was simply assumed. This is illustrated by one respondent who said ". . . the other boxes have not listed; O.K. is this an American Eskimo? Is this an American Chinese or a Chinese from China? Or an American Filipino? So why does it have to be African American? I'm an American. I mean, nobody else is—you don't say Caucasian American."

Other respondents note that some groups have the opportunity to report subgroup information (e.g., the Hispanic origin question, the American Indian write-in line, and the list of Asian and Pacific Islander categories) while other groups do not. Such respondents may write-in an ethnicity like "Italian/Irish" in an available space in the race question (Gerber et al., 1997). Thus, even at the level of answering a race question, parts of the race schema that are involved with demanding equal treatment for one's own group, are activated, and can determine responses to the survey question.

In addition to searching for "fairness," respondents often commonly react to the question as though it were part of a political dialogue about race. Being asked about race leads to associations with social problems connected with race, and responses are affected by this. One respondent replied "What is this person's race? Now, I hate that answer because I think that's what's wrong with this country, trying to divide people like that . . . It's supposed to be Americans, and what does it matter? That's the way I look at it . . . I'll say—the human race!" (Gerber et al., 1997, p. 16).

This concern with the divisiveness of the question sometimes led respondents to look for inclusive ways of responding, like writing in "human" or "American." They also told us that they might skip the question or throw the questionnaire away, indicating that these matters may have a bearing on item nonresponse and overall response rates. Thus, the question may lead to associations with wider parts of the schema, which may control response patterns in unanticipated ways.

14.3 CONCLUSIONS

This chapter has argued that cognitive testing of surveys could benefit from a wider examination of the context of our questions. We have called this wider context "ethnography" because understanding it requires a means of discovering cultural knowledge and beliefs. The concepts used in this discussion have been drawn from cognitive anthropology. There are probably other expressions of the same concepts drawn from other disciplines. We are primarily interested in how such concepts define and structure the study of the wider cultural context of survey questions.

How much wider the investigated context should be is an empirical and practical question. A complete description of American culture is beyond our resources, and would not be necessary in most instances. Some topics may not warrant a full ethnographic treatment, and some research projects may lack time or resources to include an ethnographic component. However, certain "thematic" schemas will have a wide applicability to a variety of survey questions and may be encountered commonly enough to warrant a relatively complete ethnographic treatment. The examples used in this paper are drawn from the ethnographic context of residence and race, but other topics come to mind. Surveys often involve questions about work, income, family relationships, health topics, etc.

Ideally, ethnography should be incorporated into the survey design process at a much earlier phase. I envision using in-depth interviews to give survey authors an account of the way that concepts and terminology are used by respondents. Vignettes might prove to be a useful first step towards quantifying respondents' reactions to particular situations that the survey questions are designed to tap. In this process, subsequent cognitive interviews would be concentrated solely on the questionnaire context and the response process. It does not seem likely that using ethnography in the design phase of research will soon become standard practice. We believe that cognitive researchers will do "back-up ethnography" for some time to come.

The kinds of questioning relevant to ethnography and to the cognitive interview should be regarded as separate and somewhat opposed. Ethnographic questioning leads respondents into a description of the wider schema, while the cognitive interview must stay focused on the question context. We think that survey researchers might consider using some "cognitive" respondents as ethnographic respondents, doing "in-depth" interviews with them to learn more about the ethnographic context. If this is done, the two kinds of questioning should be separated within one interview: the ethnographic questioning should be part of a debriefing after the cognitive interview is finished.

However, even if the ethnography became an early phase of the survey design, cognitive researchers would still need to have a sensitivity to the way in which respondents use the schemas which our survey questions activate. As a result, there is no way to completely separate ethnography from cognitive pretesting. Good cognitive research requires a sensitivity to ethnographic context. Developing ethnographic awareness strengthens our cognitive research and improves our questionnaires.

REFERENCES

Bently, M., Pelto, G., Strauss, W., Shuman, D., Brown, K., and Huffman, S. (1988). Rapid ethnographic assessment: Applications in a diarrhea management program. *Social Science and Medicine, 27,* 107–116.

Carlson, R. G., Seigal, H. A., and Wang, J. (1996). Attitudes toward needle "sharing" among injection drug users: Combining qualitative and quantitative research methods. *Human Organization, 5,* 361–369.

D'Andrade, R. (1992). Schemas and motivation. In R. D'Andrade and C. Strauss

(Eds.), *Human Motives and Cultural Models*, pp. 23–44. Cambridge: Cambridge University Press.

D'Andrade, R. (1995). *The Development of Cognitive Anthropology*. Cambridge: Cambridge University.

Frake, C. (1969). The ethnographic study of cognitive systems. In S. Tyler (Ed.), *Cognitive Anthropology*. pp. 28–41. New York: Holt, Rinehart, & Winston.

Friedenberg, J., Mulvihill, M., and Caraballo, L. (1993). From ethnography to survey: Some methodological issues in research on health seeking in East Harlem. *Human Organization, 52,* 151–161.

Geertz, C. (1977). The interpretation of cultures: Selected essays. *Description: Toward an Interpretive Theory of Culture*, pp. 3–30. New York: Basic Books.

Gerber, E. (1990). *Calculating residence: A cognitive approach to household membership among low income blacks*. Report for the Center for Survey Methods Research, Census Bureau.

Gerber, E. (1994). *The language of residence: Respondent understandings and census rules. Final report of the cognitive study of living situations*. Report for the Center for Survey Methods Research, Census Bureau.

Gerber, E, de la Puente, M., and Levin, M. (1997). *Race, identity, and new question options: Final report of cognitive research on race*. Draft report for the Center for Survey Methods Research, Census Bureau.

Gerber, E., and Wellens, T. (1996, July). *Perspectives on pretesting: 'Cognition' in the cognitive interview?* Paper presented at the Fourth International Social Science and Methodology Conference, Essex, England.

Gerber, E., Wellens, T., and Keeley, C. (1996). Who lives here?: The use of vignettes in household roster research. *Proceedings of the Section on Survey Research Methods, American Statistical Association*, 962–967.

Hines, A. M. (1993). Linking qualitative and quantitative methods in cross-cultural survey research: Techniques from cognitive science. *American Journal of Community Psychology, 21*, 729–746.

Holland, D. (1992). How cultural systems become desire: A case study of American romance. In R. D'Andrade and C. Strauss (Eds.), *Human Motives and Cultural Models*, pp. 61–89. Cambridge: Cambridge University Press.

Hutchins, E. (1980). *Culture and Inference: A Trobriand Case Study*. Cambridge: Harvard University Press.

Levin, P. (1997, March). *Personal communication*. Teacher Education Program, University of California, San Diego.

Massey, D. (1987). The ethnosurvey in theory and practice. *International Migration Review, XXI*, 1499–1522.

Massey, D., Alarcon, R., Durand, J., and Gonzalez, H. (1987). *Return to Aztlan: The Social Process of International Migration from Western Mexico*. Berkeley: University of California Press.

Spradley, J. (1979). *The Ethnographic Interview*. New York: Holt, Rinehart, & Winston.

Stepick, A., and Stepick, C. (1990). People in the shadows: Survey research among Haitians in Miami. *Human Organization, 49*, 64–76.

Strauss, C. (1992). Models and motives. In R. D'Andrade and C. Strauss (Eds.), *Human Motives and Cultural Models*, pp. 1–20. Cambridge: Cambridge University Press.

Strauss, C., and Quinn, N. (1994). A cognitive/cultural anthropology. In R. Borofsky (Ed.), *Assessing Cultural Anthropology*, pp. 248–297. New York: McGraw-Hill.

Suchman, L., and Jordan, B. (1990). Interactional troubles in face-to-face survey interviews. *Journal of the American Statistical Association*, *85(409)*, 232–241.

CHAPTER 15

Survey Error Models and Cognitive Theories of Response Behavior

Robert M. Groves

University of Michigan and Joint Program in Survey Methodology

15.1 INTRODUCTION

This paper describes a set of experimental results in the social and cognitive sciences over the last few years, which together have implications for the specification of statistical models describing nonsampling errors. It now seems clear that the principles underlying many sources of survey error—including coverage, nonresponse, interviewer effects, the impact of question wording and mode—will come from a variety of social science theories.

Theories of nonsampling error must explain the human behaviors that produce the error. These theories are found predominantly in the social sciences. As such, they do not have explicit links to survey estimation. This has produced a mismatch between the findings in experimental studies of nonsampling errors and their use in survey design and estimation. Advances will come when the links are made.

To understand the challenges we face, we must first note that research or sampling and nonsampling errors have different intellectual roots. In contrast to nonsampling error research, sampling statistics focuses much more on variance than bias. It exploits deductions from a well-elaborated base of probability theory rather than hypothesis generation and experimentation. Its history is found more prominently in government statistical agencies than in the academic sector.

Cognition and Survey Research, Edited by Monroe G. Sirken, Douglas J. Herrmann, Susan Schechter, Norbert Schwarz, Judith M. Tanur, and Roger Tourangeau.
ISBN 0-471-24138-5   © 1999 John Wiley & Sons, Inc.

15.2 HISTORY OF RESEARCH PRODUCTS

Figure 15.1 provides a quick overview of the timing and products of research on sampling and nonsampling errors. For each of the major types of errors, it lists early journal papers and the first book-length treatment of the field from both statistics and the social sciences. The figure shows that many of the early works appeared in the 1930's and 1940's. Noncoverage and sampling are areas that appear to be the domain of statistics exclusively. In contrast, nonresponse, interviewer effects, question effects, and mode effects have garnered the attention of both perspectives. In general, the contributions from statistics have been models that describe the variance properties of traditional estimators in the presence of a particular error source.

Over the past twenty years research methods in nonsampling errors have undergone important changes. They have moved from observational studies to experimental designs. The motivation of the research has moved from observing the differences to testing theories about the causes of the differences. In this process, it has become clear that diverse theories are useful for different nonsampling errors.

Noncoverage
    1936 Stephan paper on frame problems

Sampling
    1934 Neyman paper on allocation in stratified sampling
    1950 Deming, *Some Theory of Sampling*

Nonresponse
    1944 Hilgard and Payne on noncontacts
    1946 Hansen and Hurwitz on nonresponse
    1983 Madow *et al.*, *Incomplete Data in Sample Surveys*
    1987 Goyder, *The Silent Minority*

Interviewer Effects
    1929 Rice, effect of interviewer attitudes on responses
    1946 Mahalanobis interpenetration paper
    1954 Hyman *et al.*, *Interviewing in Social Research*

Question Effects
    1941 Rugg, experiment on wording effects
    1951 Hansen, response error paper
    1951 Payne, *The Art of Asking Questions*

Mode Effects
    1952 Larson mode effect experiment
    1979 Groves and Kahn, *Surveys by Telephone*

Figure 15.1   Early journal papers and book-length treatments in subfields of surveys.

Yet there remain great contrasts between the theories of sampling errors and those of nonsampling errors. Some of the differences arise from different uses of the term "theory." Sampling error models identify the circumstances under which error values are controlled to a specified level. They are mostly derived from probability theory; that is, the models consist of the mathematical theorems that are the basis of sampling theory. For example, the survey designer is assured that sampling error will achieve the specified level if assumed circumstances pertain. In short, the theories concern properties of statistical measures of error. In contrast the role of theory in nonsampling error is the identification of causes of behaviors. Many of these theories do not yield mathematical specifications. Hence, they identify sets of circumstances under which errors can be reduced, but not the quantitative levels to which they can be reduced. Some theories yield model-based measures of errors, but most theories provide no links to measures of error.

Why is measurability important? Design improvements in surveys inherently involve cost–error tradeoff decisions. Since cost is eminently measurable and understood (even by Congressmen), errors in surveys tend to be given less attention than costs unless they can be measured quantitatively. The framing of design decisions is quantitative in nature, and errors that are quantified receive more attention than those not quantified. Nonsampling errors need statistical measures to gain respect at the design stage.

Finally, the cognitive sciences have utilized the randomized experiment as a principal research tool. This tool, when combined with inquiries about causes of different responses to questions, tends to yield more attention to what statisticians would label "bias" terms in survey quality; that is, many of the inquiries in the CASM research tradition have tried to identify relatively ubiquitous and lawful types of reactions to question wording, order, and structure. They predict that one wording will tend to produce answers in one direction versus another and to do so systematically, repeatedly. Rarely do researchers seeking explanation of cause of behavior entertain hypotheses about instability or other properties of the behavior that might allow contact with measurement error variance models.

15.3 EMPIRICAL EVIDENCE FOR THE NEED FOR BEHAVIORAL THEORETIC BASES OF STATISTICAL ERROR MODELS

To reinforce the arguments above, this section provides an example of popular nonsampling error statistical models actually used in practice, which arose from a variance components viewpoint independent of any consideration of the social or cognitive foundations of human behavior. It offers an illustration of how models can mislead if they do not reflect well the human behavior they wish to describe.

A popular tool in government surveys is the "reinterview" as a measure of "simple response variance." Simple response variance is defined in the context of a model of the response process. If an attribute of the ith person, y_i, is sought by asking a question, the error induced in the response process is separated into systematic component, b_i and random component, e_{it}, on the tth trial of administration. Thus,

$$y_{it} = y_i + b_i + e_{it},$$

with the expected value of e_{it} being zero over all conceptual administrations of the question. Survey statisticians estimate the "simple response variance" as a function of the comparison between an answer given to y in the survey and an answer given in a later "reinterview:"

$$s_e^2 = \frac{1}{2n} \sum_i (y_{i1} - y_{i2})^2$$

where y_{i1} is the answer from the survey and y_{i2} is the answer from the same question from reinterviews with a sample of n persons. The s_e^2 quantity, the simple response variance estimator, is used as a measure of stability; the higher its value the less stable are responses to the question.

The methodology for estimating s_e^2 in many government statistical agencies entails measurement differences between the interview and reinterview. The reinterview uses a different interviewer (often a supervisor), a cheaper mode of data collection (often telephone interviews), a questionnaire containing only a subset of questions, and a relaxed respondent rule (often permitting proxy reporting). The reinterviewer approaches the sample household about two weeks after the original interview. From one perspective these are compromises for reasons of cost that move away from the ideal measurement circumstances of independent but identical measures at two points in time.

At the Census Bureau, for example, this methodology has been used for many years. Although it is not completely clear from the history of reinterview studies, it appears that high values of simple response variance are used to suggest questions in need of improvement. Alternatively, Fuller (1991) proposes to use simple response variance estimates to improve the estimation of parameters in regression models involving predictors subject to response variance.

O'Muircheartaigh (1991) observes, through empirical analysis of a reinterview survey, a weakness in the common implementation of the model. Values of simple response variance vary as a function of who responds to the interview and reinterview. His results imply that the simple response variance is lowest when the same person is reinterviewed and questioned about him/herself, next lowest when either the interview or reinterview uses a proxy respondent, and highest when two different proxy respondents are used in the two measurements.

Clearly, the findings suggest a response model acknowledging that self-reports may be based on more easily and stably accessed memory cues (and perhaps may be more contaminated by memories of the answer given in the interview) than proxy responses are. In reaction to that possibility, O'Muircheartaigh promotes a revised model, introducing both memory effects and communication effects (between persons in the same household). He suggests that values of simple response variance confound different measurement processes.

Are there behavioral theories that would have predicted O'Muircheartaigh's results? There is a fairly large literature in proxy reporting (for a review, see Moore, 1988) noting the characteristics of topics and reporters that underlie some of the self versus proxy differences. It can be argued that the utility of this literature is its identification of the influences on response processes (e.g., access to information, nature of encoding, memory structures for others versus the self) so that response variance models might incorporate some of the terms identified post hoc by O'Muircheartaigh.

15.4 CLASSES OF NONSAMPLING ERROR THEORIES THAT COULD BE INTEGRATED INTO ESTIMATION

Incorporating concerns about nonsampling errors into the design and analysis of survey data will require advances in statistical models that describe their impact on survey estimates. These models cannot be useful, however, unless they incorporate the theories that describe the human behaviors that produce the errors. That is, social science theories must guide the model specification, but statistics must integrate the models into the estimation and inference process of surveys. Obviously, insights into the causes of such behaviors can come from many theoretical bases, but they are most likely to arise from sciences that directly study human behaviors.

At this time in nonsampling–sampling error research, it appears possible to consider altering classical nonsampling error models to incorporate some findings from social science research into nonresponse and measurement errors. These include (a) theories about direction of bias, (b) theories about the causes of bias for use in existing bias adjustment, (c) theories about bias in components of error usually studied as variance properties, (d) theories related to response variance, and (e) theories about response bias *and* variance.

15.5 THEORIES ABOUT THE DIRECTION OF BIAS— FREQUENCY OF BEHAVIORS

It was common as early as the mid-50's to note survey errors in reporting numbers of doctor visits, shopping episodes, books read, and television programs watched. Sometimes these errors were biases in the direction of consistent overreporting, but sometimes, for some items, they produced underreporting. By the late 1980's we had learned much more about the causes of this phenomenon. We now understand much more about the impact of closed questions on comprehension, about the tendency to use estimation with frequent, regular events which can lead to overreports, and about the tendency to use counting for rare or irregular events, which can lead to underreports of the phenomenon in question.

First let us examine a closed question example. Table 15.1 shows results from Schwarz, Hippler, Deutsch, and Strack (1985) based on a split sample experiment embedded in a survey. Both randomly identified half samples were asked about the frequency of their television watching using a closed question. One half sample was given six response categories ranging from less than one-half hour to over two and a half hours, with the middle category being one to one and a half hours. The second half sample was given six categories ranging from less than two and a half hours to over four and a half hours, with the middle category being three to three and one half hours. As Table 15.1 displays, over 84 percent of the sample given the low frequency scale and 62 percent of the higher frequency scale cases claimed less than two and a half hours of television watching; that is, it appears that providing response categories with higher frequencies induces reporting of more television watching.

Second, consider the response formation process in survey reports of frequencies of behavior. Sometimes these may not involve deliberate enumeration of events to provide the desired report. Instead, in many circumstances, the respondent uses the response categories to make judgments. One common heuristic in assessing the response categories is the assumption that the middle category describes the central tendency of the population. With this assumption, the respondents are able to report that they are above, at, or below the average in choosing the response category. They provide such a judgment as their response. Thus, the value assigned to the middle category affects their response. [Of course, there are other strategies for answering frequency questions (see Burton and Blair, 1991).]

Table 15.1 Reported Number of Hours of TV Consumption by High- and Low-Frequency Response Categories [from Schwarz et al. (1985)]

Low-Frequency Alternatives	Percentage	High-Frequency Alternatives	Percentage
<0.5 hr	7		
0.5–1.0 hr	18		
1.0–1.5 hr	26	<2.5 hr	62
1.5–2.0 hr	15		
2.0–2.5 hr	18		
		2.5–3.0 hr	23
		3.0–3.5 hr	8
>2.5 hr	16	3.5–4.0 hr	5
		4.0–4.5 hr	2
		>4.5 hr	0
Total	100		100

Both of these phenomena may have practical implications for questionnaire designers and survey statisticians. For example, one might use these findings to avoid closed questions because of the implied population distribution in the response categories. But the research has also shown that with open questions about number of times, some respondents will estimate and some will count (without revealing their response strategy). When respondents are asked to provide a total number of events for a rare or sporadic behavior, they will tend to count the eligible episodes. The likely error in this counting process is underestimation through failure to include an episode. When respondents are asked to provide a total number of events for a common, consistently performed behavior, they will tend to estimate the number, using a variety of methods to do so. The typical error made in this response formation is overestimation, because of departures from the rule that was used to form the estimate.

How might a statistician view this situation? Let

$$Y_{ei} = \text{estimated number of times for } i\text{th person,}$$
$$Y_{ci} = \text{counted number of times for } i\text{th person, and}$$
$$Y_i = \text{actual number of times for } i\text{th person.}$$

According to the theory above, $E_t(Y_{ei}) < Y_i$ and $E_t(Y_{ci}) > Y_i$. Thus, if

$$\overline{Y}_e = \sum \frac{Y_{ei}}{n}, \quad \overline{Y}_e = \sum \frac{Y_{ci}}{n}, \quad \overline{Y}_i = \sum \frac{Y_i}{n}$$

then

$$-(\overline{Y}_e - \overline{Y}_c) \le Bias(\overline{Y}_c) \le 0$$
$$0 \le Bias(\overline{Y}_e) \le (\overline{Y}_e - \overline{Y}_c)$$

A mixed estimator that combines the two estimates can, in some circumstances, be preferable to either estimate singly. That is,

$$\theta \overline{Y}_e + (1 - \theta)\overline{Y}_c$$

can have lower bias, if the θ is chosen to give larger weight to the less biased question. Finding optimal θ thus requires some knowledge of response errors in specific measures, which might be possible in some circumstances.

Even without theoretical guidance for the mixing parameter, however, interval estimation for the mean of Y could reflect the bounding of the true value by the two question types, as in

$$[\bar{y}_c - t_{\alpha,\nu} ste(\bar{y}_c), \bar{y}_e + t_{\alpha,\nu} ste(\bar{y}_e)]$$

where α and ν, reflect the chosen risk of type 1 error and the degrees of freedom of the t-statistic.

The practical import of this might be that to improve estimation of a mean or total on Y, a survey should ask two questions, one seeking from the respondent a rate at which the behavior occurs in a week or a month, the other seeking a count of total events in the reference period. With these two questions the mixed estimator above might be constructed with more attractive bias properties than either single question estimator.

15.6 THEORIES ON CAUSES OF BIAS

Traditional post-survey adjustments for unit nonresponse in surveys include weighting class adjustments and propensity model adjustments. In both of these methods the abstract conditions for bias reduction can be deduced from statistical theory. Whether one can act with assurance that the conditions will be met needs some social science theory.

Let us examine the classical weighting class adjustment with ignorable nonresponse. Assume there are J adjustment cells within which participation is independent of (Y, I), where Y is the survey variable and I is the likelihood of inclusion. Here, the Horvitz-Thompson estimator of the mean might be written as

$$\sum \sum y_i \pi_i^{-1} r_j^{-1} \Big/ \sum \sum \pi_i^{-1} r_j^{-1}$$

where r_j is the response rate for the population in adjustment cell j.

In practice, r_j might be estimated by first estimating response propensity from sample data. Let X be a vector of variables observed for both respondents and nonrespondents and let the response propensities (R) be conditionally independent of the survey variables (Y) and the inclusion in the sample (I)

$$R \coprod (Y, I)|X$$

then $r(x_i) = Pr(R_i = 1|x_i)$.

The $r(x_i)$ can be estimated indirectly by logit models. Then J adjustment cells can be created by coarsening the estimated $r(x_i)$ into a small number of categories. When propensity models are built, we want to discover desirable X's and model specifications usable over a variety of surveys.

Research in nonresponse behavior has shown that nonresponse is multifaceted. For example, noncontact and refusal are two alternative sources of nonresponse that have very different behavioral bases. Following the logic above, let R_c refer to the probability of contact, and R_p, to the probability of participation, given contact. We want to determine if

$$R_c \coprod (Y, I)|X$$

but

$$R_p \coprod (Y, I)|Z$$

What theories might inform the specification of the X's and Z's? These might include cognitive script theory (Abelson, 1981) to describe the process by which the intentions of the interviewer seeking a survey interview would be interpreted by the householder. They might include psycholinguistic theories of comprehension, which would inform how the meaning of words used by the interviewer are interpreted in the context of the entire conversation about the survey and the interview request; and, they might include sociological theories of class and race effects (Goyder, 1987).

These theories motivate a two-step adjustment model, with the first step reflecting the process of contacting sample units and the second step reflecting the process of gaining cooperation among those contacted. The theories identify new observations to collect on respondents and nonrespondents; these involve indirect indicators of the causes of the two nonresponse phenomena. For example, the noncontact model would include as right-side variables a set of at-home influences, physical or technological impediments to access (e.g., locked apartment buildings, answering machines), and indicators of interviewer calling patterns. The cooperation model would include influences on participation, as well as correlates of key survey variables. Of particular interest are predictors motivated by theories of helping behavior, comprehension, and attitude change. In tests of these ideas thus far, these predictors have included data on householder utterances that express questions about the purposes of the interviewer's visit (regarding cognitive script errors on the part of the householder; Groves, Raghunathan, and Couper, 1995).

The first stage of weighting would estimate for the ith sample person, assigned to the jth interviewer, two propensities, using predictors on the person level (X and Z variables) and on the interviewer level (U variables):

$$R_{ci} = X_i^t \beta + U_j \delta + \alpha_{ij}$$

and the probability of interview given contact by

$$R_{pi} = Z_i^t \alpha_j + U_j \nu_j + \gamma_{ij}$$

where α_j and ν_j are random coefficients reflecting interviewer effects, sharing a mean of 0 and variances of σ_α^2 and σ_γ^2, respectively. The first propensity is used to construct the contacted group, the second propensity to adjust back to the full sample, given contact.

A completely different realm of survey measurement has been addressed through a model of response formation that is relevant to statistical bias. Smith and Jobe (1994) theorized that reports of food consumption are affected by generic knowledge about one's eating habits as well as episodic knowledge about individual meals. Their posited response formation process is a sampling from a set of items, where the tendency to respond is a function both of the base rate of consumption of the item and the episodic consumption. They posit that reporting of items from individual episodes is subject to an exponential decay function, but the reporting of base rate items is constant over time.

There are several deductions from this model. One is that "base rate" reporting dominates survey reports with long recall periods (because the episodic recall deteriorates). In contrast, the episode of eating a rarely eaten item dominates recall until decay is complete.

In this conceptual model the reporting using base rate or generic knowledge can produce response errors. The error is a function of the contrast between episodic experiences during the reference period and the base rate experiences. Thus, either underreporting or overreporting is possible.

We illustrate this by examining the probability that a single item was consumed one or more times (not the estimation of the number of items consumed). It appears that the probability of reporting the ith item is

$$p_{bi} + p_{ei}$$

the sum of a base rate of consuming the item and the probability that any episodes will be recalled. (It is the second term that declines exponentially with time.) Thus, the expression is equivalent to

$$p_{bi} + \alpha e^{-\beta t + \epsilon}$$

and thus the bias of reporting is a function of the true value r_{ei} and the report,

$$(r_{ei} - p_{bi} - \alpha e^{-\beta t + \epsilon})$$

One approach to estimating this function is to take advantage of the interplay of reference period length and recall length. The survey design would systematically vary reference period length and recall length, allowing one to estimate the relative magnitude of reports on the base rate (which should increase as a function of reference period length) and on the episodes (which should decline as a function of recall length).

Finally, there are conceptual models of reporting on attitudinal items that are akin to bias models (see Tourangeau, Rips, and Rasinski, 1999). In this case, in order to avoid the controversy of whether true values exist for an attitudinal measure, it might be better to define the true value as that report to be expected based on considerations about the object retrieved from long-term memory, excluding those considerations generated at the time the questions is asked. This itself will be controversial, but allows us to use a simple model

$$ J = \frac{\sum_L s_l + \sum_L s_n}{L + N} $$

where J is the final judgment on some response scale, s_l is the scale value assigned to a consideration retrieved from long-term memory, s_n the scale value of a newly-generated consideration, L denotes the number of considerations retrieved from long-term memory, and N the number generated at the time the question is asked. Thus, "bias" in this case might be formulated as

$$ \sum s_l/L - \sum s_n/N $$

It seems doubtful that all of these parameters can be estimated in practice, but an interpenetrated design that formally varies N, the number of considerations at the time of the measurement, might be useful. Thus, explicitly varying the number of priming questions prior to the target question on random subsets of respondents could allow one to estimate the function

$$ f(N) = J_i - \beta/N_i $$

and thereby retrieve an estimate of

$$ s_l/L - s_n/N $$

In essence, such "bias" measures would estimate the ability of context priming to alter responses to key attitudinal questions.

15.7 THEORIES ABOUT MEASUREMENT ERROR VARIANCE

As noted earlier, most of the social and cognitive theories are relevant to bias (not variance) terms in statistical error models. However, there are some exceptions. Schuman and Presser (1981, pp. 268–271) noted that respondents with weakly held attitudes (as measured by self-reports of attitude intensity) were more subject to change across waves of a panel. They interpreted this as evidence of higher response variance. Other experiments in context effects have shown that the answers to global questions can be affected by the presence of prior related questions. This is one example of the judgment process underlying survey answers being affected by cues in the interview environment. One might deduce that when a global question is not preceded by other relevant questions, the cues affecting judgment are uncontrolled by the survey design, leading to higher within-person response variance.

For theories that predict higher response variance in one format versus another, the survey statistician must give attention to attempting to measure the response variance when using the high-variance option. A common design used to estimate the response variance is either the reinterview or the multiple indicator approach, comparing relative response variance among alternative indicators of a latent variable (see Saris and Andrews, 1991).

15.8 THEORIES ON BOTH RESPONSE BIAS AND VARIANCE—EXAMPLES OF VARIABLE AND FIXED EFFECTS OF INTERVIEWERS

The statistical literature about interviewer effects on survey data contains essentially components of variance models. They model the differences that interviewers create in survey data through different ways of asking questions associated with different response errors. In contrast, the social science literature has focused on a set of fixed effects on respondent behavior from interviewer gender, race, and age. These studies have shown consistent changes in behavior of respondents on surveys dealing with topics relevant to the demographic characteristics of the interviewer.

Understanding and correctly reflecting both these variable and fixed effects of interviewers on survey data requires a modification of the specification of the interviewer effect models. For example, let $y_i = \mu_i + \epsilon_i$, for each of the i respondents. Now, let us start to acknowledge the error induced by the interviewer:

$$y_{ij} = \mu_i + b_j + e_{ij}$$

when the ith respondent is assigned to the jth interviewer. The typical assumptions made in this model, mainly to ease estimation are:

1. $\{b_1, b_2, \ldots, b_j\}$ is a random sample from an infinite population with $b_j \sim$ iid $(0, \sigma_b^2)$

2. $e_{ij} \sim$ iid$(0, \sigma_e^2)$

3. μ_i, b_j, e_{ij} are uncorrelated for all i, j

Then

$$\text{Var}(y) = (1/n)(\sigma_\mu^2 + \sigma_b^2 + \sigma_e^2)[1 + (m - 1)\rho_y] = (\text{Var}(y_{ij})/n)[1 + (m - 1)\rho_y],$$

where ρ_y is an intraclass correlation, within interviewer workloads, of response deviations. (See Biemer and Stokes, 1991, for a review of different estimators for the ρ_y parameter.)

What theories are appropriate for use in the respecification of this model? Most are theories of interpersonal influence. These theories assume that the comprehension of questions is socially constructed. They note that observable attributes of the interviewers are used to aid comprehension and judgment of appropriate answers. Finally, they observe that the race, age, and gender of interviewers have effects on answers when questions are relevant to those attributes.

The modification of interviewer variance models must acknowledge that some of the variance associated with interviewers is appropriately associated with the interviewers' demographic attributes. That is, some of the variation across interviewers is due to these fixed demographic attributes; the result may be due to attributes that are better modeled as random components. This perspective forces attention to the appropriate inferential population for the survey. For example, do we want to limit inference to essential survey conditions in which there is no variation on the interviewer attribute in question? That is, do we want to measure interviewer variability with the given mix of interviewers used in the project? (Is the demographic attribute a stratifying variable on the selection model for interviewers?) If so, then the appropriate models might be stratified variance models, reflecting "fixed" effects of interviewer attributes.

15.9 THEORIES OF RESPONSE BIAS AND VARIANCE— DATING OF EVENTS

There is much literature in survey research on the reports of dates of events within reference periods specified by the questionnaire (e.g., "In the last six months, when did you visit the doctor?"). It is common to observe errors in the dating of reported events. Generally only events occurring before the beginning of the reference period can be misdated into the reference period (forward telescoping). There is some evidence that events evoking vivid memories appear to have occurred more recently (forward telescoping). Response variance on dating appears to decline for more recent events (lower instability).

There are some models of this behavior, constructed by psychologists studying the topic. For example, Huttenlocher, Hedges, and Bradburn (1990) offer a model built on several assumptions. These include (a) the number of events is uniform over time, (b) the error rate in reporting is independent of number of events, and (c) the response standard deviation increases linearly in time. From this, one can construct a bias-adjusted estimator of the total number of events, requiring input of the response variance, the rate of increase in rate of instability over time, the length of reference period, and the uniform rate of occurrence. Such models, since they require large sets of assumptions, are not practical for use in survey estimation.

Could measurement models of more practical utility be built? The problem here is that the theory poses greater complexity than those discussed above. Both response bias and variance are found to change according to the true date of event. One approach would be to construct an estimator differentially weighting reports by response variance. This would use multiple indicators of date (e.g., free recall of date, measure of how long since event occurred, use of calendrical aid), estimation of response variance (through the covariance analysis of multiple indicators), and the use of response variance in estimation. Such an estimator, however, does not address at all the bias properties of the responses. When theories specify both bias and variance properties of response, then the survey designer must invent multiple indicators that vary in known ways on their variance and bias properties, in order to estimate parameters in a measurement model that reflect both bias and variance.

15.10 SUMMARY—MOVING NONSAMPLING ERROR THEORIES TO SURVEY ESTIMATION

Theories about error that arise in the social sciences sometimes have implications for the statistical specification of estimators. This is because they often focus on the behavior that produces the statistical error. Statistical models of error can become more useful if they reflect such theories. Unfortunately, few social scientists studying the causes of behaviors producing survey errors attend to implications of their findings for survey estimators. Their focus is on identifying principles of behavior that apply to a wide variety of contexts (including surveys). Nonresponse and response behavior, from their viewpoint, is just one example of these behaviors.

There remain, however, three barriers to a merging of these fields:

1. Most cognitive and social theories related to survey errors are at too high a level of abstraction, a few steps removed from a mathematical formulation of the processes they describe. Survey researchers need to first exploit those that seem close to the appropriate level of specification.

2. There exists within much day-to-day survey practice a reluctance to use model-based estimation of population parameters. (In contrast, those using survey data to build structural models are much more likely to incorporate such adjustments.) Outside of adjustment for nonresponse, most descriptive statistics do not utilize models in estimation. Strong examples of the utility of such models are probably needed to alter the status quo.

3. Many of the theories, as we have seen above, imply improvement in esti-

mation processes only if the survey design is altered. The examples include the use of multiple indicators, of randomized assignment of measurement procedures, etc. These changes are typically considered only by very careful investigators, who take time to deal with matters that appear secondary to the main work of getting a survey into the field. It is likely that greater success will be found in academic or government surveys than in the quick-paced world of commercial surveys.

It now seems clear that the "great leap forward" in nonsampling research will come with collaboration between cognitive and social scientists and statisticians. The collaboration will focus on a set of measurement designs and estimators that incorporate auxiliary error models. These estimators will have their error specifications motivated by the theories discovered by social scientists about behaviors producing the errors affecting the quality of survey data. However, the barriers to success in this endeavor demand direct confrontation by survey methodologists and their collaborators.

REFERENCES

Abelson, R. P. (1981). Psychological status of the script concept. *American Psychologist*, *36*, 715–729.

Biemer, P. P., and Stokes, S. L. (1991). Approaches to the modeling of measurement error. In P. P. Biemer, R. M. Groves, L. E. Lyberg, N. A. Mathiowetz, and S. Sudman (Eds.), *Measurement Error in Surveys*, pp. 487–516. New York: Wiley.

Burton, S. and Blair, E. (1991). Task conditions, response formulation processes, and response accuracy for behavioral frequency questions in surveys. *Public Opinion Quarterly*, **55**, 50–79.

Deming, W. E. (1950). *Some Theory of Sampling*. New York: Wiley.

Fuller, W. A. (1991). Regression estimation in the presence of measurement error. In P. P. Biemer, R. M. Groves, L. E. Lyberg, N. A. Mathiowetz, and S. Sudman (Eds.), *Measurement Errors in Surveys*, pp. 617–636. New York: Wiley.

Goyder, J. (1987). *The Silent Minority: Nonrespondents on Sample Surveys*. Boulder, CO: Westview.

Groves, R. M., and Kahn, R. L. (1979). *Surveys by Telephone*. New York: Academic Press.

Groves, R. M., Raghunathan, T., and Couper, M. P. (1995). *Evaluating statistical adjustments for unit nonresponse in a survey of the elderly*. Paper presented at the Sixth International Workshop on Household Survey Nonresponse, Helsinki, Finland.

Hansen, M. H. (1951). Response errors in surveys. *Journal of the American Statistical Association*, *46*, 147–190.

Hansen, M. H., and Hurwitz, W. N. (1946). The problem of non- response in sample surveys. *Journal of the American Statistical Association, 41*, 517–529.

Hilgard, E. R., and Payne, S. L. (1944). Those not at home: A riddle for pollsters. *Public Opinion Quarterly, 8(2)*, 254–261.

Huttenlocher, J., Hedges, L. V., and Bradburn, N. M. (1990). Reports of elapsed time: Bounding and rounding processes in estimation. *Journal of Experimental Psychology, Learning, Memory and Cognition, 16*, 196–213.

Hyman, H. H., Cobb, W. J., Feldman, J., Hart, C. W., and Stember, C. (1954). *Interviewing in Social Research*. Chicago: University of Chicago Press.

Larson, O. N. (1952). The comparative validity of telephone and face-to-face interviewers in the measurement of message diffusion from leaflets. *American Sociological Review, 17*, 471–476.

Madow, W. G., Nisselson, H., and Olkin, I. (1983). *Incomplete Data in Sample Surveys, Volumes 1–3*. New York: Academic Press.

Mahalanobis, P. C. (1946). Recent experiments in statistical sampling in the Indian statistical institute. *Journal of the Royal Statistical Society, 109*, 325–370.

Moore, J. C. (1988). Self-proxy response status and survey response quality: A review of the literature. *Journal of Official Statistics, 4*, 155–172.

Neyman, J. (1934). On the two different aspects of the representative methods: The method of stratified sampling and the method of purposive selection. *Journal of the Royal Statistical Society, 109*, 558–606.

O'Muircheartaigh, C. (1991). Simple response variance: Estimation and determinants. In P. P. Biemer, R. M. Groves, L. E. Lyberg, N. A. Mathiowetz, and S. Sudman (Eds.), *Measurement Errors in Surveys*, pp. 551–574. New York: Wiley.

Payne, S. L. (1951). *The Art of Asking Questions*. Princeton, NJ: Princeton University Press.

Rice, S. A. (1929). Contagious bias in the interview: A methodological note. *American Journal of Sociology, 35*, 420–423.

Rugg, D. (1941). Experiments in wording questions. *Public Opinion Quarterly, 5*, 91–92.

Saris, W., and Andrews, F. (1991). Evaluation of measurement instruments using a structural modeling approach. In P. P. Biemer, R. M. Groves, L. E. Lyberg, N. A. Mathiowetz, and S. Sudman (Eds.), *Measurement Errors in Surveys*, pp. 575–598. New York: Wiley.

Schuman, H., and Presser, S. (1981). *Questions and Answers in Attitude Surveys: Experiments in Question Form, Wording, and Context*. New York: Academic Press.

Schwarz, N., Hippler, H. J., Deutsch, B., and Strack, F., (1985). Response categories: Effects on behavioral reports and comparative judgments. *Public Opinion Quarterly, 49*, 388–395.

Smith, A. F., and Jobe, J. B. (1994). Validity of reports of long-term dietary memories: Data and a model. In N. Schwarz and S. Sudman (Eds.), *Autobiographical Memory and the Validity of Retrospective Reports*, pp. 121–140. New York: Springer-Verlag.

Stephan, F. F. (1936). Practical problems of sampling procedure. *American Sociological Review, 1*, 569–580.

Tourangeau, R., Rips, L., and Rasinski, K. A. (1999). *The Psychology of Survey Responding*. London: Cambridge.

CHAPTER 16

New Connectionist Models of Mental Representation: Implications for Survey Research

Eliot R. Smith
Purdue University

Theoretical conceptions of the nature of mental representations (of all sorts, including autobiographical memories, general beliefs, and attitudes) have changed over the years. Consider first the naive or person-in-the-street view of mental representation. In this view, which we term "view 0" because of its pre-scientific character, memory is understood using metaphors involving a camera or tape recorder, and a file cabinet.

> **View 0:** Mental representations are *faithful copies* of the stimulus information that was available on a past occasion (like photographs), and are stored as *discrete records* that can be retrieved separately (like pages in a file cabinet).

In this view, retrieving a memory involves searching in the file cabinet to find the correct representation; the original event can then be reexperienced. For survey researchers, this view would imply that to obtain accurate memory reports, one simply has to motivate the respondent to search carefully in the file cabinet for the correct record.

Cognition and Survey Research, Edited by Monroe G. Sirken, Douglas J. Herrmann, Susan Schechter, Norbert Schwarz, Judith M. Tanur, and Roger Tourangeau.
ISBN 0-471-24138-5 © 1999 John Wiley & Sons, Inc.

16.1 THROWING OUT THE CAMERA: REPRESENTATIONS ARE CONSTRUCTED

Psychologists now know that memory functions nothing like a camera or tape recorder. Mental representations are *constructed* by the perceiver rather than being direct copies of available stimulus information. This insight leads to:

> **View 1:** Mental representations are *constructed* based on the perceiver's interpretation of the available information, and are stored as *discrete records* that can be retrieved separately (like pages in a file cabinet).

The construction process involves selection, inference, and interpretation. It draws on the person's general knowledge and prior expectations as well as the stimulus information that was present. Conceptualizations of this construction process often invoke the concept of a schema (Fiske and Taylor, 1991), an organized body of knowledge and expectations that is used to interpret and make sense of incoming information. The result of the construction process is considered to be an associative structure formed by connecting nodes representing individual objects or concepts with links. Just as meaningful words are combined to make sentences, nodes representing "John" and "honest" can be linked to represent the idea or interpretation that John is honest. Many current models in cognitive psychology (e.g., Anderson, 1983) and social cognition (Wyer and Srull, 1989) use assumptions like these (see Smith, 1998). For survey researchers, View 1 implies that potential sources of inaccuracy in memory reports include not only the respondent's locating and using the wrong representation in memory, but also defects in the perceiver's interpretive and constructive processes. Clearly, View 1 is much more sophisticated and commands much more empirical support than the naive View 0. Still, though the camera metaphor is discarded, View 1 keeps the filing cabinet metaphor. Associative structures that represent the perceiver's interpretation of an event or situation are still assumed to be stored as discrete entities and retrieved independently. However, new connectionist models of mental representation question even this assumption.

16.2 OVERTURNING THE FILE CABINET: REPRESENTATIONS ARE SUPERPOSED

Connectionism developed in the 1970s and 1980s, and great interest was sparked by the 1986 publication of two "Parallel Distributed Processing (PDP)" volumes (McClelland, Rumelhart, and PDP Research Group, 1986; Rumelhart, McClelland, and PDP Research Group, 1986). Today, connectionist models are influential in many areas of psychology. Space limitations permit only the briefest of introductions to be given in this chapter; accessible presentations can be found in the first few chapters of Rumelhart, McClelland, and PDP Research Group (1986) or Smith (1996).

A connectionist model involves a large number of simple processing units (modeled in an abstract way on neurons) each possessing an activation level that can change rapidly over time. The units are interconnected and send activation to each other over weighted links. In a simple model, the output from each unit j equals its activation level a_j, and this signal is multiplied by the weight w_{ij} (which may be a positive or negative number) on the connection from unit j to unit i (see Figure 16.1). The total input received by unit i, which affects its activation level at the next instant, is the sum of its input from all units j that send connections to it, $\Sigma w_{ij}\alpha_j$. The weights are assumed to change only slowly as the network learns, and therefore serve as the repository of the network's memory. An important aspect of most connectionist models is the use of distributed representations. A semantically meaningful representation is a pattern of activation across many processing units (McClelland and Rumelhart, 1986). Activity of a single unit has no fixed meaning independent of the pattern of which it is a part. As an analogy, think of the individual dots or pixels on a TV screen: by taking on different patterns of illumination, the set of dots can represent many meaningful patterns, but the brightness or color of any individual dot has no fixed meaning. A connectionist memory operates in several stages.

1. A set of input units (which might receive their activation from sensory receptors or some other external source) feed a pattern of activation into the network. Activation flows through the connections, producing a pattern of activity across the units in the network. This pattern depends not only on the current inputs but also on the existing connection weights. The pattern corresponds to the person's interpretation of those inputs, and may not be an identical copy of the input pattern because it may involve selection, inference, and distortion.

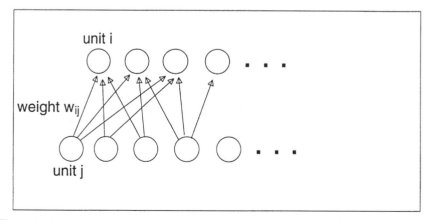

Figure 16.1   A portion of a connectionist network showing connections along which activation signals flow between units. (Note: Not all units or connections are shown.)

2. Changes in the connection weights then take place, which represent the storage of a "memory trace" of the given stimulus or experience. Different models use learning rules that differ in detail. However, generally speaking, the weights on connections between concurrently active units are increased in strength, while weights on connections between an active unit and an inactive one are decreased.

3. At a later time, if a subset of the same input pattern or a similar pattern is again input, some of the units that formed part of the activation pattern will again be turned on. As activation flows through the connections, the modifications produced by the learning rule will tend to result in the recreation of the entire original pattern.

However, since all learning involves modifying the same matrix of connection weights, the recreation will be imperfect and will be affected by other experiences as well. In a sense, the entire set of connection weights constitutes a single representation in which representations of all learned patterns are superposed or "mushed together" (Carlston and Smith, 1996). An individual memory trace is the change in the connection weights in the network produced by the learning algorithm when one input pattern is processed on one occasion. Each memory trace is embodied in changes in many weights, just as each weight is a part of many memory traces (van Gelder, 1991).

Distributed representations require new ways of thinking about the nature and function of memory. In the file cabinet metaphor, discrete representations are thought of as being inscribed on separate sheets of paper that can be independently accessed. Even the very terms "memory storage," "search," and "retrieval" invoke this familiar metaphor. With connectionist models that use distributed representations, there is no discrete location for each representation. Instead, accessing one representation necessarily accesses all, because all representations are encoded in the same set of connection weights. Similarly, adding a new experience changes many weights and therefore alters (perhaps minimally) the representation of all (van Gelder, 1991, p. 45). Instead of "search" and "retrieval," better metaphors involve similarity and "resonance." Thus, we might say that a new stimulus resonates with, and activates, representations in memory that resemble it. In a connectionist memory the reconstruction of a memory will often be imperfect and subject to influence from the person's other knowledge (such as schemas and scripts), but this characteristic is typical of actual human memory performance (van Gelder, 1991; Carlston and Smith, 1996). This new picture of mental representations leads beyond the now-familiar View 1 and its filing cabinet to:

View 2: Mental representations are *constructed* based on the perceiver's interpretation of the available information, and are stored in a common set of weights, so that no one representation can be stored, accessed, or changed independent of all others.

16.3 MAJOR IMPLICATIONS OF CONNECTIONIST MODELS

16.3.1 Buildup of General Knowledge from Exemplars

Abandoning the idea of discrete representations means reconsidering the way general knowledge is constructed, but this may actually be an advantage, for accounts of this process in traditional symbolic models are often vague and incomplete. In such models, the construction of an item of general knowledge (or the creation of a schema) is assumed to reflect generalization or abstraction from many experiences (e.g., Fiske and Taylor, 1991, p. 98). However, it is unclear when and how this abstraction or summarization process works—when do people decide that they have enough experience with a domain to stop storing individual exemplar representations and compute and store a summary instead? Schema theories generally deal in detail with the activation and use of schemas but tend to give little attention to such questions involving the origins and construction of schemas.

In distributed connectionist models, the construction of general representations from a number of related experiences occurs automatically, as a function of the learning rule. Recall that learning involves adjusting the connection weights in a network after a particular pattern is processed, in a way that facilitates the reconstruction of that pattern in the future. Suppose a network is exposed to a number of related patterns that amount to variations on a common theme, such as multiple members of a specific category involving both general similarities and idiosyncratic differences. When the representations of all the instances are stored in a single set of connections, the repeated weight changes produced by the learning rule will reinforce the representation of those respects in which they are similar, while weight changes produced by their unique and variable aspects will tend to cancel out. The resulting representation will emphasize the central features shared by most of the instances. In effect, an average or prototype representation for the category is created automatically (McClelland and Rumelhart, 1986).

The automatic abstraction property of connectionist representations has applications beyond the case of multiple exemplars of a category. Many real domains of knowledge are what Plaut, McClelland, Seidenberg, and Patterson (1996) termed "quasi-regular," involving both general rules and specific exceptions. There are many examples in the area of language, where the formation of past tenses of verbs or the spelling of words involve both important regularities and multiple special cases. Much of our general world knowledge is also "quasi-regular;" for example, knowledge about President Kennedy's assassination is partly based on general knowledge about rifles, motorcades, etc., but also involves unique and specific details (Dallas, November 22, etc.).

McClelland, McNaughton, and O'Reilly (1995) demonstrated how a connectionist network could learn such information. They trained a network on a number of facts about living things, plants, animals, and so on. Sample facts are "a robin can fly," "a rose can grow," "an oak is a tree," "a bird is an animal." After considerable training on a set of these facts, the network constructed distributed representations that reflected a hierarchy of category memberships. The patterns representing canaries and robins were highly similar, as were the patterns for oak and pine; at a higher level the patterns for plants and animals were quite different. This structure reflects the similarities among the facts and inferences that the network learned for each pattern. Moreover, the network was able to answer questions by making inferences and generalizations based on its prior knowledge. For example, if the single new fact "a sparrow is a bird" is learned, the network will use its other knowledge to infer that a sparrow can fly, can grow, is an animal, etc.

Hierarchical knowledge representations contain a mixture of systematic, regular information with unique and idiosyncratic information. For example, every bird can fly, every animal can grow, and so on. These facts are sufficiently predictable that they can be stored with the higher level concepts of bird and animal, rather than separately with every individual type of creature. However, canaries are yellow, and this information must be specifically stored for it cannot be predicted or reconstructed from more general knowledge. When such a mixture of information is learned, the predictable and regular aspects will of course be learned faster than the arbitrary and unique aspects, because the former benefit from a greater number of repetitions. If learning is not continued until performance is error free, the network's errors will tend to be overregularizations or schematizations. A familiar example is the way children who are just learning to speak will often regularize irregular verbs, saying "goed" instead of "went." Another consequence of the same process is the "schematic" quality of long-term memory (McClelland et al., 1995, p. 438). As memory fades into the distant past, unique and arbitrary details tend to be lost in favor of common and typical experiences.

16.3.2 Retrieval as Multiple Simultaneous Constraint Satisfaction

With multiple representations all stored in a single set of connection weights, retrieval can no longer be regarded as the process of "searching" for the single desired representation and "retrieving" it. Instead, retrieval is a reconstructive process that seeks to satisfy multiple constraints simultaneously. Neisser (1967) likened memory to the work of a paleontologist, who reconstructs a dinosaur from some fragments of bones (data) and a body of preexisting knowledge about dinosaurs (theory). Similarly, what is retrieved as a "memory" is constrained not only by the cues that trigger a recollection and by the specific details of the to-be-remembered event or object itself, but also by related general knowledge, memory traces of similar events, etc. Many stored representations necessarily contribute to each reconstructed memory, because all representations are stored in the same set of connection weights, and the flow of activation through all the connections contributes to the final output.

Specific examples of the operation of memory as multiple constraint satisfaction have been provided in several domains. Rumelhart, Smolensky, McClelland, and Hinton (1986) gave a connectionist memory information about typical features of different rooms. A living room is large and usually has a sofa. A bathroom is small and has a tub and sink, etc. Once the network had learned these facts, it could be questioned about various rooms by providing a single feature as a cue. So given "oven" as a cue, the output would be the typical set of features that had been associated with the kitchen description. This could be interpreted as simple retrieval of the previously-learned pattern. Thus, when people have learned specific facts through multiple repetitions (such as 3 + 2 = 5 or "my birthday is Feb. 4"), an appropriate cue (like 3 + 2 = ? or "what is your birthday?") will quickly and reliably elicit retrieval of the well-learned answer.

However, connectionist memories can go beyond retrieval of single patterns. The "room" network could be given more than one cue simultaneously, such as "bed" and "sofa." Even though these two features never occurred together during the learning process, the network came up with a description that could be regarded as a large, fancy bedroom, complete with easy chair, floor lamp, and fireplace. The network in effect used the known bedroom and living room patterns as constraints, and combined mutually consistent aspects of them to come up with a new type of room description. In other words, it constructed its output by simultaneously satisfying multiple learned constraints as well as the input cues, rather than by finding a single representation that matched the input.

16.3.3 Failure of Source Memory

When information from multiple experiences is combined into a single representation, it may be impossible at a later time to tell what was actually experienced and what was inferred. Suppose you learn that Ann is taller than Barbara and Barbara is taller than Charlie. If you combine these items into a single representation (an ordered list of all three), you can later answer questions about who is taller than whom. But you may not be able to say whether each fact was learned or inferred. This inability is a near-inevitable cost of the increased inferential power given by a unified representation (compared to the alternative of storing each given fact separately).

The same principle holds true when connectionist memories maintain multiple representations in a single memory structure (a single set of connection weights). The network that learned about properties of living things, described earlier, can serve as an example (Rumelhart, Smolensky, McClelland, and Hinton, 1986). The network created a structured knowledge representation, and accessed this to answer questions. However, there is no way to tell whether a particular answer is based on a specifically learned fact (and hence represents "remembering") or is a generalization or inference from learned facts ("knowing"). Again, our ability to generalize and make inferences is a crucial cognitive skill that permits us to "go beyond the information given." But then one may not be able to say what specific information was, in fact, "given": there is no clear way to distinguish what is remembered from what is known, judged, or inferred.

In general, the source of any information that we report—whether it derives from a memory formed on a specific past occasion or represents an inference—is an attributional judgment (see Carlston and Smith, 1996, pp. 202–203). Studies of "reality monitoring" examine how people decide whether a memory is based on perception or inference. For example, the more concrete perceptual details are part of the memory (e.g., colors, sounds), the more likely it is to be labeled perceptual rather than inferential (Johnson and Raye, 1981). However, these judgments are fallible. Work by Loftus (1979) and others shows that people can be misled (for example, by leading questions) into reporting as "memories" details that were not present in the original scene. Young children and the elderly are less able to accurately monitor source information than are nonelderly adults (Squire, 1995). These groups should therefore have particular difficulty in distinguishing accurately between what has been seen or heard and what has been imagined or inferred.

16.4   CATASTROPHIC INTERFERENCE AND DUAL MEMORY SYSTEMS

16.4.1   The Problem of Catastrophic Interference

It is now time to emphasize some limitations of a connectionist network's learning capabilities. In general, when multiple patterns are to be learned by a network, they must be presented repeatedly in an interleaved order. If instead one takes a network that has already learned a set of patterns and repeatedly presents a new pattern, it will be learned—but the preexisting knowledge will be destroyed in the process. This phenomenon has been termed catastrophic interference (McCloskey and Cohen, 1989). It has been used to argue that connectionist models are inadequate accounts for human memory performance.

16.4.2   Dual Memory Models

A major article by McClelland et al. (1995) points to one way to escape these limitations. (Several other theorists have advanced basically similar ideas.) The authors argue that catastrophic interference reflects the functional incompatibility of two conflicting demands that an adequate human memory system must meet. One demand is to record information slowly and incrementally (i.e., to change connection weights only a small amount after each experience) so that the total configuration of information in memory reflects a large sample of experiences. This is important so that general expectancies and long-term schematic knowledge can be based on the average, typical properties of the environment. It is a function that is served very well by the types of connectionist networks that we have described earlier. A second function calls for rapid learning of new information (i.e., large changes in connection weights) so that a novel experience can be remembered after a single occurrence. After all, people can at least sometimes learn things by being told once; however, the demands for slow and fast learning are incompatible.

The McClelland et al. (1995) model explains how humans can both quickly form memory traces of unique events and also integrate many experiences over time. These two forms of memory correspond roughly to "episodic" and "schematic" memory, and McClelland and his colleagues argue that they are handled in separate memory systems. This idea may seem unparsimonious, but many types of psychological and neuropsychological evidence support the existence of two memory systems with the postulated properties. For example, this model makes sense of the particular patterns of memory impairment exhibited by patients with damage in specific brain areas (Squire, 1995). The two memory systems appear to be anatomically mediated by the hippocampus and the neocortex. The system involving the hippocampus rapidly changes its synaptic strengths to quickly acquire new memories. This system is responsible for one-shot learning about novel events and stimuli, in a way that depends on selective attention and thus on conscious awareness. The other system, located in the neocortex, learns only gradually and incrementally with experience. (Note that both of these postulated systems are distinct from "working memory" or "short-term memory," which is generally assumed to consist of the activated subset of representations held in long-term memory rather than being an independent memory store.)

In the process of "consolidation," a newly formed memory is transferred by repeated presentations from the hippocampal to the neocortical system. Consolidation is known to take much time in humans and the authors suggest that it is necessarily slow so that new knowledge can be integrated nondisruptively into the stably structured representations maintained in the neocortical system. Estimates of the time course of consolidation differ (ranging from weeks to years) depending on the research methods used and the content of the memories that are studied. Survey research might help shed light on this question by indicating when respondents asked about the frequencies of events stop recalling and counting specific incidents (held in the fast-learning system) and turn instead to giving general estimates based on schematic knowledge in the slow-learning system (see Burton and Blair, 1991).

16.4.3 Implications of Dual Memory Systems

If it is true that people have separate fast- and slow-learning memory systems, there are important implications for the concerns of survey researchers. One is that people have *separate* mechanisms that perform the functions that many current theories attribute to associative networks and schemata (Smith, 1998). The schematic function of interpreting input information in terms of stable, general world knowledge may be performed by neocortical systems that learn slowly, extracting regularities in the environment and using them to interpret and process further inputs. In contrast, the rapid construction of new associative structures that bind together information about different aspects of an object or experience in its context, seems to take place in hippocampal systems that exhibit one-shot learning and mediate conscious, explicit recollection. The fast-learning system is closely tied to language, for its ability to rapidly construct and store novel representations is essential for both the production and comprehension of language (Smolensky, 1988). Thus, people's verbal reports about what they know may draw more directly on the episodic memory system's linguistically encoded representations, although of course, information in schematic memory (even if it is not linguistically encoded in the first place) can often be translated or expressed in verbal forms.

In addition to these differences in learning speed and conscious accessibility, the two systems are predicted to differ in the type of information to which they attend. Schematic learning is chiefly concerned with regularities, so it records primarily what is typical and expected. In contrast, episodic memories should record the details of events that are novel and interesting: in other words, this system attends more to the unexpected and unpredicted. Social psychological studies show that under many circumstances, people attend to and recall mostly information that is inconsistent with their expectations (Higgins and Bargh, 1987). This empirical finding may reflect the more basic differences between two underlying memory systems: one that learns quickly and emphasizes novelty, one that accumulates information slowly and emphasizes regularities. For example, if someone has had mostly positive experiences with their health maintenance organization (HMO) but is treated rudely on one occasion, they may hold inconsistent representations in the two memory systems. The schematic system can maintain a generally positive opinion of the HMO because when a large number of experiences are summarized, the positive ones far outnumber the negative. However, in episodic memory the unexpected negative encounter may be most readily accessible (Higgins and Bargh, 1987). Depending on which memory system is tapped by a survey question, very different answers may be generated.

The independence of these two memory systems may help explain many seemingly puzzling observations. For example, "intuitive" emotional reactions such as a fear of flying are often stubbornly independent of our conscious and explicit knowledge (Kirkpatrick and Epstein, 1992). In problem-solving studies using the "conjunction fallacy" and other types of problems as well, logical reasoning and intuition give completely different answers, each seeming compelling in its own way (Sloman, 1996). A common-sense assumption is embodied in many current psychological theories: that all our knowledge and beliefs are represented in a single memory system—so that, for example, the beliefs we can consciously access and verbally report are the same ones that guide our preconscious interpretation of our experiences and reconstruction of our explicit memories. This assumption now seems highly questionable (McClelland et al., 1995; Squire, 1995).

The two memory systems appear to differ in the ease and automaticity with which they operate. There is ample evidence that the use of well-learned schemas for interpretation and inference is effortless and automatic, occurring preconsciously so that the perceiver is not even aware that he or she is making interpretations rather than just seeing "what is out there" (Bargh, 1994). In contrast, everyday experience tells us that retrieving a specific fact or autobiographical memory takes time (perhaps a second or two) and some mental effort. Thus, both motivation and cognitive capacity (e.g., availability of ample time, freedom from distraction) should influence the relative impact of the two memory systems on people's overt judgments and memory reports. Low levels of motivation or capacity will lead people to give their general schematic or stereotypic impressions rather than specific recollections. Consistent with this idea, Smith and Jobe (1994) found that when asked questions like "How often do you eat bananas?" people tend to give general estimates (based on schematic knowledge) rather than counting specific episodes. Still, if respondents are highly motivated and the recall task is made easy enough (for example, by asking "How many times did you eat bananas yesterday?") they can probably be induced to make reports based on episodic memory.

16.5 CONCLUSIONS AND SUMMARY OF IMPLICATIONS FOR SURVEY PRACTICE

Connectionist representations maintain information in memory as distributed patterns superposed in a single set of connection weights. They have several desirable properties, including the ability to build up general knowledge from repeated similar experiences, and the ability to use multiple applicable representations simultaneously as constraints in the process of memory retrieval or judgment. However, connectionist memories cannot quickly learn new information within the same system as preexisting information. One potential solution involves the use of two separate memory systems, one slow-learning and responsible for accumulating general "schematic" knowledge, and the other fast-learning and responsible for learning new or unexpected information based on a single exposure. This model solves the problem of catastrophic interference and is consistent with many types of psychological as well as neuropsychological evidence.

What direct implications do these new developments in connectionist models of memory have for survey research?

1. If an individual can have different representations in multiple memory systems, they may sometimes differ in content. Indeed, such differences will typically occur because the role of the fast-learning episodic memory system is precisely to store information that is novel and unexpected from the viewpoint of the slow-learning schematic system. Survey researchers must ask themselves which of these potentially different types of representation their questions will access. Factors such as motivation and cognitive capacity, as well as the nature

of the cues offered by the questions, may influence which system is used. Obviously the answers that will be obtained may differ drastically depending on which system is tapped.

2. A more fundamental question is closely related: if people can simultaneously hold multiple, potentially inconsistent representations in different memory systems, which *should* survey researchers attempt to elicit? The answer must depend on the survey's purpose, for example, whether the goal is to gather detailed recollections of specific facts or events, or overall impressions of the kind that are more likely to guide other evaluations, judgments, and behavior.

3. Information that is initially stored in the fast-learning episodic system is transferred over time into the slow-learning schematic system. This consolidation process qualitatively changes the nature of information that the person has available and can report, for the fast-learning system maintains information about unexpected or novel details as well as the context in which the information was encountered, while the schematic system emphasizes regularities, and the gist of the information, preserving context less effectively. For survey researchers interested in respondents' reports of specific past events, qualitatively different types of information may be obtained at different points in time. If relatively specific details are required, it is important to ask questions before too many weeks have passed.

4. Survey researchers who are interested in reports of specific facts must abandon the idea that a respondent can access a discrete representation formed on a unique occasion. Such representations do not exist as distinct entities. Memories (even episodic traces of novel events) are always superposed, and therefore influenced by the content of other similar memory traces. Retrieval is always reconstructive, influenced by the context of the occasion on which retrieval is attempted as well as by the person's interpretations and judgments. Researchers have recently come to realize that judgments (such as attitudes) constructed on-line are intrinsically variable and context sensitive (Wilson and Hodges, 1992). We must now realize that memory has exactly the same properties. In fact, as Carlston and Smith (1996) argue, it is very difficult to draw a principled line between memory and judgment.

5. Connectionist memories store multiple representations in a single combined (superposed) structure. At a later time, appropriate cues can elicit not only information that was specifically learned and stored, but also additional information that was inferred. This automatic inference process obviously has advantages, but from the survey researcher's point of view it means that people are often unable to subjectively distinguish between memory and inference. In other words, the source of a given item of information is not automatically available to the respondent; source is an attribution—a judgment—and therefore fallible. Other theoretical treatments of memory for the source of information have made similar points. As noted above, young children and the elderly may be important populations that have special difficulties with source monitoring.

Connectionist models of mental representation are still under active development today, and some issues remain controversial. Still, the widespread popularity of these models in various forms, as well as their consistency with multiple types of evidence (including neuropsychological as well as behavioral) suggest that their implications for those of us who work with human memory should be taken seriously.

ACKNOWLEDGMENTS

Preparation of this chapter was supported by NIMH grants R01 MH46840 and K02 MH01178. Thanks to Jamie Decoster, Lynda Mae, Roger Tourangeau, and Michaela Waenke for comments on earlier drafts.

REFERENCES

Anderson, J. R. (1983). *The Architecture of Cognition*. Cambridge, MA: Harvard University Press.

Bargh, J. A. (1994). The four horsemen of automaticity: Awareness, intention, efficiency, and control in social cognition. In R. S. Wyer and T. K. Srull (Eds.), *Handbook of Social Cognition*, 2nd ed., Vol. 1, pp. 1–40. Hillsdale, NJ: Erlbaum.

Burton, S., and Blair, E. (1991). Task conditions, response formulation processes, and response accuracy for behavioral frequency questions in surveys. *Public Opinion Quarterly, 55*, 50–79.

Carlston, D. E., and Smith, E. R. (1996). Principles of mental representation. In E. T. Higgins and A. Kruglanski (Eds.), *Social Psychology: Handbook of Basic Principles*, pp. 184–210. New York: Guilford Press.

Fiske, S. T., and Taylor, S. E. (1991). *Social Cognition, 2nd ed*. New York: McGraw-Hill.

Higgins, E. T., and Bargh, J. A. (1987). Social cognition and social perception. *Annual Review of Psychology, 38*, 369–426.

Johnson, M. K., and Raye, C. L. (1981). Reality monitoring. *Psychological Review, 88*, 67–85.

Kirkpatrick, L. A., and Epstein, S. (1992). Cognitive-experiential self-theory and subjective probability: Further evidence for two conceptual systems. *Journal of Personality and Social Psychology, 63*, 534–544.

Loftus, E. F. (1979). *Eyewitness Testimony*. Cambridge, MA: Harvard University Press.

McClelland, J. L., McNaughton, B. L., and O'Reilly, R. C. (1995). Why there are complementary learning systems in the hippocampus and neocortex: Insights from the successes and failures of connectionist models of learning and memory. *Psychological Review, 102*, 419–457.

McClelland, J. L., and Rumelhart, D. E. (1986). A distributed model of human learning and memory. In J. L. McClelland, D. E. Rumelhart, and PDP Research Group

(Eds.), *Parallel Distributed Processing: Explorations in the Microstructure of Cognition*, Vol. 2, pp. 170–215. Cambridge, MA: MIT Press.

McCloskey, M., and Cohen, N. J. (1989). Catastrophic interference in connectionist networks: The sequential learning problem. In G. H. Bower (Ed.), *The Psychology of Learning and Motivation*, Vol. 24, pp. 109–165. New York: Academic Press.

Neisser, U. (1967). *Cognitive Psychology*. New York: Appleton-Century-Crofts.

Plaut, D. C., McClelland, J. L., Seidenberg, M. S., and Patterson, K. (1996). Understanding normal and impaired word reading: Computational principles in quasi-regular domains. *Psychological Review*, *103*, 56–115.

Rumelhart, D. E., McClelland, J. L., and PDP Research Group (Eds.) (1986). *Parallel Distributed Processing*, Vol. 1. Cambridge, MA: MIT Press.

Rumelhart, D. E., Smolensky, P., McClelland, J. L., and Hinton, G. E. (1986). Schemata and sequential thought processes in PDP models. In J. L. McClelland, D. E. Rumelhart, and PDP Research Group (Eds.), *Parallel Distributed Processing*, Vol. 2, pp. 7–57. Cambridge, MA: MIT Press.

Sloman, S. A. (1996). The empirical case for two systems of reasoning. *Psychological Bulletin*, *119*, 3–22.

Smith, A. F., and Jobe, J. B. (1994). Validity of reports of long-term dietary memories: Data and a model. In N. Schwarz and S. Sudman (Eds.), *Autobiographical Memory and the Validity of Retrospective Reports*, pp. 121–140. New York: Springer-Verlag.

Smith, E. R. (1998). Mental representation and memory. In D. Gilbert, S. Fiske, and G. Lindzey (Eds.), *Handbook of Social Psychology*, 4th ed., Vol. 1, pp. 391–445. New York: McGraw-Hill.

Smith, E. R. (1996). What do connectionism and social psychology offer each other? *Journal of Personality and Social Psychology*, *70*, 893–912.

Smolensky, P. (1988). On the proper treatment of connectionism. *Behavioral and Brain Sciences*, *11*, 1–74.

Squire, L. R. (1995). Biological foundations of accuracy and inaccuracy in memory. In D. L. Schacter (Ed.), *Memory Distortion*, pp. 197–225. Cambridge, MA: Harvard University Press.

van Gelder, T. (1991). What is the "D" in "PDP"? A survey of the concept of distribution. In W. Ramsey, S. P. Stich, and D. E. Rumelhart (Eds.), *Philosophy and Connectionist Theory*, pp. 33–59. Hillsdale, NJ: Erlbaum.

Wilson, T. D., and Hodges, S. D. (1992). Attitudes as temporary constructions. In L. L. Martin and A. Tesser (Eds.), *The Construction of Social Judgments*, pp. 37–65. Hillsdale, NJ: Erlbaum.

Wyer, R. S., and Srull, T. K. (1989). *Memory and Cognition in its Social Context*. Hillsdale, NJ: Erlbaum.

SECTION D

CHAPTER 17

Potential Contributions of the CASM Movement Beyond Questionnaire Design: Cognitive Technology and Survey Methodology

Douglas J. Herrmann
Indiana State University

As discussed elsewhere in this book, the process by which data collection instruments were developed and tested changed swiftly and dramatically in the early 1980's. For decades, questionnaire design had been guided by behaviorism. In very simple terms, the behaviorist approach viewed questions as stimuli that elicited responses much in the same way as a bell elicited saliva from Pavlov's dog. Through such exposure to information, respondents associated ideas with certain terms. When asked a question, respondents supposedly generated the answer that was most strongly associated with the term or terms in the question (Sirken and Herrmann, 1996). Traditionally, some survey methodologists recognized that the response process was a complicated function of questions (Cantril, 1944; Payne, 1951), but question answering was seen fundamentally as a stimulus–response process.

In the early 1960's and 1970's, several forces revealed weaknesses in the behaviorist approach. Linguists found that language acquisition could not be explained in terms of stimulus–response learning. Artificial intelligence (AI) experts sought to model mental processes and, like the linguists, found behav-

Cognition and Survey Research, Edited by Monroe G. Sirken, Douglas J. Herrmann,
Susan Schechter, Norbert Schwarz, Judith M. Tanur, and Roger Tourangeau.
ISBN 0-471-24138-5 © 1999 John Wiley & Sons, Inc.

iorism inadequate as an explanation of how people perceive, remember, and solve problems. In the place of behaviorism, an increasing number of scientists proposed a cognitive approach. Answers were no longer responses to stimuli. Instead, answers were derived from mental processes which resembled computer operations that shaped the content of experience.

Within psychology, the cognitive approach became popular for three reasons. First, after five decades of extensive use, the behaviorist concept of stimulus and response had reached the limits of its usefulness. Researchers had demonstrated that Skinner's stimulus–response approach to language could not account for the diverse and creative way that children acquire language. Second, many psychologists stopped investigating the learning of rats and instead began to investigate human learning. This research demonstrated that people recalled what they learned not only on the basis of stimulus–response associations but also with inferences about the topic at hand. Third, the development of computer models to study how the mind worked brought a new way of conceptualizing human behavior. Such models did not rely just on stimulus–response associations but also employed processes that were in essence cognitive (i.e., rehearsal, reasoning).

In the early 1980's, leading survey methodologists and cognitive psychologists met at a seminar concerning cognitive aspects of survey methodology (CASM I) to determine whether cognitive psychology might provide a better account of how people answer survey questions. Many of the participants to this discussion agreed that the behaviorists were incorrect when they assumed that respondents answered survey questions on the basis of previously established habits elicited by questions. In contrast to the behaviorists, these participants concluded that answers to questions were derived, not elicited, through a series of cognitive processes: perception, encoding, comprehension, memory retrieval, thought, editing of potential answers, and expression of the answer (Jobe and Mingay, 1991). Moreover, the cognitive approach enabled a principled explanation of measurement error, something that eluded behaviorism. The cognitive approach showed that respondent error was a function of the kinds of ideas embraced by a question and the knowledge and/or the cognitive skills that a respondent did or did not possess.

The conclusions drawn at CASM I led to a dramatic change in the methods of survey questionnaire development and pretesting (Jabine, Straf, Tanur, and Tourangeau, 1984; Tanur, 1984). In addition to using large-sample field tests to identify defects in question and questionnaire design, cognitive methods that relied on small samples of respondents proved to be very useful for questionnaire development and pretesting (Sirken, Mingay, Royston, Bercini, and Jobe, 1988). For example, having a small sample of subjects ($n < 30$) "think aloud" as they attempted to answer questions revealed cognitive glitches in questions that were less likely to be detected by field tests and survey-design experts. Thus, the research emphasis focused on data collection instruments and, in particular, improving our understanding and use of cognitive methods to pretest survey questions.

In the fifteen years since the first CASM Seminar and other conferences concerned with cognitive issues, numerous research projects have been conducted and reported in the journals of survey methodology and cognitive psychology. This is not to suggest that the cognitive revolution in surveys is over. Much work remains to be done on the elements of respondent error (i.e., the wording of the question, inappropriate or unclear use of skip patterns, and so forth). Indeed, chapters in the earlier sections of this book demonstrate that new discoveries continue to be made all the time. In particular, these sections present important advances in understanding respondent error. Willis, DeMaio, and Harris-Kojetin's (Chapter 9) discussion on improving validity and on the evaluation of cognitive techniques, as well as Schober's account (Chapter 6) of the interaction between interviewer and respondent, are examples of some of the riches contained in the earlier sections.

17.1 NEW DEVELOPMENTS IN COGNITIVE PSYCHOLOGY

The survey research field has worked hard to develop useful applications of cognitive psychology to survey research. Most of these developments have involved the fundamental theories and findings of cognitive psychology that inspired, in part, the first CASM Seminar. Survey researchers remained cognizant of new developments in cognitive psychology but, understandably, they have focused first on applying the early discoveries in cognitive psychology. For example, cognitive research demonstrated that what a person recalls is often dependent on whether directions to recall contain the same cues that were present when the memories were formed (Tulving, 1983). Accordingly, respondent error was evaluated in terms of cognitive errors based on cues that are present or absent in a question. Eventually, a wider battery of a variety of methods was developed from cognitive psychology to identify questionnaire problems (Forsyth and Lessler, 1991).

Today there are many more sources of ideas about how to solve cognitive problems. Many of the theories and findings available now were not available at the beginning of the cognitive movement. Indeed, new theories and findings are reported each year. Basic research programs in cognitive psychology have been developed at most major colleges, universities, public research institutes, and government labs. Additionally, because so much basic cognitive research has accumulated, many applied researchers have come to apply basic cognitive theories not only to questionnaire issues but also to many other common problems in government, the military, law, education, advertising, publishing, industry, business, and medicine (Barber, 1988; Berger, Pezdek, and Banks, 1987; Gruneberg, Morris, and Sykes, 1978, 1988; Herrmann, McEvoy, Hertzog, Hertel, and Johnson, 1996; Hoffman and Deffenbacher, 1992). Applied researchers in these fields have solved real world problems, and devised remedies for situations in which people have difficulty in perceiving comprehension, learning, remembering, reasoning, and problem solving (Nickerson, 1997; Marsh and Gorayska, 1997). Besides survey methodology, cognitive psychology has been applied with considerable success to psychotherapy for depression, rehabilitation for victims of brain damage, the operation of machinery, the study of history, and more.

In the years since the beginning of the cognitive movement, basic cognitive researchers have continued to investigate the phenomena that originally elicited the interest of survey research. For example, cues present at input and output have been analyzed and categorized in a variety of ways. It is recognized now that some questions about past experience can be inferred from other knowledge and, hence, are not recalled in the true sense of the term. Also, recall is now seen as dependent on both inferential processes (Beatty, Herrmann, Puskar, and Kerwin, 1998), and the respondent's health and emotional state (Herrmann and Parente, 1994).

17.2 ACCESS TO THE NEW COGNITIVE PSYCHOLOGY

These days, it is much more challenging to keep up with the newest developments in cognitive psychology than in the early days of the cognitive movement in survey research. Several new basic cognitive journals and applied cognitive journals have been established. Numerous books have been developed from conferences devoted to specific theoretical cognitive issues and cognitive applications (e.g., Benjafield, 1997; Searleman and Herrmann, 1994).

Some of the recent cognitive research has been published in out of the way places, such as conference texts or journals not commonly examined by survey researchers. Other research contributions have appeared in publications frequented by survey researchers, but there has not yet been sufficient time to consider how these developments might be used in the survey process. The purpose of this section is to call attention to some of the frontiers that remain to future survey research.

Many new cognitive research areas have arisen, bringing with them new bodies of knowledge and methods that may be applied not only to coping with respondent error but also to all of the components of the survey process. For example, cognitive psychology has been successfully applied to guiding the interviewing process through software development, improving visual graphics to use in data analysis, and graphing the data collected by surveys and created by data analyses. These topics are examined in the following chapters and were selected because they are especially important to the process of gathering and analyzing survey data. However, there are other areas of cognitive psychology that could have been selected, and some of these are discussed below.

The first wave of applications of cognitive psychology to survey research and development focused on reducing respondent error from survey data. The efforts of that first wave made creative and comprehensive use of cognitive findings that pertain to respondent error. However, there are many other phases of the survey process that have not been, or barely have been, subjected to cognitive analyses. For example, the conceptualization of a sponsor's measurement goals, the transmission of a sponsor's goals to those involved in the survey design, and the design of computer applications for data collection, data analysis, and data dissemination are only now just beginning to be addressed from a cognitive perspective. Since the cognitive approach has been so successful when applied to respondent error, it seems reasonable to extend investigations to the other phases of the survey measurement process involving different survey tasks that are subject to different kinds of measurement error (Schechter, Sirken, Tanur, Martin, and Tucker, 1997). For example, the cognitive approach that successfully identified the stages of the survey response process has been extended to situations in which people extract information from statistical maps (Pickle and Herrmann, 1994; Herrmann and Pickle, 1996) or graphical user interface (Norman, 1997).

17.3 CRITICAL EXAMPLES OF THE NEW COGNITIVE PSYCHOLOGY

The chapters that follow report on the latest cognitive work in some of the key phases of the survey process mentioned above. It will be seen that their findings suggest how survey methodology in the future may adopt a comprehensive use of cognitive psychology, past and present. The work reported in this section does not by any means exhaust new developments in cognitive theory and findings that would be useful to survey researchers. Rather, the work of these researchers were seen by the editors of this text as especially informative to current research in survey methodology. In particular, these chapters address new cognitive developments in the administration of surveys, in the use of new technologies to aid the survey process, in the analysis of survey data, and in the presentation of such data to other survey researchers.

In Chapter 18, Mick Couper reports on recent cognitively-oriented research that has investigated different computer methods for collecting data. His chapter demonstrates how computer assisted data collection greatly facilitates collecting of data and eliciting answers to questions. Chapter 19 addresses how survey content can guide survey design. Responses to open ended surveys are typically coded by analysts; similarly, administrative data often requires coding. Fred Conrad reports in his chapter how AI systems have been developed specifically to perform many of the decisions previously made by coders, with the coders mental activity reserved for classifications too difficult for the AI system. In particular, his findings illustrate how respondent error is sometimes due to designer or sponsor error. In Chapter 20, Michael Friendly illustrates how the visualization of data can influence an analyst's cognition about his or her data. Graphical presentation of data may reveal patterns in data that are not apparent any other way. Three-dimensional presentations and data presented through visual filters can reveal hidden patterns and identify outliers. These methods enable analysts to understand data in ways that would otherwise not be possible. In Chapter 21, Steve Lewandowsky demonstrates important new points about data presentation in the form of statistical maps and graphs. For example, maps that present data by some graphic symbol, such as color, shading, or dot density, facilitate the detection of geographic trends in the data. This chapter reports on many investigations that have shown that cognitive factors sometimes lead people to misread maps and graphs. Alternatively, findings are discussed that show how graphs can enable us to grasp patterns in data. Finally, Chapter 22, by Elizabeth Martin and Clyde Tucker, identifies several targets for future development and application of cognitive sciences to surveys. This chapter reports the outcome of the CASM II Seminar working groups that examined different survey problems that the organizers felt could benefit from a research agenda that takes a cognitive and interdisciplinary approach. Martin and Tucker insightfully review the conclusions about the kinds of research that emerged from a series of discussions about the survey process. Some of the problems examined in this chapter are generic to surveys (data collection, interviewing) and some are specific to particular survey goals (measurement of disability, income, establishment surveys, household surveys). If research is conducted as Martin and Tucker propose, both the survey methodology field and cognitive psychology will benefit.

17.4 WHAT COGNITIVE RESEARCH ON SURVEY METHODS MAY BECOME IN THE FUTURE

The topics addressed by the chapters in this section cut across the phases of the survey process. The chapters themselves demonstrate that cognitive psychology can yield valuable insights and contribute to the development of better cognitive survey methodologies. Computer methods of data collection facilitate cognitive processes of respondents and interviewers. AI systems allow survey analysts to avoid the cognitive burden of tasks from which they otherwise would have to do. Cognitive approaches facilitate detection of patterns in the data through visualization and presentation by maps and graphs. Cognitive concepts help in analyses of surveys and help define a variety of research questions. These chapters highlight how basic cognitive psychology can be applied in an innovative fashion to help the survey research field progress at the rate necessary for future societal demands for data (Herrmann and Raybeck, 1997; Sirken, Herrmann, and White, 1993). If the survey field can establish routine channels of communication between basic cognitive researchers and applied cognitive researchers, it will help ensure that the excellent pieces of research reported here will become common and will lead to greater public support of survey research (Herrmann, 1998; Herrmann and Raybeck, 1997; Payne and Conrad, 1997).

We are living in exciting times. Cognitive psychology has been extremely helpful to the investigation of response error, but the usefulness of cognitive psychology to survey research has only just begun. In this chapter we have considered future applications of cognitive psychology to the survey process. As the cognitive survey methods field continues to evolve, it will probably come to share more and more generic problem areas with cognitive ergonomics, cognitive engineering, human engineering, and human factors. Nevertheless, the field of the cognitive aspects of survey methodology will continue to stand apart from related disciplines. The cognitive survey field has a stronger allegiance to basic cognitive theory than these other disciplines, partly because surveys are devices for communication, and because cognitive processes are central to communication. Thus, the cognitive aspects of survey methods takes on a unique set of challenges that define the direction of research, methods, and theory in the survey process.

Cognitive research has revealed many of the bases of respondent error, but cognitive research into other phases of the survey process has just begun. Cognitive psychology holds the promise of solving a wide variety of survey problems, only a few of which have yet been developed. The future of the cognitive aspects of survey methodology is likely to be as exciting, or more so, as its past.

ACKNOWLEDGMENTS

I thank the following people for influencing me in the writing of this chapter, directly or indirectly: Paul Beatty, Linda Pickle, Monroe Sirken, and Susan Schechter.

REFERENCES

Barber, D. (1988). *Applied Cognitive Psychology*. London: Methuen.

Beatty, P., Herrmann, D. J., Puskar, C., and Kerwin, J. (1998). "Don't know" responses in surveys: Is what I know what you want to know, and do I want you to know it? *Memory, 4*, 407–426.

Benjafield, J. G. (1997). *Cognition*, 2nd ed. Upper Saddle River, NJ: Prentice Hall.

Berger, D. E., Pezdek, K., and Banks, W. P. (1987). *Applications of Cognitive Psychology: Problem Solving, Education, and Computing*. Hillsdale, NJ: Erlbaum.

Cantril, H. (Ed.). (1944). *Gauging Public Opinion*. Princeton, NJ: Princeton University Press.

Forsyth, B. H., and Lessler, J. T. (1991). Cognitive laboratory methods: A taxonomy. In P. P. Biemer, R. M. Groves, L. E. Lyberg, N. A. Mathiowetz, and S. Sudman (Eds.), *Measurement Errors in Surveys*, pp. 393–418. New York: Wiley.

Gruneberg, M. M., Morris, P. E., and Sykes, R. N. (Eds.) (1978). *The Practical Aspects of Memory Research*. New York: Academic Press.

Gruneberg, M. M., Morris, P. E., and Sykes, R. N. (Eds.) (1988). *The Practical Aspects of Memory*, Vols. 1 and 2. Chichester: Wiley.

Herrmann, D. J. (1995). Reporting current, past, and changed health status: What we know about distortion. *Medical Care, 33*, AS89–AS94.

Herrmann, D. J. (1998). The relationship of basic and applied cognitive research. In C. P. Thompson, D. J. Herrmann, D. Bruce, D. G. Payne, J. D. Read, and M. P. Toglia, M. P. (Eds.). *Autobiographical Memory*, pp. 13–27. Hillsdale, NJ: Erlbaum.

Herrmann, D. J., McEvoy, C., Hertzog, C., Hertel, P., and Johnson, M. (Eds.) (1996). *Basic and Applied Memory*, Vols. 1 and 2. Hillsdale, NJ: Erlbaum.

Herrmann, D. J., and Parente, R. (1994). A multi-modal approach to cognitive rehabilitation. *NeuroRehabilitation, 4*, 133–142.

Herrmann, D. J., and Pickle, L. W. (1996). A cognitive subtask model of statistical map reading. *Visual Cognition, 3*, 165–190.

Herrmann, D. J., and Raybeck, D. (1997). The relationship between basic and applied research cultures. In D. G. Payne and F. G. Conrad (Eds.), *Intersections in Basic and Applied Memory Research*, pp. 25–44. Mahwah, NJ: Erlbaum.

Hoffman, R. R., and Deffenbacher, K. A. (1992). A brief history of applied cognitive psychology. *Applied Cognitive Psychology, 6*, 1–48.

Jabine, T., Straf, M., Tanur, J., and Tourangeau, R. (Eds.) (1984). *Cognitive Aspects of Survey Methodology: Building a Bridge Between Disciplines*. Washington, DC: National Academy Press.

Jobe, J. B., and Mingay, D. J. (1991). Cognition and survey measurement: History and overview. *Applied Cognitive Psychology, 5*, 175–192.

Marsh, J., and Gorayska, B. (1997). Cognitive technology: What's in a name? *Cognitive Technology, 2*, 40–43.

Nickerson, R. (1997). Cognitive technology: Reflections on a long history and promising future. *Cognitive Technology, 2*, 6–20.

Norman, K. L. (1997). Cognitive impact of graphical user interfaces. *Cognitive Technology, 2*, 22–30.

Parente, R., and Herrmann, D. J. (1996). *Retraining Cognition*. Gaithersburg: Aspen.

Payne, D. G., and Conrad, F. G. (Eds.) (1997). *Intersections in Basic and Applied Memory Research*. Mahwah, NJ: Erlbaum.

Payne, S. L. (1951). *The Art of Asking Questions*. Princeton, NJ: Princeton University Press.

Pickle, L. W., and Herrmann, D. J. (1994). The process of reading statistical maps: The effect of color. *Statistical Computing and Graphics Newsletter, 5*, 1, 12–15.

Schechter, S., Sirken, M., Tanur, J., Martin, E., and Tucker, C. (1997). CASM II: Current and future directions in interdisciplinary research. 1997 *Proceedings of the Section on Survey Research Methods of the American Statistical Association*, 1–10.

Searleman, A., and Herrmann, D. J. (1994). *Memory from a Broader Perspective*. New York: McGraw Hill.

Sirken, M., and Herrmann, D. J. (1996). Relationships between cognitive psychology and survey research. *Proceedings of the Section on Survey Research Methods, American Statistical Association*, 245–249.

Sirken, M., Herrmann, D. J., and White, A. (1993). The cognitive aspects of designing statistical maps. *Proceedings of the Section on Survey Research Methods, American Statistical Association*, 586–591.

Sirken, M. G., Mingay, D. J., Royston, P., Bercini, D., and Jobe, J. B. (1988). Interdisciplinary research in cognition and survey measurement. In M. M. Gruneberg, P. E. Morris, and R. N. Sykes (Eds.), *Practical Aspects of Memory: Current Research and Issues*, Vol. 1, pp. 531–536. Chichester: Wiley.

Tanur, J. M. (1984). Preface. In T. Jabine, M. Straf, J. Tanur, and R. Tourangeau (Eds.). *Cognitive Aspects of Survey Methodology: Building a Bridge Between Disciplines*, pp. ix–xi. Washington, DC: National Academy Press.

Tulving, E. (1983). *Elements of Episodic Memory*. Oxford, England: Oxford University Press.

CHAPTER 18

The Application of Cognitive Science to Computer Assisted Interviewing

Mick P. Couper
University of Michigan and Joint Program in Survey Methodology

18.1 INTRODUCTION

The move from paper and pencil to computer assisted interviewing is well underway (see Couper and Nicholls, 1998). While computer assisted telephone interviewing (CATI) has been an accepted technology for over two decades, with the relatively recent development of computer assisted personal interviewing (CAPI) and associated methods such as computer assisted self-interviewing (CASI) in the late 1980's, the shift to computers has gained momentum. Computer assisted interviewing (CAI) has the potential for fundamentally changing the entire data collection process. While we are in the midst of a revolution in the way survey data are created, collected and processed, developments in CAI to date have been somewhat uneven. Great strides have been made in the technological aspects of CAI (both hardware and software), but less progress has been made on issues relating to the design and use of CAI systems. In other words, the dominant focus of CAI has been on issues of feasibility, while less attention has been paid to issues of usability.

At the same time, the CASM movement has advanced our understanding of the respondent's cognitive processes. Unfortunately, the cognitive activities of the interviewer have been largely neglected. The increasing sophistication of CAI instruments is beginning to raise awareness of the cognitive demands placed on the interviewer.

Cognition and Survey Research, Edited by Monroe G. Sirken, Douglas J. Herrmann, Susan Schechter, Norbert Schwarz, Judith M. Tanur, and Roger Tourangeau.
ISBN 0-471-24138-5 © 1999 John Wiley & Sons, Inc.

This chapter attempts to redress these imbalances. We argue that usability is an important but heretofore largely neglected element of computer assisted interviewing. In interviewer-administered surveys the focus of usability is more on the interviewer than the respondent. We provide a framework in which the usability aspects of CAI can be examined, by linking survey research applications to the broader field of human-computer interaction (HCI) research. We highlight the contributions that cognitive science, through HCI, can make to computer assisted survey data collection. We also discuss methods and approaches for advancing the design of computer assisted survey instruments, and the challenges and opportunities we face in applying usability principles and procedures to the survey world.

18.2 WHY IS USABILITY IMPORTANT IN CAI?

If relatively little attention has been paid to usability issues in survey research, the obvious question is, why *should* we be concerned about usability? There is a growing body of literature on CAI pointing to the potential importance of design issues. Three sets of examples are offered that suggest the need to attend to design and interface issues in computer assisted surveys.

First, there is evidence from a number of studies that while interviewers react favorably, and even enthusiastically, to the introduction of CAPI (e.g., Baker, 1992; Edwards, Bittner, Edwards, and Sperry, 1993; Weeks, 1992), there are still a number of complaints expressed about the systems used (Couper and Burt, 1994). These range from complaints about weight (e.g., National Center for Health Statistics and Bureau of the Census, 1988; Couper and Groves, 1992), screen visibility (e.g., Bradburn et al., 1991), loss of eye contact or rapport with the respondent (Bernard, 1989), and speed or response time of the computer (e.g., Statistics Sweden, 1990). The fact that interviewers express positive attitudes toward computer assisted interviewing does not mean that they do not also experience problems with particular operations in a CAI environment (see also Sperry, Edwards, Dulaney, and Potter, 1998). This apparent paradox does not lead one to question the *feasibility* of CAI systems, but it does point to potential *usability* problems.

A second set of examples come from mode comparisons of CAI versus paper and pencil interviewing. While generally few substantive mode differences have been found, many of the larger effects can be attributed to changes in the design of the instrument (e.g., Baker, Bradburn and Johnson, 1995; Bergman, Kristiansson, Olofsson, and Säfström, 1994). While many of these are seemingly innocuous changes of layout rather than wording, they nonetheless convey different expectations of what kinds of information are required.

The third set of pointers to potential usability problems come from studies of interviewer use of various CAPI or CATI functions in interviewing, whether from self-reports (e.g., Baker, 1992) or from keystroke or trace files (Couper, Hansen, and Sadosky, 1997; Couper, Horm, and Schlegel, 1996;

Sperry, Edwards, Dulaney, and Potter, 1998). These studies generally find that interviewers use certain CAI functions (e.g., help and backup) less than might be expected. While this does not necessarily imply problems with the instrument, there is evidence of interviewer variation in the use of various CAI functions that may suggest potential usability concerns.

Collectively, these diverse findings suggest that design does matter. The way in which a CAI instrument is presented to the interviewer (or to the user in CASI) may affect the data that are collected. Similarly, the design of various functions and utilities (navigation aids, help screens) may also affect the likelihood of interviewers identifying and correcting such errors as may occur.

This is by no means meant to imply that CAI is worse than paper and pencil, or that such design problems occur only in CAI. Indeed, there is evidence that some design features of CAI systems serve to increase data quality (see Nicholls, Baker, and Martin, 1997, for a review). Given this, we argue that cognitive psychology can, through HCI research, inform design in CAI.

CAI is qualitatively different from paper and pencil interviewing, and hence, demands special attention. Some aspects of this difference include limited screen real estate (especially in CAPI), limited design options (especially in DOS-based interviewing systems), and the relative unfamiliarity of interviewers with computers. Further, both the input and output tasks are different in CAI. While it is clear that the computer has brought important benefits in certain areas, there may be other things that are harder to do on computer than paper. Again, we should recognize that CAI is different from paper-based survey instruments, and these differences bring additional challenges to design.

While a great deal has been learned from cognitive and communicative sciences regarding the content or wording of survey questions, much less attention has been paid to the design or form of the question on the printed page or computer screen. This paper is about design rather than content or wording (the traditional domain of CASM). It is about computer assisted interviews rather than paper and pencil, but much of the discussion could be applied to the latter. Furthermore, the paper focuses primarily on the interviewer as user of CAI. Usability issues are even more important in CASI, especially with the increased interest in methods such as audio-CASI, e-mail and Internet surveys, etc. Finally, there are other classes of users (e.g., instrument authors, managers and supervisors, and end users or analysts) that will not be focused on here (see Couper, 1994).

18.3   INTRODUCTION TO HCI AND USABILITY

CAI design, as with question wording, needs to be regarded more as a science than an art. There is an existing body of knowledge, containing both theoretically-based principles and empirical findings, that can be readily applied to the CAI world. This field, HCI or usability research, has its intellectual roots in cognitive psychology.

Marchionini and Sibert (1991, p. 18) define HCI research as being "concerned with the design of interfaces that allow easy and efficient use of computer systems." This encompasses a field of research and application that goes under various names, including usability, cognitive ergonomics, and human factors, and was originally termed "software psychology." In Shneiderman's book of the same title, he described software psychology as "a new science which applies the techniques of experimental psychology and the concepts of cognitive psychology to the problems of computer and information science" (1980, p. 3). HCI or usability research emerged from a blend of cognitive psychology and computer science (see Carroll, 1993, 1997). Since these early days the field has expanded rapidly to incorporate a variety of other disciplines and perspectives, including sociology, social psychology, anthropology, ethnography, communication research, information science, and so on. The notion of usability is used in the broader context here, considering contributions beyond cognitive science.

The field of HCI has grown rapidly from its uncertain beginnings in the late 1970's, and until quite recently was still struggling for recognition within the traditional disciplines that gave it life (cognitive psychology and computer science). The field has only recently emerged as an accepted discipline in its own right, and can now boast graduate programs at several universities, as well as a number of professional journals and conferences devoted to this area of research and application. During this time the field has succeeded in breaking its narrow disciplinary bounds, and now encompasses a wide variety of theoretical and applied approaches focusing on the human–computer interface. However, the tensions between theory and practice, between research and application, have not diminished. The field ranges widely from the more theoretical to the highly applied, with different associations representing various perspectives. In fact, Cooper and Bowers (1995, p. 49) characterize the field of HCI as "fragmented, contested, and dynamic." Thus, HCI can by no means be described as a well-established field with a long tradition and with a unified theoretical base and shared sense of purpose. Rather, it is represented by diverse perspectives and interest groups. In some respects, HCI research has many similarities to CASM, in addition to sharing the same theoretical underpinnings of cognitive psychology.

The first significant work in HCI was on the development of analytic or cognitive models to predict human performance using a computer. An influential example of this early work was the Model Human Processor proposed by Card, Moran, and Newell (1980, 1983; see also Newell and Card, 1985). The goal of this work was analytic rigor in decomposing the interactions of the human and computer into measurable components, at the level of predicting performance on individual keystrokes. Carroll (1993) argues that the impact of the models has been relatively narrow, being more useful in domains where relatively low-level user performance efficiency matters.

The second wave of the HCI movement began by critiquing the limits of this early "mentalist" tradition (see Cooper and Bowers, 1995; Frohlich, Drew, and Monk, 1994), arguing that useful theory is impossible because of the complexity of human–computer systems. Landauer (1991, p. 61) goes so far as to say that "the theory of human cognition is now and forever may be too weak to be the main engine driving HCI." The newer approaches were less cognitive and more social, and placed increasing emphasis on the user and on interaction (rather than interface). Despite the rhetoric, developments continue in the field of cognitive modeling, and the various approaches coexist under the broad HCI rubric, resulting in the broadening of perspectives.

More recently, these developments have evolved further into the "contextualist" or "interactionist" perspective, based largely on qualitative analysis of user activity in situ (see Carroll, 1993; Frohlich et al., 1994; Greatbatch, Heath, Campion, and Luff, 1995a; Greatbatch, Heath, Luff, and Campion, 1995b). Much of this is attributed to the pioneering work of Lucy Suchman (1987). This trend has broadened HCI to focus on social as well as cognitive aspects (Anderson, Heath, Luff, and Moran, 1993), and has led to the application of both conversational analytic and ethnographic methods to the study of human–computer interaction (see, e.g., Greatbatch, Luff, Heath, and Campion, 1993; Cooper, Hine, Rachel, and Woolgar, 1995).

Regardless of the perspective one takes, there appears to be agreement that a key element of HCI research is interface design. As Cooper and Bowers (1995) note, the human interface is the site of HCI knowledge and practice. While in the early days the interface was conceived of more narrowly as the input/output part of the system, over time the view of the interface has broadened to include almost any activity related to the interaction between human and machine. As Norman (1986) argues, the notion of the interface cannot be separated from the system. He writes (p. 61) "From the point of view of the user, the interface *is* the system." From the CAI perspective, this would involve everything the user does in interacting with the computer and everything the computer does in interacting with the user. In the CAI world, this extends to human–computer–human interaction. Thus, the focus is not only on interviewer–computer interaction and interviewer–respondent interaction, but also on their mutual influences.

In summary, very much like the CASM movement, the field of HCI research encompasses a broad range of theoretical perspectives and methods. Similarly, there are tensions between the experimental or theoretical world (deriving principles, testing theories, etc.) on the one hand, and the applied world (evaluating questionnaires, designing instruments, etc.) on the other. In the balance of this chapter, we attempt to demonstrate how these approaches can be applied to the study of CAI. First, some of the methods employed in HCI research are reviewed, then we present a few brief examples of common threads between HCI and CAI.

18.4 METHODS OF USABILITY RESEARCH AND TESTING

Methods are the tools of the trade for any discipline. While usability research and testing have two different goals, they generally share the same methods.

HCI or usability research is focused on exploration and understanding of cognitive and interactional aspects of computer use. Usability testing is aimed at evaluating the quality of particular products or systems with the goal of improving design. Clearly, the results of either approach could inform the other. The latter focus has tended to dominate in recent years, particularly in the commercial sector. This distinction has parallels in the CASM world, where much of the work is focused on testing and improving particular survey instruments.

There are a variety of techniques used to conduct usability research or testing, and these can be classified in a number of ways. These range from the more formal to the less formal methods. Macleod and Rengger (1993) distinguish between theoretical methods, expert methods, and user-based methods. Theoretical or formal methods involve the application of various formal models (e.g., keystroke level models, task action grammars, cognitive complexity theory) to build predictive models of user performance (see, e.g., Dix, 1995; Drury and Hoffman, 1992; Gillan, Holden, Adam, Rudisill, and Magee, 1992; Howes, 1995; Lansdale and Ormerud, 1994). We will not discuss these further here. In the remainder of this section we focus on the two major usability approaches in use today, expert evaluation methods, and end-user evaluation methods.

18.4.1 Expert Evaluation Methods

Expert methods generally involve the evaluation of a system by one or more HCI experts. This is closely analogous to expert review in the evaluation of survey questionnaires (see Presser and Blair, 1994). These are also called usability inspection methods. One of the key proponents of this approach is Jakob Nielsen who, with various colleagues, has developed methods of heuristic evaluation (e.g., Nielsen, 1992; Nielsen and Mack, 1994; Nielsen and Molich, 1990; see also Jeffries, Miller, Wharton, and Uyeda, 1991). This approach involves experts being provided with a small set of usability heuristics (e.g., be consistent, provide feedback, minimize user memory load) distilled from a larger number of interface design principles. The experts then test the system and note violations of particular heuristics. In general, from three to five experts are recommended for this approach. Lansdale and Ormerud (1994) note that heuristic evaluation can be highly unreliable, but can find many problems relatively cheaply.

Heuristic evaluation has been used in the survey world to evaluate prototype instruments for the Consumer Price Index (CPI) Housing survey at the U.S. Bureau of Labor Statistics (BLS) (see Bosley, Conrad, and Uglow, 1998). A set of eight specific heuristics for the pen-based CAPI system was designed. They recruited three usability specialists with knowledge of pen and graphical-user interfaces, but without knowledge of the CPI-Housing instrument. After a set of individual sessions in which the evaluators recorded problems with the system, the problem reports were consolidated into a single list, and all items were rated for severity by each of the experts. Levi and Conrad (1995) conducted a similar heuristic evaluation of a web site for dissemination of BLS data.

Researchers at the Census Bureau (Sweet, Marquis, Sedivi, and Nash, 1997) undertook expert reviews of two electronic computerized self-administered questionnaire (CSAQ) instruments. The experts used a similar set of heuristics to those proposed by Nielsen and Molich (1990) and Levi and Conrad (1995). The most problems uncovered across both instruments related to aesthetic design (37 out of 143 unique problems), while navigation and organization (20), data entry (19), and consistency (18), were the next most frequently violated heuristics.

Both the BLS and Census Bureau examples yielded a large number of design problems at the early stages of instrument development, and these were used to influence subsequent design of the instruments. As Bosley, Conrad, and Uglow (1998) note, while there is no substitute for end-user performance data, the heuristic evaluation provides valuable feedback to the developers.

Another set of expert methods are called "cognitive walkthroughs." These generally involve experts, singly or in groups, walking through an interface in the context of core tasks a typical user will need to accomplish (Jeffries et al., 1991). Walkthroughs are intended to be used by the designers of the software, rather than outside experts. The approach is useful in that it forces designers to define user goals and explicitly state assumptions.

18.4.2 End-User Evaluation Methods

This category encompasses a wide variety of different approaches, with almost the only common element being the involvement of the people who will ultimately use the systems. End-user evaluations can be laboratory based or field based. They can involve observational methods, empirical collection of performance data, or direct questioning of users, or combinations of all three. They can include structured tasks (scripted activities) or "natural" interactions (free exploration of the system). Finally they can be more or less obtrusive, ranging from instances where the evaluator is present and interacting with the user, to the collection of automatic performance data without the knowledge of the user. After mentioning some examples from the CAI world, we review some of the design issues in end-user testing.

In work on CAI usability, a variety of approaches are being adopted. Some of these involve the automatic collection of keystroke data from live interviews in the field (see Couper et al., 1996, 1997). We have also used keystroke data to evaluate interviewer performance and identify instrument problems in scripted mock interviews (Couper, Sadosky, and Hansen, 1994). While the collection of keystroke data is relatively cheap, they provide only a limited window into what happens in the survey interview, and should be complemented by other methods involving end-users at early stages of the design process.

A variety of observation techniques are currently in use in surveys, and there is no reason why these could not be expanded to include usability issues. Generally, monitoring (in CATI) and behavior coding (in CAPI) have been used to evaluate interviewer performance or explore respondent problems with the questions. Typically, interviewer difficulties with the computerized instrument have not been the focus of these approaches. Similarly, observation is used to evaluate interviewer performance in the field, but no systematic data have been collected on interaction or usability problems. These methods provide rich qualitative details on how interviewers interact with the computer, and are useful both for hypothesis generation and design of experiments (see Frese et al., 1991; Vaske and Grantham, 1990). However, as with keystroke file analyses, these approaches are more useful for evaluation of existing systems than for testing and development.

Direct end-user evaluations of CAI instruments have been slow in coming to the survey world, despite arguments for the incorporation of such methods (see Couper, 1994). Bosley, Conrad, and Uglow (1998) mention using a variety of informal end-user evaluations of the CPI-Housing instrument, and Hansen (1996) reports on initial development of CAI usability methods on a CATI instrument, but there have been few reports of formal end-user testing of CAI systems.

The Survey Research Center (SRC) at the University of Michigan is currently conducting a series of usability evaluations of different instruments. The early work led to the development of a fully-equipped laboratory for conducting such evaluations (see Hansen, Fuchs, and Couper, 1997). While similar to cognitive laboratories, additional equipment facilitates the usability evaluation. This includes a scan converter for capturing video images of the computer screen, and ceiling-mounted cameras to capture both the interviewer's hands on the keyboard, and the respondent–interviewer interaction. Dumas and Redish (1993) and Rubin (1994) provide guidance on setting up a usability laboratory.

Two studies currently in progress will illustrate use of the usability laboratory. The first is a small-scale experiment of two alternative versions of rostered questions. We developed rough prototypes of a grid- or screen-based version and an item-based version of a set of questions. A small number of interviewers conducted scripted mock interviews on each of the prototype instruments, with the order of presentation randomized. We are examining a variety of objective performance measures (time, accuracy, etc.) and subjective reactions (from debriefing interviews) to evaluate the relative effectiveness of the two methods of collecting rostered data.

The second study involves observation of "real" interviews in a laboratory setting, using the National Health Interview Survey, with Census Bureau interviewers and respondents especially recruited for this purpose. The objectives of this work are two-fold. First, the observations provide real data on the usability of a production CAPI system. Second, we are using the videotapes to explore issues of respondent–interviewer interaction, and the role that the presence of the computer plays in such interaction.

A further goal of these and other activities is to evaluate the relative effectiveness of a variety of end-user methods for evaluating CAI instrument design. Issues of what kind of video and audio data should be captured and how these data are to be coded and analyzed are key concerns if we hope to make usability testing an integral part of the survey design process.

Think-alouds are a useful cognitive laboratory technique that work well with self-administered surveys. The method is often used in HCI to get users to articulate their thoughts and reactions in interacting with a computer. We are planning on using think-alouds for an evaluation of interviewer use of a new sample management system. However, this approach makes less sense for interviewer-administered surveys, where the conversation between interviewer and respondent may reveal much of what is happening with the interviewer–computer interface.

Conversation analytic methods have also been used in HCI. This method seems to be particularly suited to the survey world. Greatbatch and colleagues (1995a, 1995b) analyzed doctor–patient interactions before and after the introduction of a computerized record system. We are using similar approaches at SRC to explore respondent–interviewer interactions in both CAI and paper surveys. Many of the techniques used in usability research and testing have their analogs in the survey world, and are familiar to those who use cognitive laboratory and related methods to evaluate survey questions.

18.4.3 Issues in Usability Research and Testing

Finally, we review some of the key issues in usability testing or research. A critical decision is, of course, the choice of method. This depends on a variety of factors, including the time and money available for testing, the availability of experts or end-users, the equipment and facilities available, the stage of development or purpose of the evaluation, and so on. While certain methods are better suited to some studies than others, given the current state of play in the survey world, at this point we would concur with Bosley, Conrad, and Uglow (1998) that any usability testing is better than none. However, there are a number of other issues to be faced in usability testing, particularly involving end-users.

The first is whether to evaluate designs using novice interviewers, those with some experience, or expert interviewers with broad experience. For example, using interviewers familiar with the CAI system may help them to focus on design issues rather than focus on the basic operation of the system. On the other hand, this does not reveal how inexperienced interviewers will react to the system. In addition, testing new or alternative designs with experienced interviewers may produce biases in favor of the methods they currently use. Often the choice of subjects will be dictated by the design and the target of inference, and constrained by the available resources.

Another issue in usability testing already alluded to is the massive amount of qualitative data that can be produced. In addition to the various video images we obtain in our tests (computer screen, interviewer's hands, and respondent–interviewer interaction), we also collect information from debriefing interviews of both interviewer and respondent, time stamps from the system, trace files, and substantive data from the interview (Hansen et al., 1997).

To this could be potentially added reaction time measurement, eye movement data, and so on. Our goal is in part to determine what information from the laboratory is most useful to study the human–computer interface. Furthermore, we are using the same tests to focus on multiple issues (e.g., the CAPI interface and the interviewer–respondent interaction). Even if we narrow the focus, the information gathered still has the capacity to rapidly overwhelm the analyst.

There are two general approaches to the reduction and analysis of this wealth of data. The first is to code the information, and the other is to review and subset the data. There are a variety of products and tools to facilitate coding and analysis of videotaped interactions (e.g., Jordan and Henderson, 1995; Macleod and Rengger, 1993; Sweeney, Maguire, and Shackel, 1993), but this still remains a laborious task. We are developing a restricted set of codes (much like behavior coding) that will provide a relatively cheap but superficial overview of the entire interaction, and provide quantitative data on the incidence and frequency of various events. The second approach is to review the videotapes and create a subset of instances where difficulty was observed. This approach is often used in conversation analysis, and is particularly useful for in-depth analysis of a smaller set of interactions.

A further set of related issues is whether to use experimental or observational methods, and whether to observe "natural" interactions or to create scripts designed to simulate known difficulties. Observations run the risk of revealing few difficulties, especially if the incidence of problems is low, or dependent on certain responses. Scripted interviews or scenarios could be used to ensure that the interactions contain sufficient difficult situations, but they would enhance the artificiality of the testing environment. A final issue is when to do usability testing. Some interface problems can (and are) identified as part of the testing process, but we suspect from our own work that asking evaluators to focus both on functionality and usability may be too burdensome a task. We would argue for separate processes focusing on debugging and usability testing.

We are in the early stages of developing and evaluating tools for studying computerized survey instruments. But this work shows a great deal of promise. Not only can survey research borrow the methods of HCI research, but many of the theoretical perspectives are also relevant to our field. Some of these contributions are in the next section.

18.5 APPLICATION OF USABILITY RESEARCH TO SURVEYS

Examples of HCI research with potential applicability to survey research abound; for example, navigation (e.g., Dillon, 1994), online help (e.g., Duffy, Palmer, and Mehlenbacher, 1992), menus (e.g., Norman, 1991), screen layout (e.g., Staggers, 1993; Tullis, 1983), input devices (e.g., Sears, 1991), and so on. Much of the HCI research is focused on users interacting with complex computer systems. The current state of play in CAI is not at this level, and generally involves a limited range of features and actions. While the translation may not be automatic, there are some areas of HCI research that are clearly relevant to CAI. In this section we offer a selection of findings and perspectives from the usability literature to show how they may apply to the CAI world.

18.5.1 Cognition and Beyond

Given its roots in cognitive psychology, it is not surprising that much of the focus of HCI research is on the cognitive aspects of interacting with computers. One area of cognition important to usability is the notion of mental models. A mental model is a user's cognitive representation of the system's structure and function (Brown, 1988; Davis and Bostrom, 1992; Norman, 1986). Mental models are a broad class of concepts, including declarative and procedural knowledge, scripts, mental maps, and so on, that affect the degree of cognitive activity required to interact with a system.

Ideally, systems should be designed around the mental models users have (Norman, 1988). What are the mental models of interviewers? Do CAI instruments in any way reflect these models? How different are the mental models of interviewers using (or familiar with) paper and pencil surveys to those using CAI instruments, and how can these differences affect the transfer of knowledge from one method of data collection to the other?

Errors or difficulties in interaction often arise through incompatibilities between the user's mental model and the computer program. Mental maps or models can be developed in three ways, through training, usage, and/or analogy. Since we expect interviewers to be competent users shortly after training, and we want to minimize the costs of training interviewers to use CAI systems, designing such systems to facilitate the development and use of appropriate mental models is important. Many of the design principles and guidelines proposed in HCI (e.g., Galitz, 1993; Hix and Hartson, 1993; Mayhew, 1992; Powell, 1990) are based on this assumption. Some of these will be mentioned briefly here (for a fuller discussion, see Couper, 1994).

For example, consistency is one of the most universally endorsed design principles, aimed at facilitating the transfer of knowledge across parts of a system. Examples of the effect of inconsistency on interviewer error abound in the CAPI world (see, e.g., Couper et al., 1996). Another principle is that recognition is an easier cognitive task than recall, and efforts should be made to reduce the memory load on users. Again, many CAI systems violate this principle, in that interviewers are expected to remember the function keys, rather than having them provided on the screen or template. Another tool for facilitating the match between user and system is the use of physical analogies (e.g., representation of a paper form, natural mapping of function keys, etc.). Similarly, informative feedback is a means of facilitating user understanding of a system. The interviewer does not need to know what the internal workings of the system are, but does need to know how a function is operated on by the system. While these and other design guidelines all seem self-evident, they have received remarkably little attention in CAI design, despite the strong theoretical underpinnings and empirical support.

Many of the problems of navigation or segmentation, first noted by Groves and Mathiowetz (1984), but still recognized as a problem today (e.g., Sperry, Edwards, Dulaney, and Potter, 1998), could be attributed to deficiencies in the mental representations interviewers have of the survey instrument. A common problem noted in CAI training is how to represent the complexity of the typical survey instrument to the interviewer (see Wojcik and Hunt, 1998).

While the cognitive aspects of interaction are important, it has long been recognized that mental activity is not the only element of human–computer interaction. Both perceptual and motor activities are involved (see Card et al., 1983). One of the most frequently cited models is Norman's (1986) seven stages of human computer interaction. These stages are: (1) forming the goal, (2) forming the intention, (3) specifying an action, (4) executing the action, (5) perceiving the system state, (6) interpreting the system state, and (7) evaluating the outcome. It is clear that the process involves cognitive (e.g., stages 1 and 2), motor (stage 4) and perceptual (stage 5) activities. However, the domain of CASM has generally been that of cognitive (and communicative) aspects of surveys. Aside from some limited work on forms design (e.g., Lessler and Holt, 1987; Rothwell, 1985), and the conceptual work of Jenkins and Dillman (1994, 1997), perceptual issues have not received much attention in surveys. The increased use of the computer, with its enhanced visual features (e.g., color, typeface, font, etc.) and limitations (e.g., screen size) may increase our awareness of perceptual issues in CAI. There are a number of HCI studies on these issues whose findings can be readily applied to CAI instrument design.

Various studies have shown that user preferences for screen complexity follow an inverted U (e.g., Coll and Wingertsman, 1992; Staggers, 1993; Tullis, 1983). Not only are users uncomfortable with screens that are too complex or too simple, but their performance will be reduced when using such screens. This is especially true of more experienced users. This runs contrary to the "white space is a virtue" dictum of self-administered surveys (Dillman and Salant, 1994, p. 120). In CAI, we have heard many interviewers express frustration at the single question per screen approach, and say they would prefer a grid approach (as they are accustomed to in paper instruments).

Finally, sensorimotor activity is a new area to consider in CAI. The physical demands of interacting with a computer (especially a laptop in CAPI) may lead to errors and reduced efficiency. For example, we found that a large number of function key errors in a CAPI interview could be attributed to the physical layout (both size and relative position) of the function keys (Couper et al., 1997). This can especially be an issue in CAI where the interviewer is dividing attention between the computer and the respondent.

More recent work has extended the domain beyond the single user and computer to include all interactants and the environment in which the activity takes place. As Greatbatch and colleagues (1993, p. 210) note, ". . . designers of computer systems should be sensitive to the orientations of all participants in the socio-interactional environment." HCI research suggests that we should extend our focus beyond the cognitive activities of the respondent, to include the cognitive, perceptual, and motor activities of the interviewer in interacting with the computer, as well as the interaction between interviewer and respondent. Finally, we should also consider the environment in which this interaction takes place (see Couper, 1997).

18.5.2 Human Error

While the study of human error extends beyond HCI, it plays a key role in usability research. A major contribution of this field has been an effort to understand and classify the different types of errors that can be made in interacting with computers (e.g., Norman, 1983; Rasmussen, Duncan, and Leplat, 1987; Reason, 1990). Understanding the level of activity at which errors occur (see Zapf, Brodbeck, Frese, Peters, and Prümper, 1992) is important for distinguishing between mental errors (e.g., forgetting, errors of judgment), perceptual errors (e.g., failing to notice cues, misinterpreting system feedback), and sensorimotor errors (including typographical or keying errors). By understanding the types of errors interviewers make, exploring the reasons why they may make them, and examining the consequences of these errors, we can design systems to better manage such errors.

It is widely recognized that errors can never be entirely eliminated, and the focus should be on error management, which includes minimizing the incidence of errors, maximizing the discovery of errors, and facilitating the recovery from errors (Bagnara and Rizzo, 1989; Lewis and Norman, 1986). Errors can be reduced both on the system side (through good design) and user side (through training). Frese and Altmann (1989) suggest that training should focus on dealing with errors rather than simply avoiding them. Frese et al. (1991) note that users' mental models are enhanced when errors are made and successfully corrected. Most CAI training sessions are quite structured, allowing only limited time for free exploration of the system. Often these sessions require trainees to work through a series of scripts in tandem, not deviating from the prescribed sequence of actions.

Error detection generally occurs in one of three ways (see Norman, 1993; Reason, 1990; Rizzo, Bagnara, and Visciola, 1987) through (a) self-monitoring, either by the user or the system; (b) the occurrence of an unintended consequence; or (c) discovery by a third party. This suggests that CAI systems should not be too automatic. If the interviewer has to pay little attention to the system and receives little feedback from the system, there may be a tendency to turn control over to the system. This may reduce the likelihood of error detection through self-monitoring.

Finally, error recovery proceeds from the assumption that errors will inevitably occur. System features such as informative feedback, reversible actions, forcing functions (e.g., a confirmation dialogue) for actions with serious consequences, and so on, all make the task of error recovery easier for the user.

18.5.3 Knowing the User

Early consensus in the HCI field was reached on the key issue of recognizing the importance of the user. This is manifested in a number of dictums or design principles such as "know the user," or "design for the user." The notion of user-centered design was first promoted by Donald Norman (1983, 1986; see also Landauer, 1995). A critical assumption behind this perspective is that the person who programs the system or instrument has different skills, knowledge, and goals than those who will ultimately use the system. For example, Thimbleby (1990, p. 134) writes: "Designers design for themselves. Designers often think of themselves as exemplary users and evaluators." It is safe to say that the user-centered design perspective has been slow to take hold in CAI.

A variety of efforts have been undertaken to identify various types of users and differences in how they interact with computer systems (e.g., Bødker, 1991; Cuff, 1980; Shneiderman, 1992). For example, experts and novices appear to differ in the mental models they have of computer systems, and how they use these mental models in interacting with the systems. Experts have more elaborate mental models, permitting greater generalization across functions, and improved learning from errors. When interacting with systems, novices are more likely to use declarative knowledge (discrete chunks of information) while experts are more likely to use procedural knowledge (sequences of actions).

The types of errors made by experts and novices also appear to differ. For example, Prümper, Zapf, Brodbeck, and Frese (1992) found that experts do not make fewer errors than novice users, but they do take less time handling the errors. Thus, their error management skills are enhanced through practice or usage.

Santhanam and Wiedenbeck (1993) describe discretionary users as those for whom the software is simply one tool among many which they use to facilitate their professional work. Such users are likely to use one software package all the time for a particular task and to have little or no exposure to other alternative software. This seems similar to interviewers' use of CAI systems. They found that on a small set of frequently performed tasks, the performance of discretionary users resembles that of expert users, in that they know a method to accomplish the task, and they do so with few errors and with very little hesitation or need for help or feedback. However, for tasks that the user has never done before, as well as tasks done so rarely that the user has forgotten how to do them, different types of knowledge are required. In performing these tasks, they found discretionary users to be similar to novices. We have found similar behavior among survey interviewers (Couper et al., 1997).

The key implications of this work for CAI are in identifying and understanding the likely users of the systems, and designing systems appropriately. Frequent or expert users have different needs and expectations of the system than infrequent or novice users. With the relatively high turnover of survey interviewers and the periodic nature of much interviewing work, survey interviewers probably lie closer to the novice than the expert end of the continuum.

Hollnagel (1993) notes that every set of specifications for a system includes a set of assumptions about the user. Some of these assumptions may be implicit or hidden. It is the goal of usability research to make these assumptions explicit.

18.5.4 Measurable Outcomes

Another lesson we can learn from usability research is the focus on measurable outcomes. The popular notion of "user friendliness" is unproductive. Users do not want friendly systems, they want useful systems. A classic example of a friendly system that had an extremely limited lifespan is the recent "Microsoft Bob" interface. Many early evaluations of CAPI systems measured only the interviewers' subjective reactions to the systems. User preference or satisfaction is only one outcome that can be quantified and measured in evaluation of computer systems. Other outcomes include time taken to learn the system, retention of skills and knowledge over time, the number and types of errors made, and the time taken to complete tasks (see Brown 1988; Shneiderman, 1992). This means that alternative systems can be directly compared to each other on a variety of dimensions such as these, and empirical data can be obtained on the relative merits of each alternative. Also, measurable outcomes mean that design can be more of a science than an art. As in questionnaire design, subjective reactions and programmer preferences for a certain style can be refuted with good evidence on what works and what does not. Unfortunately, much current design of CAI systems is based on the personal tastes of the designer.

18.5.5 Research Needs in CAI

The kinds of research needs for CAI range from the mundane to the highly complex. Considering how little systematic research has been done on the design of survey instruments, even fairly basic work on screen design and layout would be useful. This is not to say we need an experiment to test every alternative format that could be used, but observational data on how these factors may affect interviewer performance may be useful. Similarly, we do not need to mount costly experiments to verify that the HCI findings on the use of color, upper or lower case text, and so on, apply equally to the survey world. We can build on an existing body of knowledge, but should not uncritically import all HCI findings to CAI. It is important to understand how human–computer interaction in CAI is different from that in other applications, and how this can affect the translation of findings from other research areas.

At the other end of the spectrum, there are a wide variety of issues facing designers of CAI instruments, both present and future. Some of these are issues the field has been grappling with for some time, such as multiple item screens, rosters or grids, and navigation issues. These are ripe for usability research and testing on existing instruments, or with current versions of CAI software. An ongoing tension in CAI is the degree of control exercised over the interview, or the level of flexibility offered the interviewer (and respondent). Because CAI systems are based on highly structured linear programs, they tend to dictate the order in which the questions are asked, how the responses are entered, and what actions the interviewer can take. But this need not necessarily be so, and indeed the HCI literature argues for giving the user more (not less) control over the system. Research is now focusing on how respondents provide household roster information (for instance), and how CAI systems can be designed to accommodate respondent (and interviewer) preferences without sacrificing the error-checking, skip functions and other features that make CAI powerful (Hansen et al., 1997; Moore, 1996).

Looking more toward the future, there are a number of design challenges facing the CAI field. Some of these are developments external to CAI over which we have little control (e.g., the move from text-based applications to graphical user interfaces), while others are exciting new developments within CAI (e.g., audio-CASI). Whatever the source of these new opportunities, we should be conducting research now on the optimal ways to harness the new features of computers to extend the power of surveys.

Turning to the interactional components of the survey interview, there are a number of research areas where contributions can be made. We know remarkably little about the dynamics of respondent–interviewer interaction, particularly as it is shaped by the presence of the computer. Greatbatch and colleagues (1995a, 1995b) found that doctor–patient interactions have been changed by the introduction of computers into doctors' offices. We are currently exploring similar issues in CAI.

We have attempted to demonstrate a number of parallels between the broader HCI field and the more limited CAI applications. Despite these similarities, we should also recognize that there may be differences. As noted earlier, much HCI work focuses on a single user and a single computer. CAI typically involves both an interviewer and a respondent. However, this is not the same as computer-mediated communication in which the computer serves as intermediary or interface between two users (as in e-mail or online chat rooms). It is also not the same as computer-supported collaborative work in which multiple users are working together on the computer. While CAI is distinct from many of these newer developments in HCI, it may share similarities with computer use in service interactions, in which one user interacts with the computer to assist another user (e.g., directory assistance, catalog shopping, reservations, 911 operators, etc.). The challenge is to determine how these similarities and differences affect the applicability of HCI findings to the CAI realm, and vice versa.

18.6 THE FUTURE OF USABILITY IN SURVEY RESEARCH

Finally, we summarize both the potential promise of usability research for survey applications, as well as the challenges facing the field in the application of HCI principles and procedures to surveys. In terms of promise, we believe the contributions of HCI to computer assisted interviewing are likely to be more rapid than those of cognitive psychology to question wording. This is based on the following reasons:

1. The field of HCI research is more applied and practice-driven than cognitive psychology. In some sense, much of the translation (from cognitive theory to design principles) has already been done. What remains is to further adapt or translate this knowledge to the survey world.

2. The link between general computer interface design and CAI screen design is much stronger than the basic material of cognitive science research and the applied field of questionnaire design. This further facilitates the translation from HCI findings to CAI practice.

3. Many of the research and evaluation methods are already familiar to practitioners in survey research. Usability laboratories, for example, can be seen as an extension of cognitive laboratories. Much of what we have learned about conducting research in the latter can be applied to the former. Many of the other techniques used in usability research and testing (whether experimental design, think-alouds, expert review or conversation analysis) are part of the survey culture, in part, due to the success of the CASM movement.

4. There is much room for improvement in the design of CAI instruments. The survey world is so far behind the rest of the computing field that design improvements can be made almost immediately, even without extensive testing.

In summary, much of the groundwork for the application of HCI to CAI has already been done, partly by the contributions of CASM to survey research, and partly within the HCI field itself. Some of these benefits might also be seen as the downside of usability for survey research, in that the HCI work might not be seen as theoretically interesting enough. It may be viewed as too applied to justify basic research, and too specific to generalize to other systems and applications. We believe there is enough interest, to both theory and practice, to generate research activities on CAI.

Another possible drawback of CAI usability is that the gains (in terms of data quality) are not likely to be as big as, say, changes in question wording. For example, the use of color to enhance comprehension and operation of a CAPI instrument by an interviewer is unlikely to produce huge gains in efficiency or reductions in error. However, usability research may inform the design and deployment of new technologies and approaches that may fundamentally alter the way in which we collect survey data.

In conclusion, we offer a set of challenges that we see for the emerging field of CAI usability:

1. Demonstrate real payoffs in terms of costs and/or quality. Usability testing should be seen as not just another costly and time-consuming step in the development process, but as a critical element in the design and development of a high-quality survey instrument.

2. Produce generalizable findings. The results of usability research should extend across multiple surveys and different software systems. To do so, we need to derive broad principles and theoretically-grounded findings from our research.

3. Identify and tackle interesting research problems in usability. To be a vibrant area of research activity, CAI usability should use a variety of methods, both observational and experimental. It should conduct both applied and basic research. The field should strive to achieve a synergy of application-driven and hypothesis-driven research.

4. Advance our understanding of the process of human–computer interaction and human–computer–human interaction, especially as it applies to the survey realm. The research findings for CAI usability can be extended beyond the survey realm to other settings in which two (or more) humans use a computer to accomplish a task.

5. Extend the power of CAI methods by developing and testing new methods of survey data collection. This is already being done with audio-CASI (for example), but can be extended to other areas. The real payoff comes when we use cognitive science and HCI not simply to improve what we are already doing, but to design, test, and evaluate entirely new ways of doing things.

It is clear that we believe CAI usability to have a bright future. In large part this is because it can build on the foundation laid by the cognitive movement, and extend this work to the newer methods of survey data collection.

REFERENCES

Anderson, R. J., Heath, C. C., Luff, P., and Moran, T. P. (1993). The social and the cognitive in human-computer interaction. *International Journal of Man-Machine Studies*, *38*, 999–1016.

Bagnara, S., and Rizzo, A. (1989). A methodology for the analysis of error processes in human-computer interaction. In M. J. Smith and G. Salvendy (Eds.), *Work with Computers: Organizational, Management, Stress and Health Aspects*, pp. 605–612. Amsterdam: Elsevier.

Baker, R. P. (1992). New technology in survey research: computer assisted personal interviewing (CAPI). *Social Science Computer Review*, *10*(2), 145–157.

Baker, R. P., Bradburn, N. M., and Johnson, R. A. (1995). Computer-assisted personal interviewing: an experimental evaluation of data quality and costs. *Journal of Official Statistics*, *11*(4), 415–431.

Bergman, L. R., Kristiansson, K.-E., Olofsson, A., and Säfström, M. (1994). Decentralized CATI versus paper and pencil interviewing: Effects on the results in the Swedish Labor Force Surveys. *Journal of Official Statistics*, *10*(2), 181–195.

Bernard, C. (1989). *Survey data collection using laptop computers*. Paris: INSEE (Report No. 01/C520).

Bradburn, N. M., Frankel, M., Hunt, E., Ingels, J., Wojcik, M., Schoua-Glusberg, A., and Pergamit, M. (1991). A comparison of computer assisted personal interviews (CAPI) with personal interviews in the National Longitudinal Survey of Labor Market Behavior—youth cohort. *Proceedings of the Annual Research Conference*, 389–397. Washington, DC: U.S. Bureau of the Census.

Bødker, S. (1991). *Through the Interface: A Human Activity Approach to User Interface Design*. Hillsdale, NJ: Erlbaum.

Bosley, J., Conrad, F. G., and Uglow, D. (1998), "Pen CASIC: Design and usability. In M. P. Couper, R. P. Baker, J. Bethlehem, C. Z. F. Clark, J. Martin, W. L. Nicholls II, and J. O'Reilly (Eds.), *Computer Assisted Survey Information Collection*, pp. 521–541. New York: Wiley.

Brown, C. M. (1988). *Human-Computer Interface Design Guidelines*. Norwood, NJ: Ablex.

Card, S. K., Moran, T. P., and Newell, A. (1980). The keystroke-level model for user performance with interactive systems. *Communications of the ACM, 23*, 396–410.

Card, S. K., Moran, T. P., and Newell, A. (1983). *The Psychology of Human-Computer Interaction*. Hillsdale, NJ: Erlbaum.

Carroll, J. M. (1993). Creating a design science of human-computer interaction. *Interacting with Computers, 5(1)*, 3–12.

Carroll, J. M. (1997). Human-computer interaction: Psychology as a science of design. *Annual Review of Psychology, 48*, 61–83.

Coll, R., and Wingertsman, J. C. (1992). The effect of screen complexity on user preference and performance. *International Journal of Human-Computer Interaction, 2(3)*, 255–265.

Cooper, G., and Bowers, J. (1995). Representing the user: Notes on the disciplinary rhetoric of human-computer interaction. In P. J. Thomas, (Ed.), *The Social and Interactional Dimensions of Human-Computer Interaction*, pp. 48–66. Cambridge: Cambridge University Press.

Cooper, G., Hine, C., Rachel, J., and Woolgar, S. (1995). Ethnography and human-computer interaction. In P. J. Thomas (Ed.), *The Social and Interactional Dimensions of Human-Computer Interaction*, pp. 11–36. Cambridge: Cambridge University Press.

Couper, M. P. (1994). Discussion: What can CAI learn from HCI? In Office of Management and Budget Seminar on New Directions in Statistical Methodology, Statistical Policy Working Paper 23, pp. 363–377. Washington, D.C.: Statistical Policy Office.

Couper, M. P. (1997). Changes in interview setting under CAPI. *Journal of Official Statistics, 12(3)*, 301–316.

Couper, M. P., and Burt, G. (1994). Interviewer attitudes toward computer assisted personal interviewing (CAPI). *Social Science Computer Review, 12(1)*, 38–54.

Couper, M. P., and Groves, R. M. (1992). Interviewer reactions to alternative hardware for computer assisted personal interviewing. *Journal of Official Statistics, 8(2)*, 201–210.

Couper, M. P., Hansen, S. E., and Sadosky, S. A. (1997). Evaluating interviewer use of CAPI technology. In L. Lyberg, P. Biemer, M. Collins, E. de Leeuw, C. Dippo,

N. Schwarz, and D. Trewin (Eds.), *Survey Measurement and Process Quality*, pp. 267–285. New York: Wiley.

Couper, M. P., Horm, J., and Schlegel, J. (1996). *The use of trace files for evaluation of questionnaire and instrument design*. Paper presented at the International Conference on Computer Assisted Survey Information Collection, San Antonio, TX.

Couper, M. P., and Nicholls II, W. L. (1998). The history and development of computer assisted survey information collection. In M. P. Couper, R. P. Baker, J. Bethlehem, C. Z. F. Clark, J. Martin, W. L. Nicholls II, and J. O'Reilly (Eds.), *Computer Assisted Survey Information Collection*, pp. 1–21. New York: Wiley.

Couper, M. P., Sadosky, S., and Hansen, S. E. (1994). Measuring interviewer behavior using CAPI. *Proceedings of the Section on Survey Research Methods, American Statistical Association*, 845–850.

Cuff, R. N. (1980). On casual users. *International Journal of Man-Machine Studies, 12*, 163–187.

Davis, S., and Bostrom, R. (1992). An experimental investigation of the roles of the computer interface and individual characteristics in the learning of computer systems. *International Journal of Human-Computer Interaction, 4*(2), 143–172.

Dillman, D. A., and Salant, P. (1994). *How to Conduct Your Own Survey*. New York: Wiley.

Dillon, A. (1994). *Designing Usable Electronic Text: Ergonomic Aspects of Human Information Usage*. London: Taylor & Francis.

Dix, A. J. (1995). Formal methods. In A. F. Monk and G. N. Gilbert (Eds.), *Perspectives on HCI: Diverse Approaches*, pp. 9–43. London: Academic Press.

Drury, C. G., and Hoffman, E. R. (1992). A model for movement time on data-entry keyboards. *Ergonomics, 35*(2), 129–147.

Duffy, T. M., Palmer, J. E., and Mehlenbacher, B. (1992). *On-Line Help: Design and Evaluation*. Norwood, NJ: Ablex.

Dumas, J. S., and Redish, J. C. (1993). *A Practical Guide to Usability Testing*. Norwood, NJ: Ablex.

Edwards, B., Bittner, D., Edwards, W. S., and Sperry, S. (1993). CAPI effects on interviewers: A report from two major surveys. *Proceedings of the Annual Research Conference*, pp. 411–428. Washington, DC: U.S. Bureau of the Census.

Frese, M., and Altmann, A. (1989). The treatment of errors in learning and training. In L. Bainbridge and S.A. Ruiz Quintanilla (Eds.), *Developing Skills with Information Technology*, pp. 65–86. Chichester: Wiley.

Frese, M., Brodbeck, F., Heinbokel, T., Mooser, C., Schleiffenbaum, E., and Thiemann, P. (1991). Errors in training computer skills: On the positive function of errors. *Human-Computer Interaction, 6*, 77–93.

Frohlich, D., Drew, P., and Monk, A. (1994). Management of repair in human-computer Interaction. *Human-Computer Interaction, 9*, 385–425.

Galitz, W. O. (1993). *User-Interface Screen Design*. Boston: QED.

Gillan, D. J., Holden, K., Adam, S., Rudisill, M., and Magee, L. (1992). How should Fitt's law be applied to human-computer interaction? *Interacting with Computers, 4*(3), 291–313.

Greatbatch, D., Luff, P., Heath, C., and Campion, P. (1993). Interpersonal communication and human-computer interaction: An examination of the use of computers in medical consultations. *Interacting with Computers, 5(2)*, 193–216.

Greatbatch, D., Heath, C., Campion, P., and Luff, P. (1995a). How do desk-top computers affect the doctor-patient interaction? *Family Practice, 12(1)*, 32–36.

Greatbatch, D., Heath, C., Luff, P., and Campion, P. (1995b). Conversation analysis: Human-computer interaction and the general practice consultation. In A. F. Monk and G. N. Gilbert (Eds.), *Perspectives on HCI: Diverse Approaches*, pp. 199–222. London: Academic Press.

Groves, R. M., and Mathiowetz, N. A. (1984). Computer assisted telephone interviewing: Effect on interviewers and respondents. *Public Opinion Quarterly, 48*, 356–369.

Hansen, S. E. (1996). *Instrument usability testing.* Paper presented at the International Conference on Computer Assisted Survey Information Collection, San Antonio, TX.

Hansen, S. E., Fuchs, M., and Couper, M. P. (1997). *CAI instrument and system usability testing.* Paper presented at the annual conference of the American Association for Public Opinion Research, Norfolk, VA.

Hix, D., and Hartson, H. R. (1993). *Developing User Interfaces: Ensuring Usability Through Product and Process.* New York: Wiley.

Hollnagel, E. (1993). The design of reliable HCI: The hunt for hidden assumptions. In J. L. Alty, D. Diaper, and S. Guest (Eds.), *People and Computers VIII, Proceedings of the HCI Conference*, pp. 3–15. Cambridge: Cambridge University Press.

Howes, A. (1995). An introduction to cognitive modelling in human-computer interaction. In A. F. Monk and G. N. Gilbert (Eds.), *Perspectives on HCI: Diverse Approaches*, pp. 97–119. London: Academic Press.

Jeffries, R., Miller, J. R., Wharton, C., and Uyeda, K. M. (1991). User interface evaluation in the real world: A comparison of four techniques. *Proceedings of the ACM Conference on Human Factors in Computing Systems*, pp. 119–124. New York: ACM Press.

Jenkins, C. R., and Dillman, D. A. (1994). The language of self-administered questionnaires as seen through the eyes of respondents. In Office of Management and Budget Seminar on New Directions in Statistical Methodology, Statistical Policy Working Paper 23, pp. 470–516. Washington, DC: Statistical Policy Office.

Jenkins, C. R., and Dillman, D. A. (1997). Towards a theory of self-administered questionnaire design. In L. Lyberg, P. Biemer, M. Collins, E. de Leeuw, C. Dippo, N. Schwarz, and D. Trewin (Eds.), *Survey Measurement and Process Quality*, pp. 165–196. New York: Wiley.

Jordan, B., and Henderson, A. (1995). Interaction analysis: Foundations and practice. *Journal of the Learning Sciences, 4(1)*, 39–103.

Landauer, T. K. (1991). Let's get real: A position paper on the role of cognitive psychology in the design of humanly useful and usable systems. In J. M. Carroll (Ed.), *Designing Interaction: Psychology at the Human-Computer Interface*, pp. 60–73. New York: Cambridge University Press.

Landauer, T. K. (1995). *The Trouble with Computers: Usefulness, Usability, and Productivity*. Cambridge, MA: MIT Press.

Lansdale, M. W., and Ormerud, T. C. (1994). *Understanding Interfaces: A Handbook of Human-Computer Interaction*. London: Academic Press.

Lessler, J. T., and Holt, M. (1987). Using response protocols to identify problems in the U.S. Census long form. *Proceedings of the Section on Survey Research Methods, American Statistical Association*, 262–266.

Levi, M. D., and Conrad, F. G. (1995). A heuristic evaluation of a world wide web prototype. *Interactions, 3(4)*, 50–61.

Lewis, C., and Norman, D. A. (1986). Designing for error. In D. A. Norman and S. W. Draper (Eds.), *User Centered System Design*, pp. 411–432. Hillsdale, NJ: Erlbaum.

Macleod, M., and Rengger, R. (1993). The development of DRUM: A software tool for video-assisted usability evaluation. In J. L. Alty, D. Diaper and S. Guest (Eds.), *People and Computers VIII; Proceedings of the HCI Conference*, pp. 293–309. Cambridge: Cambridge University Press.

Marchionini, G., and Sibert, J. (Eds.) (1991). An agenda for human-computer interaction: Science and engineering serving human needs. Report of an invitational workshop sponsored by the National Science Foundation. *SIGCHI Bulletin, 23(4)*, 17–32.

Mayhew, D. J. (1992). *Principles and Guidelines in Software User Interface Design*. Englewood Cliffs, NJ: Prentice-Hall.

Moore, J. C. (1996). *Person- vs. topic-based design for computer assisted household survey instruments*. Paper presented at the International Conference on Computer Assisted Survey Information Collection, San Antonio, TX.

National Center for Health Statistics and U.S. Bureau of the Census. (1988). *Report of the 1987 Automated National Health Interview Survey Feasibility Study*. National Center for Health Statistics, Working Paper No. 32.

Nicholls, W. L., Baker, R. P., and Martin, J. (1997). The effect of new data collection technologies on survey data quality. In L. Lyberg, P. Biemer, M. Collins, E. de Leeuw, C. Dippo, N. Schwarz, and D. Trewin (Eds.), *Survey Measurement and Process Quality*, pp. 221–248. New York: Wiley.

Nielsen, J. (1992). Finding usability problems through heuristic evaluation. *Proceedings of CHI '92, Human Factors in Computing Systems*, pp. 373–380. New York: ACM Press.

Nielsen, J., and Mack, R. (Eds.) (1994). *Usability Inspection Methods*. New York: Wiley.

Nielsen, J., and Molich, R. (1990). Heuristic evaluation of user interfaces. *Proceedings of CHI '90, Human Factors in Computing Systems*, pp. 249–256. New York: ACM Press.

Newell, A., and Card, S. K. (1985). The prospects for psychological science in human-computer interaction. *Human-Computer Interaction, 1*, 209–242.

Norman, D. A. (1983). Design principles for human-computer interfaces. *Proceedings of CHI '83, Human Factors in Computing Systems*, pp. 1–10. New York: ACM Press.

Norman, D. A. (1986). Cognitive engineering. In D. A. Norman and S.W. Draper (Eds.), *User Centered System Design*, pp. 31–61. Hillsdale, NJ: Erlbaum.

Norman, D. A. (1988). *The Design of Everyday Things*. New York: Doubleday.

Norman, D. A. (1993). Design rules based on analyses of human error. *Communications of the ACM, 26(4)*, 254–258.

Norman, K. L. (1991). *The Psychology of Menu Selection: Designing Cognitive Control of the Human/Computer Interface*. Norwood, NJ: Ablex.

Powell, J. E. (1990). *Designing User Interfaces*. San Marcos, CA: Microtrends Books.

Presser, S., and Blair, J. (1994). Survey pretesting: Do different methods produce different results? *Sociological Methodology, Vol. 24*, pp. 73–104. Washington, DC: American Sociological Association.

Prümper, J., Zapf, D., Brodbeck, F.C., and Frese, M. (1992). Some surprising differences between novices and experts in computerized office work. *Behaviour and Information Technology, 11(6)*, 319–328.

Rasmussen, J., Duncan, K., and Leplat, J. (Eds.). (1987). *New Technology and Human Error*. Chichester: Wiley.

Reason, J. (1990). *Human Error*. Cambridge: Cambridge University Press.

Rizzo, A., Bagnara, S., and Visciola, M. (1987). Human error detection processes. *International Journal of Man-Machine Studies, 27*, 555–570.

Rubin, J. (1994). *Handbook of Usability Testing*. New York: Wiley.

Rothwell, N. D. (1985). Laboratory and field response research studies for the 1980 census of population in the United States. *Journal of Official Statistics 1*, 137–157.

Santhanam, R., and Wiedenbeck, S. (1993). Neither novice or expert: The discretionary user of software. *International Journal of Man-Machine Studies, 38*, 201–229.

Sears, A. (1991). Improving touchscreen keyboards: Design issues and a comparison with other devices. *Interacting with Computers, 3(3)*, 253–269.

Shneiderman, B. (1980). *Software Psychology: Human Factors in Computer and Information Systems*. Boston: Little Brown.

Shneiderman, B. (1992). *Designing the User Interface: Strategies for Effective Human-Computer Interaction*, 2nd. ed. Reading, MA: Addison-Wesley.

Sperry, S., Edwards, B., Dulaney, R., and Potter, D. E. B. (1998). Evaluating interviewer use of CAPI navigation features. In M. P. Couper, R. P. Baker, J. Bethlehem, C. Z. F. Clark, J. Martin, W. L. Nicholls II, and J. O'Reilly (Eds.), *Computer Assisted Survey Information Collection*, pp. 351–365. New York: Wiley.

Staggers, N. (1993). Impact of screen density on clinical nurses' computer task performance and subjective screen satisfaction. *International Journal of Man-Machine Studies, 39(5)*, 775–792.

Statistics Sweden (1990). *Computer assisted data collection production test in the labour force surveys August 1989–January 1990*. Stockholm/Örebro: Statistics Sweden (R&D Report 1990:11 E).

Suchman, L.A. (1987). *Plans and Situated Actions: The Problem of Human-Machine Communication*. Cambridge: Cambridge University Press.

Sweeney, M., Maguire, M., and Shackel, B. (1993). Evaluating user-computer interaction: A framework. *International Journal of Man-Machine Studies, 38(4),* 689–712.

Sweet, E., Marquis, K., Sedivi, B., and Nash, F. (1997). *Results of expert review of two internet R&D questionnaires.* Washington DC: U.S. Bureau of the Census (unpublished memorandum).

Thimbleby, H. (1990). *User Interface Design.* New York: ACM Press.

Tullis, T. S. (1983). The formatting of alphanumeric displays: A review and analysis. *Human Factors, 25(6),* 657–682.

Vaske, J. J., and Grantham, C. E. (1990). *Socializing the Human-Computer Environment.* Norwood, NJ: Ablex.

Weeks, M. F. (1992). Computer-assisted survey information collection: A review of CASIC methods and their implications for survey operations. *Journal of Official Statistics, 8(4),* 445–465.

Wojcik, M. S. and Hunt, E. (1998). Training field interviewers to use computers: Past, present and future trends. In M. P. Couper, R. P. Baker, J. Bethlehem, C. Z. F. Clark, J. Martin, W. L. Nicholls II, and J. O'Reilly (Eds.), *Computer Assisted Survey Information Collection,* pp. 331–349. New York: Wiley.

Zapf, D., Brodbeck, F. C., Frese, M., Peters, H., and Prümper, J. (1992). Errors in working with office computers: A first validation of a taxonomy for observed errors in a field setting. *International Journal of Human-Computer Interaction, 4(4),* 311–339.

CHAPTER 19

Customizing Survey Procedures to Reduce Measurement Error

Frederick G. Conrad
Bureau of Labor Statistics

Because surveys involve people, it can be hard to get to the truth—at least as "truth" is defined by survey designers and statistical organizations. Despite their best efforts, people taking part in surveys make errors and behave in unexpected ways—at least with respect to the plans of survey designers and their organizations. Respondents may misunderstand what they are being asked, answering a question that differs from what the question author had in mind; coders and reviewers may forget to apply relevant knowledge, misclassifying an open ended response or overlooking problematic data; end users of a survey's findings may be unaware of the context surrounding particular statistics and consequently misinterpret them; and so on.

The challenge to methodologists is to develop procedures that will help survey participants be more accurate and consistent from the perspective of the survey designers. Different kinds of inaccuracies and inconsistencies warrant different procedures. Researchers working in the CASM arena have, in my opinion, made the most progress addressing "mental slips," that is, errors that people make even though they are otherwise competent to carry out the task at which they have erred. A hallmark of slips is that the action is unintended, as is the outcome. For example, when a respondent is asked how frequently she has engaged in an activity over a recent time period she might forget the current date and instead bound the recall period with yesterday's instead of today's or last month instead of this month. Certainly she did not intend to do this; she just slipped. The CASM-based solution to these problems tends to involve

Cognition and Survey Research, Edited by Monroe G. Sirken, Douglas J. Herrmann, Susan Schechter, Norbert Schwarz, Judith M. Tanur, and Roger Tourangeau.
ISBN 0-471-24138-5 © 1999 John Wiley & Sons, Inc.

revising the survey instrument so that it is easier for people to do what they are capable of doing, in this case filling the current date into the question when the interviewer reads it to the respondent.

A source of error, potentially more pervasive and costly than mental slips, becomes evident if we think about survey tasks as a combination of plans and actions. In this case, plans are created by survey designers and actions are intentionally taken by survey participants in order to follow the designer's plan. The difference between the designer's plan and the participant's action can be thought of as a kind of measurement error. In principle, the more closely the survey participant follows the designer's plan, the smaller the associated error. We will advocate an approach to more closely align designers' plans and participants' actions by modeling the different ways that different participants perform a task, and then adapting procedures to these models. This approach is inspired by the practice of "user modeling" in software development where, to varying degrees, designers customize the plans they intend users to follow based on what can be determined about the users.

In the first section, I will explore the idea that the cognitive origins of measurement error are related to the degree to which designers' plans and participants' actions coincide. In the following section, I will briefly discuss some of the user modeling techniques designed to tailor human–computer interaction (HCI) to the preferences and abilities of the user (see Couper, Chapter 18, for further discussion of HCI research in the survey context). Then, I will illustrate how some of these techniques can be used to help close the gap between ideal and actual performance by survey participants. The particular tasks I will explore are (a) answering survey questions whose meanings may not be clear, (b) reviewing and coding complex data after they have been collected, and (c) retrieving published survey data using electronic dissemination tools. I will conclude by discussing the implications of this approach for theories of measurement error.

19.1 PLANS, ACTION, AND MEASUREMENT ERROR

People plan to act in particular ways but stray from their plans in response to the situation. Consider the following example from Suchman (1987, p. 52). When planning to run a stretch of rapids in a canoe, the canoeist is likely to formulate a goal (probably to get downstream without capsizing) and a plan for accomplishing that goal (for example, stay to the left to pass between the two big rocks and then backwater hard to the right to move around the next set of rocks). However, once the run is underway, the canoeist tries to follow this plan but inevitably improvises at certain points to get downstream safely and dry. In this case, the canoeist developed the plan herself. However, she could just as easily have tried to follow, but ultimately strayed from, a plan formulated by someone else, say a canoeing coach. By analogy, survey designers play the role of the coach—they establish the goals of survey tasks and provide the participants with plans for reaching those goals (where the plans are expressed as survey questions, human–computer interfaces, coding procedures, etc.); participants, on the other hand, play the role of the canoeist—they try to implement those plans while reacting and adapting to circumstances.

The important point is that plans and actions are not the same; people do what they have to do, or can do, or think they should do, to achieve a goal, even if this is not what they set out to do. This idea is at the core of a general theory of action developed by Lucy Suchman (1987, 1993). "Situated action," as the idea is known, separates plans in people's heads from the way they enact those plans in particular contexts.

Suchman (1987) developed the idea in the context of human–machine interaction, in particular, people using a copying machine. Essentially, she argues that designers create machines for generic users working in abstractly defined situations. But she observed that the *actual* situations in which machines are used are defined, in part, by the beliefs and knowledge of *particular* users. Because the machine does not have access to the users' beliefs and knowledge (except to the extent that these are reflected by user actions like latching the document cover of the copying machine), the machine may "assume" users are in certain mental states at certain times, when, in fact, they are not. When this happens, users can depart from the plan that the designers intended them to follow, without the machine detecting it. This can reduce the chance that the users will meet their objectives, or at least that they will be satisfied with how they have been met.

For example, Suchman (1987, pp. 149–150) describes a pair of users making five copies of a bound document. This is a two-stage process. First, the users manually produce a single, unbound copy of the bound document one page at a time; this is the master copy. In the second stage, the machine automatically reproduces the master copy four times. The designer's plan for the user, inherent in the design of the machine, requires the user to signal that he has finished the first stage by latching a particular cover; this is a prerequisite for the machine to begin the second stage.

Because these particular users were unaware of this requirement, they left the cover unlatched after making the master copy. From the perspective of the machine, the user was still working on the first stage. As a result, it did not start the second stage. Because the machine was inactive, the users concluded that they had misunderstood the procedure and that what they were really intended to do was repeat the manual copying four more times. They had unnecessarily revised the designers' otherwise workable plan because they lacked a critical piece of knowledge (the cover needs to be latched when done). Had the designers anticipated this, they could have taken measures to enable the users to follow the plan, such as making the instructions about closing the latch more salient. Alternatively, they could have adjusted the design so that the user would not have had to explicitly signal completing the first stage, for example, treating a long user pause as a signal that the master copy was made.

19.1.1   What Does This Have to Do With Surveys?

The idea of separating plans from action can be a productive way to think about measurement error in surveys. When the survey designer formulates a task for a survey participant to perform, he has, in effect, provided a plan for carrying out that task, much like the designers of a machine have. If a survey participant follows the plan exactly as its designer intended, measurement error (due to the participants, at least) would vanish. The way to reduce this sort of measurement error is, therefore, to help survey participants closely follow the designers' plans.

This is easier said than done. Moreover, there are cases in which a designer's plan, even when followed closely by the participant, does not lead to the designer's goal; in such cases, the participant's faithful adherence to the designer's plan does not reduce measurement error, and may actually increase it.

19.1.2   No Official Plan

It is possible the designers have not established an official plan for participants to follow in executing a particular task. Consider the respondent who is asked how many films he has watched in the last six months. There are potentially many ways the respondent could answer this question (e.g., counting remembered episodes, reporting the usual rate of film watching, converting a qualitative impression like "all the time" into a numerical estimate) and each can lead to very different estimates (Conrad, Brown, and Cashman, 1998). Without committing to particular plans and, therefore, measurement goals (e.g., estimates of moderate precision), the designer leaves the respondent at sea. Moreover, if there is no official plan, it is hard for a methodologist to evaluate the quality of the respondent's performance which makes it virtually impossible to help the respondent improve his performance.

19.1.2   Misinterpreted Plan

It is possible that the survey designers may have specified how a particular task should be carried out but that the participant does not interpret the plan as the designer intended. The participant might come to understand the plan differently than the designer conceived of it if their goals differ. For example, a respondent's goal might be to complete the interview as quickly as possible while the designer's goal might be to collect thoughtful responses. Under these circumstances the respondent might come to understand the plan as limiting the time devoted to each question while the designer would never have intended for this to happen.

Alternatively, the participant may simply have misunderstood what she is being asked to do. Consider a respondent who interprets her task as checking the one answer category with which she most strongly agrees. She may carry this out exactly as she planned but if the survey designer intended her to check all that apply, the response is not what the designer believes it to be.

In both cases, the participants intended to act as they did prior to acting: they had a plan in mind, albeit not the designer's plan. These are *planning errors* as opposed to *execution errors* (Reason, 1990) or *mistakes* as opposed to *slips* (Norman, 1981, 1993). It may be possible to help participants correctly interpret the designer's plan by clarifying the instructions. However, in other situations, instructions may not be practical because (aside from the extra time involved) the participant may lack the knowledge to understand the designer's plan. I will discuss a version of this problem in the section on data retrieval where information seekers do not necessarily know enough about the sponsoring organization to navigate their way to the desired statistics.

19.1.4 Defective Plan

In other cases, the survey designer may have developed a plan for the participant to follow that, as it turns out, does not always lead to the designer's goal. Like misinterpreted plans, defective plans lead to mistakes or planning errors because the participants intend to act as they do but intend to produce a different outcome. For example, according to the philosophy of standardized interviewing, if the interviewer reads a pretested and refined question exactly as worded, the respondent will understand it exactly as intended. This approach implies that, in general, a speaker communicates to a listener by encoding his thoughts as words, and then uttering those words; the words are heard by the listener and create meaning in the listener's head that corresponds to the speaker's meaning (e.g., Akmajian, Demers, Farmer, and Harnish, 1990). By implicitly adopting this view of communication, survey designers are treating it as the plan by which interviewers and respondents are to transfer information from the respondent's head or records to the survey's data base.

In many interview circumstances this works just fine. However, if the respondent is confused about what the question means, he has no way to clarify its meaning and if the interviewer suspects the respondent misunderstands the question, she cannot correct a potential misunderstanding (Schober, Chapter 6; Schober and Conrad, 1997). In these cases, the interviewer and respondent must be "licensed" to clarify concepts and confirm that they understand each other. This is a case in which the official plan, based on a controversial model of communication, may actually be defective under certain circumstances—even if the participants try, in good faith, to follow it. The solution may involve revising the official plan. We will explore this problem more in the section on an experimental, self-administered questionnaire.

19.1.5 Mental Slips

Finally, there are situations in which there is a sanctioned plan, where the participants correctly understand the plan, and where, if they closely follow the plan, they should be able to satisfy the goal. People still make errors, however, because they forget facts or routines, because their working memory is overworked, because they do not correctly hear or correctly access the meanings of particular words, because their eyes land in the wrong place when reading, and so on. These errors are essentially "mental slips," errors that would not occur as often if the amount of information involved in the task were reduced, or its presentation improved, or the participant were better rested, etc. These errors are not due to the ease with which respondents can interpret the plan. Slips are implicated when a plan prescribes routine actions but these routines are not perfectly reliable, occasionally leading to unintended outcomes (Norman, 1981, 1993). In the section on data review, we will discuss a software tool that helps reviewers perform a classification task, in part, because the tool is not subject to slips.

As was asserted earlier, such mental slips are the kind of error on which CASM researchers have concentrated and the area in which they have enjoyed the greatest success. Laboratory pretesting, for example, has proven effective in identifying question characteristics that are likely to promote this type of error in response tasks (for example, Esposito and Rothgeb, 1997; Willis, DeMaio, and Harris-Kojetin, Chapter 9; Willis, Royston, and Bercini, 1991).

While such mental lapses undoubtedly contribute to measurement error, so do the other deviations from ideal behavior that have been sketched above. In fact, it seems that if the participant fails to grasp or adhere to the designer's plan, the results can be more persistent and costly than incorrectly recalling the current date. Resolving this latter class of problems seems a matter of refining the survey question; addressing the former type of problem may require redesigning the plan.[1]

In the copy machine example presented earlier, the two users seemed hopelessly lost compared to a user who might, for example, have failed to notice that two-sided copying was turned on. The latter problem might be solved by fine-tuning the interface, for example, repositioning the two-sided copying indicator so that it is harder to miss. In contrast, the problem that Suchman's users displayed in the copy machine example speaks to a more fundamental incompatibility between the design and the user's actions. If we look at survey tasks as plans and actions, I believe that we can see sources of measurement error whose reduction requires more than "interface adjustments," but rather better underlying plans. We turn now to user modeling, an approach taken by various cognitive scientists to reduce this type of error in human-computer interaction.

19.2 USER MODELING

Suchman's ideas have helped increase awareness that software users do not always use software the way it was designed to be used, and that when this happens, users perform less accurately and are less satisfied with the interaction.

[1] See Reason (1990) for a related discussion on why "mistakes" are costlier than "slips."

Certain researchers in artificial intelligence have responded by embedding "user models" in larger software systems in order to simulate the user's mental states as he is interacting with the computer. When the user model is in different states, the software behaves differently, in effect, adapting the designer's plan to the user's behavior.

19.2.1 Generic User Models

In their most general form, user models are no more than a description of proto-typical user behavior in a particular domain, without differentiating among users (Kay, 1995). This description informs the design of the software and almost certainly leads to a product that is more compatible with its users' behavior than would be its counterpart, developed without such analysis (Benyon, 1993).

19.2.2 User Class Models

A more concrete type of user model, based on the notion of stereotypes (e.g., Rich, 1989), represents different classes of users, each as a kind of schema. Each stereotype (schema) lists the characteristics and preferences of typical users in that class. In addition, the stereotype includes a triggering condition, a particular user action (e.g., typing a certain word or choosing a menu option) that indicates the current user is a member of that class. In this way, based on a simple user action, the system can attribute the various features in the schema to an individual user. While the stereotype rarely matches an individual user perfectly, it can be an adequate (and computationally simple) model.

Rich (1989) describes a travel agent-like system that includes stereotypes for various different kinds of travelers. If the system infers that the user is traveling as part of her job, then, based on the business traveler stereotype, it might reserve an expensive hotel room (having assumed that the trip is covered by an expense account) with office-like facilities, and will reserve a luxury sedan. If the system can determine that the user is planning a family vacation, it will reserve more modest accommodations with kitchen facilities and will reserve a large vehicle with child seats.

In one version, the system modifies particular attributes of the model if they are contradicted by the users' behavior. Assuming this occurs for a minority of the attributes, this can be an efficient way to produce relatively individualized interactions.

19.2.3 Individual User Models

Another, more customized approach to user modeling is to construct and update the model over the course of each user session with the computer. Based on the state of the model at a given time, the program can adjust the plan that the user must follow. Programs that embody this kind of adaptability have been called "self-modifying" systems (Benyon, 1993). The idea has been effectively used in so called intelligent tutoring systems to model the student's state of knowledge at any point in the curriculum.

A good example of student modeling is the work by Anderson and his colleagues (e.g., Anderson, Corbett, Boyle, and Lewis, 1990; Corbett and Anderson, 1995; Corbett, Anderson, and O'Brien, 1995). The basic approach is to think of the skill being learned, for example introductory computer programming, as a set of rules. At any given point, a particular rule is either learned or not learned. The student models in Anderson's tutoring systems consist of a list of the component programming rules and, for each rule, the probability that it has been learned. The probability associated with each rule is updated based on a combination of the student's most recent experience using the rule and her history using it. If the student's actions differ from the actions of an "ideal student" (a simulation program that simultaneously solves the same problems as the actual student), the tutoring system provides some instructional feedback, designed to help the student use a particular rule correctly on the next try. The tutoring system assigns exercises to the student which involve rules that, according to the student model, are currently not learned.

There are many variations on these themes across the research in the user modeling community (Kay, 1995). The important point for our purposes is that when people follow procedures with the help of computers, the procedures—the designer's plans—can be tailored to what the users know and do not know, to what they prefer, and to what they need to accomplish, while at the same time, helping them perform consistently and accurately.

19.3 ADAPTING SURVEY TASKS TO PARTICULAR PARTICIPANTS

Survey designers can help participants perform their tasks more accurately by developing plans that can be adapted to the participants' knowledge, beliefs, and behaviors much like user models are designed to promote more accurate human–computer interaction. In this section, I will describe three research efforts that illustrate this approach. While all of these involve survey tasks that are performed with the help of computers, the approach does not require computers. For example, the first illustration involves a computer administered questionnaire that is designed to clarify confusing concepts when the respondent's behavior indicates this could be helpful. Not only is this something that can be done by interviewers without a computer but interviewers can take into account many more types of cues about respondent confusion than can computers.

19.3.1 Answering Self-Administered Questions

There is a debate about the best way to conduct survey interviews that can be thought of as a contest between two plans for obtaining information from respondents, standardized interviewing and conversational interviewing. Standardized interviewing, which is almost universally embraced, seeks to reduce error due to interviewers by restricting them to reading each question as it appears on the survey script and to using "non-directive probes" to elicit additional information from the respondent (Fowler and Mangione, 1990; see Beatty, 1995 for an historical view). This means, for example, that the interviewer cannot respond substantively to a respondent's request for clarification. The rationale is to eliminate bias by exposing every respondent to exactly the same stimulus. By doing this, the argument goes, differences in responses are not due to differences between interviewers but rather to actual differences between respondents.

Conversational interviewing has been advocated in order to improve the communication between interviewer and respondent, and therefore to assure that the data are valid (e.g., Suchman and Jordan, 1990, 1992). Under this approach, interviewers are licensed to use their ordinary conversational skills in order to establish that the respondent understands the question as it was intended. The interviewer can say whatever she believes is necessary to achieve this. For example, if the respondent indicates that he is unsure about the meaning of the question, either by asking for help or through less direct linguistic cues, the interviewer can provide substantive help.

From the perspective of plans and actions, standardization holds the designer's plan constant across respondents while conversational interviewing allows interviewers to adapt the plan to particular respondents' circumstances. The debate has centered around which of these plans better satisfies the goal of producing high-quality data. Michael Schober and I conducted an experiment in which we compared response accuracy under one version of standardized interviewing and two versions of conversational interviewing using a computer assisted self interviewing (CASI) instrument.

By administering the "interview" with the computer, we were able to model certain characteristics of the interaction in standardized interviews and certain characteristics in conversational interviews, each as a distinct version of the CASI instrument. Moreover, the CASI instrument allowed us to (1) evaluate just those features of the interaction that we believed were central to each approach and (2) compare perfectly reliable instances of the two techniques.

Our implementations of standardized and conversational interviewing were distinguished by the type of clarification available to respondents. We simulated standardized interviewing by displaying the question without providing respondents access to any information (such as definitions) that could resolve possible confusion about the question's meaning (No Clarification). We modeled a version of conversational interviewing by enabling the respondents to obtain definitions for potentially confusing concepts when needed, by positioning the mouse cursor over the confusing portion of the question and clicking the mouse (Clarification on Demand)[2]. We also simulated a second form of

---

[2]We recognize that in many survey organizations, the interviewers who are considered to be standardized are allowed to provide definitions when respondents request help. However, this is not, strictly speaking, standardized (Fowler and Mangione, 1990, pp. 20–21). If some respondents are given the definition and some are not, then not all respondents are exposed to the same stimulus. Moreover, long definitions are susceptible to being read inconsistently.

conversational interviewing by both enabling the respondent to obtain help by clicking on the problematic text and by designing the CASI instrument so that after a long period of inactivity, the instrument would "offer" to clarify confusing concepts (Unsolicited Clarification). The evidence that the respondent was confused was prolonged inactivity. In other words, if the respondent neither answered the question nor asked for help after a certain amount of time, the instrument offered to help.

The questions were actual items in ongoing government surveys. All of them involved technical concepts that could differ from everyday notions of those concepts. The survey programs from which the items were borrowed all published definitions of these technical concepts. These definitions formed the content of the clarification text in both Clarification on Demand and Unsolicited Clarification.

So we could assess the accuracy of the responses, we asked the respondents to answer on the basis of fictional scenarios rather than on the basis of their own lives (Schober and Conrad, 1997). We constructed a packet of scenarios, one per question, for which we knew the correct answer based on the official definitions. We designed half of the scenarios so that they corresponded to the questions in a straightforward way and half so that this mapping was complicated. For example, a "yes"–"no" question about recent purchases of household furniture was answered on the basis of a retail receipt for either an end table or a floor lamp. We thought respondents would be reasonably certain to consider an end table an instance of household furniture but expected their decision about a lamp to be difficult. As it turns out, lamps are excluded from household furniture under the official definition but, without access to the definition, respondents could be confused.

In an initial study of the CASI tool, respondents answered the questions presented by the instrument based on the content of the scenarios. Because the number of observations is quite small, these results are really intended to illustrate our approach and suggest follow-up questions.

All of the respondents were very accurate when the mappings were straightforward, regardless of whether definitions were available (94 percent accurate with No Clarification, 100 percent with Clarification on Demand, and 97 percent with Unsolicited Clarification). When the mappings were complicated, the respondents were far less accurate with No Clarification (33 percent) but still quite accurate with Clarification on Demand (86 percent) and Unsolicited Clarification (83 percent). Clearly, under some circumstances (complicated mappings), the standardization plan (No Clarification) did not meet the quality goal. Alternatively, the two ways that we implemented conversational interviewing showed large accuracy advantages over standardization under these circumstances. These results begin to point to situations in which the standardization plan is unworkable for respondents.

How might one redesign the plan? We experimented with two implementations of conversational interviewing. Our initial thought was that accuracy would be highest under Unsolicited Clarification because it allows either party to initiate help sequences. In fact, respondents receiving Unsolicited Clarification requested help and accepted it from the instrument about equally often. However, this did not improve accuracy beyond its high level in Clarification on Demand. This could suggest that the effects of conversational interviewing are the same whether or not interviewers are licensed to volunteer information, however, the respondents using both "conversational" versions of the instrument were instructed to seek clarification when needed. It is possible that without this instruction, respondents would not request definitions as often, leading to somewhat lower accuracy in Clarification on Demand but comparably high accuracy in Unsolicited Clarification.

19.3.2 Classifying Complex Data

Survey organizations sometimes employ classification experts to code open-ended or complex data after the data have been collected. One example of this is found in the Consumer Price Index (CPI) program at BLS. We (Conrad, 1997; Conrad, Kamalich, Longacre, and Barry, 1993) have developed an experimental expert system to help commodity analysts (CA's) review and classify the comparability of products sampled in the CPI. The expert system, known as COMPASS (COMParability ASSistant), helps CA's follow official plans by catching their mental slips. Moreover, COMPASS customizes the official plan for individual CA's by modeling their unique expertise in particular product areas.

The quality of the CA's classification decision has a direct bearing on the quality of the data in the CPI. The CPI measures inflation in the United States by comparing monthly price change for a fixed set of goods and services. If a product that was priced in the previous month is not currently available, the data collector can sometimes substitute a similar product and treat it as if it were the original. Each product substitution is reviewed by a CA; should the CA decide the substitute is comparable to the original, then their prices can be compared in computing the index.

If the CA approves a substitute that is actually not comparable, then the price of the substitute product does not measure what it is intended to measure; this directly contributes to the survey's measurement error.[3] On the other hand, if the CA excludes a substitute that is actually comparable, this disrupts the series for the particular item, making the sample less representative and contributing to sampling error. The CA can commit classification errors of both sorts by mentally slipping, for example, misreading the product descriptions, forgetting to apply a comparability rule, applying a "rule" that is actually invalid (Brown and VanLehn, 1980),[4] and so forth.

COMPASS is intended to reduce this type of error by checking the CA's reasoning. When a CA and COMPASS disagree with each other, the CA must

---

[3]Groves (1991) considers such classification errors to be *processing* as opposed to *measurement* error. For current purposes we will consider them to be measurement error.

[4]Reason (1990, p. 56) describes the use of inappropriate rules as a rule-based mistake, not a slip.

resolve the discrepancy. In some of these cases the discrepancy is due to incomplete or incorrect knowledge in COMPASS but in other cases it is due to a slip by the CA that COMPASS does not make. Noting and correcting these deviations is one way to improve data quality.

COMPASS is made up of individual knowledge bases for different product areas. When different CA's evaluate comparability, they exhibit surprisingly uniform decision strategies across the various product types so each knowledge base is structured around the same skeletal reasoning strategies. In general, CA's tend to look for evidence that products are not comparable. If they cannot find such evidence, they look for indications that the substitute could be made comparable by statistically adjusting its price for quality differences (Armknecht and Weyback, 1989). Only after the analyst has not turned up either kind of evidence is the substitute deemed comparable. This reasoning process was modeled in COMPASS as a set of hypotheses which are tested in sequence. These hypotheses and the sequence in which they are tested represent a kind of generic user model which is, in effect, a blueprint of the plan for evaluating comparability.

Each CA's knowledge about particular product areas is added to this skeletal knowledge and thus constitutes a kind of individualized user model in each knowledge base. A CA's specific product knowledge concerns the details of how product specifications (like screen size and sound quality for televisions) combine to affect quality and price in that product area. For example, the CA's rule might be to disqualify a substitute if it differs from the original product on one particular specification (for example, color versus black and white television sets). But in other cases, the CA will allow mismatches for certain specification values although not for others (for example, a cotton and linen garment may be comparable but a cotton and nylon garment may not be comparable). In our experiment (Conrad et al., 1993), we developed individual knowledge bases for 13 product areas out of about 350 potential areas.

COMPASS was implemented as a check on the CA's reasoning, not as part of producing the CPI, so after a CA had evaluated a substitution, COMPASS made the same evaluation. By requiring CA's to *analyze* the discrepancies between their decisions and COMPASS', as opposed to *acting* on the discrepancies, it is more their thinking than their action that is brought into line with the designers' plans. Nonetheless, detecting mental slips may help CA's avoid committing them in the future.

The approach successfully highlighted CA error. For a representative knowledge base, Women's Pants and Shorts, COMPASS reached a substantive decision (either not comparable, adjust or comparable) for 81 percent of the substitutions. The rest were referred to the analyst.[5] COMPASS and the analyst agreed on 68 percent of the substantive decisions and disagreed on 13 percent. The average correlation over the four data sets was 0.79.

Of course, COMPASS is doing its job when it disagrees with the CA, assuming this is due to CA error. Many of the disagreements were, in fact, due to CA

---

[5]COMPASS can conclude that a decision is beyond its scope and "refer" it to the CA.

error. For example, when looking at data for one month, the CA and expert system disagreed on 19 out of 126 substitutions, and of these, 15 were attributed to some kind of analyst oversight. All of these were cases in which the CA judged the substitute comparable and COMPASS judged it not comparable. Had COMPASS been used prior to producing the CPI for that month, 15 erroneous substitutes would have been excluded and the overall measure would have been that much more accurate.

19.3.3 Seeking Statistical Information

Surveys are done to create statistical products. As survey organizations increasingly disseminate their products electronically, in particular via web sites, the conflict between site designers' plans and data users' actions is a central issue, even at the end of the survey process.

The way a web site is designed implies to someone visiting the site how he should retrieve the information he is seeking. Because many organizations structure their sites to reflect the structure of the organization, this implies that an information seeker should work his way to the program or office that produces the statistics of interest. Of course, many people know the topic of the statistics they are interested in, but not how the entity that produced the statistics is structured. If the site design requires data users to follow plans for which they lack critical information, they are almost sure to misinterpret the plan. This is akin to the problem faced by the people reproducing the bound document who simply did not know that the cover needed to be latched to start the second stage.

This is a problem in the Bureau of Labor Statistics' (BLS) web site as well as in other large statistical sites. By analyzing the logs automatically maintained by web site servers (Hert and Marchionini, 1997; Marchionini and Hert, 1997), it has become clear that the plan implied by the site structure was not workable for many people. It was evident from these logs that a surprisingly large proportion of visitors were using the search facility, for example, typing in a string like "inflation" or "unemployment" instead of navigating to the page for the appropriate office.

In addition, by analyzing e-mail messages sent to the BLS help desks and interviewing help desk staff, it has become clear that many frustrated data seekers are unable to find specific statistics that are, in fact, available at the site. This is true whether users are seeking a specific number like the unemployment rate or a time series like monthly unemployment over the last year. Clearly, the designers' intent was to provide easy access to statistics, not hide them or make them difficult to find.

A related problem concerns users' difficulty obtaining textual descriptions that provide context for numerical data, so-called meta-data. Data and meta-data are mostly separate from each other in the BLS web site. This suggests that users seeking one need to follow a different plan than users seeking the other; unfortunately, both plans appear to be unclear to a large number of people.

One way to better tailor the site and the way it is used to meet the needs of data seekers is to create different "views" of the site for different data retrieval tasks. This is a version of the user class model idea, however, instead of creating different stereotypes for different users, the proposal (Hert and Marchionini, 1997) is to create different front pages to the site for different user tasks. Someone visiting the site would be asked to choose from a list of frequent tasks (for example, retrieving a specific number, a time series, or a document). Their choice would select a page containing the links that the designers believe are most relevant to the chosen task. The success of this approach will depend on how well the designers can characterize the diversity of tasks that people try to accomplish at the web site.

19.4 CONCLUSION

I have discussed the idea that a serious and under-addressed type of measurement error results when survey participants cannot follow the plans of the designer, or when following those plans does not have the intended outcome. I then mentioned how user models have made it possible to create computer programs that better accommodate the way people actually behave. Finally, I reported three research activities that were aimed at bringing survey participants' actions closer to survey designers' goals. Each relied on (or would rely on) some user modeling ideas to achieve this. I believe these ideas can extend well beyond their embodiment in software to various survey tasks.

So if survey designers can only get people to follow their plans then measurement error will be significantly reduced. Of course this is an ideal, something to work toward in trying to characterize truth for a population. But truth is a thorny concept. As Groves (1991, p. 24) points out, a quantity as seemingly straightforward as the weight of a potato may not be so straightforward because "measured weight varies by humidity [and because of] problems defining where the potato begins and ends (do you count dirt and what is dirt?). Keep in mind that gravity varies over parts of the earth." Survey measurement is at least as elusive.

By emphasizing the steps taken by participants in performing survey tasks, we may be able to avoid what Groves (1991) calls the "true value morass." Instead of trying to estimate the difference between observed and true values we might instead measure the degree to which the participants followed the designer's plans in producing those observed values. This shifts the burden from the participant to the designer to specify accurate plans that are workable for a diversity of people in a diversity of situations.

O'Muircheartaigh (1997, p. 1) makes a similar point. "[Error is] *work purporting to do what it does not do*. Rather than specify an arbitrary (pseudo-objective) criterion, this redefines the problem in terms of the aims and frame of reference of the researcher. It immediately removes the need to consider *true value* concepts in any absolute sense, and forces a consideration of the needs for which the data are being collected. Broadly speaking, every survey operation has an objective, an outcome, and a description of that outcome. Errors . . . will be found in the mismatches among those elements."

We can reduce mismatches between stated survey objectives (goals) and actual outcomes (the result of the participants' plan-following behavior) by focusing on what comes in-between the two—what designers *ask* people to do and how they actually *do* it. In survey measurement, as in many walks of life, actions speak louder than words.

ACKNOWLEDGMENTS

I am grateful to Michael Schober for helpful comments on an earlier draft of this manuscript. In addition, many thanks to Al Corbett, Carol Hert, Irv Katz, and Gary Marchionini, for helping clarify some of the ideas presented here. I am solely responsible for the views expressed in the final product. They do not necessarily reflect the positions or opinions of the Bureau of Labor Statistics.

REFERENCES

Anderson, J. R., Boyle, C. F., Corbett, A. T., and Lewis, M. W. (1990). Cognitive modeling and intelligent tutoring. *Artificial Intelligence*, *42*, 7–49.

Akmajian, A., Demers, R. A., Farmer, A. K., and Harnish, R. M. (1990). *Linguistics: An Introduction to Language and Communication*, 3rd ed. Cambridge, MA: The MIT Press.

Armknecht, P., and Weyback, D. (1989). Adjustments for quality change in the U.S. consumer price index. *Journal of Official Statistics*, *5*, 107–123.

Beatty, P. (1995). Understanding the standardized/non-standardized interviewing controversy. *Journal of Official Statistics*, *11*, 147–160.

Benyon, D. (1993). Adaptive systems: A solution to usability problems. *User Modeling and User-Adapted Interaction*, *3*, 65–87.

Brown, J. S., and VanLehn, K. (1980). Repair theory: A generative theory of bugs in procedural skills. *Cognitive Science*, *4*, 379–426.

Conrad, F. (1997). Using expert systems to model and improve survey classification processes. In L. Lyberg, P. Biemer, M. Collins, E. de Leeuw, C. Dippo, N. Schwarz, and D. Trewin (Eds.), *Survey Measurement and Process Quality*, pp. 393–414. New York: Wiley.

Conrad, F., Brown, N., and Cashman, E. (1998). Strategies for estimating behavioral frequency in survey interviews. *Memory*, *6*, 339–366.

Conrad, F., Kamalich, R. Longacre, J., and Barry, D. (1993). An expert system for reviewing commodity substitutions in the consumer price index. *Proceedings of the Ninth Conference on Artificial Intelligence for Applications*, pp. 299–305. Los Alamitos, CA: IEEE Computer Society Press.

Corbett, A. T., and Anderson, J. R. (1995). Knowledge tracing: Modeling the acquisition of procedural knowledge. *User Modeling and User-Adapted Interaction, 4*, 253–278.

Corbett, A. T., Anderson, J. R., and O'Brien, A. T. (1995). Student modeling in the ACT programming tutor. In P. Nichols, S. Chipman, and B. Brennan (Eds.), *Cognitively Diagnostic Assessment*, pp. 19–41. Hillsdale, NJ: Erlbaum.

Esposito, J. L., and Rothgeb, J. M. (1997). Evaluating survey data: Making the transition from pretesting to quality assessment. In L. Lyberg, P. Biemer, M. Collins, E. de Leeuw, C. Dippo, N. Schwarz, and D. Trewin (Eds.), *Survey Measurement and Process Quality*, pp. 541–571. New York: Wiley.

Fowler, F. J., and Mangione, T. W. (1990). *Standardized Survey Interviewing: Minimizing Interviewer-Related Error*. Newbury Park, CA: Sage.

Groves, R. M. (1991). Measurement error across the disciplines. In P. Biemer, R. Groves, L. Lyberg, N. Mathiowetz, and S. Sudman (Eds.), *Measurement Errors in Surveys*, pp. 1–25. New York: Wiley.

Hert, C., and Marchionini, G. (1997). *Seeking statistical information in federal websites: Users, tasks, strategies, and design recommendations.* Unpublished report to the Bureau of Labor Statistics.

Kay, J. (1995). Vive la difference! Individualized interaction with users. In C.S. Mellish (Ed.), *Proceedings of the Fourteenth International Joint Conference on Artificial Intelligence*, pp. 978–984. San Mateo, CA: Morgan Kaufmann Publishers.

Marchionini, G., and Hert, C. (1997). *Usability testing for large institutional web sites.* Position paper for workshop on website usability testing at SIGCHI Human Factors in Computing Systems Conference, Atlanta, Georgia.

Norman, D. A. (1981). Categorization of action slips. *Psychological Review, 88*, 1–15.

Norman, D. A. (1993). *Things that Make us Smart: Defending Human Attributes in the Age of the Machine*. Reading, MA: Addison-Wesley.

O'Muircheartaigh, C. A. (1997). Measurement error in surveys: A historical perspective. In L. Lyberg, P. Biemer, M. Collins, E. de Leeuw, C. Dippo, N. Schwarz, and D. Trewin (Eds.), *Survey Measurement and Process Quality*, pp. 1–25. New York: Wiley.

Reason, J. (1990). *Human Error*. New York: Cambridge University Press.

Rich, E. (1989). Stereotypes and user modeling. In A. Kobsa and W. Wahlster (Eds.), *User Models in Dialogue Systems*, pp. 35–51. Berlin: Springer-Verlag.

Schober, M. F., and Conrad, F. G. (1997). Reducing survey measurement error through conversational interaction. *Public Opinion Quarterly, 61*, 576–602.

Suchman, L. A. (1987). *Plans and Situated Actions: The Problem of Human-Machine Communication*. New York: Cambridge University Press.

Suchman, L. A. (1993). Response to Vera and Simon's situated action: A symbolic interpretation. *Cognitive Science, 17*, 71–75.

Suchman, L., and Jordan, B. (1990). Interactional troubles in face-to-face survey interviews. *Journal of the American Statistical Association, 85(409)*, 232–241.

Suchman, L., and Jordan, B. (1992). Validity and the collaborative construction of meaning in face-to-face surveys. In J. M. Tanur (Ed.), *Questions About Questions: Inquiries into the Cognitive Bases of Surveys*, pp. 241–267. New York: Sage.

Willis, G. B., Royston, P., and Bercini, D. (1991). The use of verbal report methods in the development and testing of survey questionnaires. *Applied Cognitive Psychology*, *5*, 251–267.

CHAPTER 20

Visualizing Categorical Data

Michael Friendly
York University

For some time we have wondered why graphical methods for categorical data are so poorly developed and little used compared with methods for quantitative data. For quantitative data, graphical methods are commonplace adjuncts to all aspects of statistical analysis, from the basic display of data in a scatterplot, to diagnostic methods for assessing assumptions and finding transformations, to the final presentation of results. In contrast, graphical methods for categorical data are still in its infancy. There are not many methods, and those that are available in the literature are not accessible in common statistical software, consequently they are not widely used.

What has made this contrast puzzling is the fact that the statistical methods for categorical data are, in many respects, discrete analogs of corresponding methods for quantitative data: log-linear models and logistic regression, for example, are such close parallels of analysis of variance and regression models that they can all be seen as special cases of generalized linear models.

Several possible explanations for this apparent puzzle may be suggested. First, it may be that those who have worked with and developed methods for categorical data are just more comfortable with tabular data, or that frequency tables representing sums over all cases in a dataset are more easily apprehended in tables than quantitative data. Second, it may be argued that graphical methods for quantitative data are easily generalized; for example, the scatterplot for two variables provides the basis for visualizing any number of variables in a scatterplot matrix; available graphical methods for categorical data tend to be more specialized.

However, a more fundamental reason may be that quantitative and categorical

Cognition and Survey Research, Edited by Monroe G. Sirken, Douglas J. Herrmann, Susan Schechter, Norbert Schwarz, Judith M. Tanur, and Roger Tourangeau.
ISBN 0-471-24138-5   © 1999 John Wiley & Sons, Inc.

319

data display are best served by different visual metaphors. Quantitative data rely on the natural visual representation of magnitude by length or position along a scale; for categorical data, it will be seen that a count is more naturally displayed by an area or by the visual density of an area. Graphical methods, therefore, can foster a clearer cognitive model of the data than numeric methods alone.

To make the contrast clear, we include a section on graphical methods for categorical data that describes and illustrates both some old and some relatively novel methods, with particular emphasis on methods designed for large, multi-way contingency tables. Some methods (sieve diagrams, mosaic displays) are well-suited for detecting patterns of association in the process of model building; others are useful in model diagnosis, or as graphical summaries for presentation of results. A final section describes some ideas for effective visual presentation. But first we will outline a general framework for data visualization methods in terms of communication goal (analysis vs. presentation), display goal, and the psychological and graphical design principles for which graphical methods for different purposes should adhere.

20.1 GOALS AND DESIGN PRINCIPLES FOR VISUAL DATA DISPLAY

Designing good graphics is surely an art, but it is also one that ought to be informed by science. In constructing a graph, quantitative and qualitative information is encoded by visual features, such as position, size, texture, symbols, and color. This translation is reversed when a person studies a graph. The cognitive representation of numerical magnitude and categorical grouping, and the apperception of patterns and their *meaning* must be extracted from the visual display in order that a person may make use of these cognitions.

There are many views of graphs, of graphical perception, and of the roles of data visualization in discovering and communicating information. On the one hand, one may regard a graphical display as a stimulus—a package of information to be conveyed to an idealized observer. From this perspective certain questions are of interest: which form or graphic aspect promotes greater accuracy or speed of judgment (for a particular task or question)? What aspects lead to greatest memorability or impact? Cleveland (Cleveland and McGill, 1984, 1985; Cleveland, 1993a), Jenkins and Dillman (1997), Lewandowsky and Spence (1989; Spence, 1990), and Tufte (1990, 1993) have made important contributions to our understanding of the cognitive aspects of graphical display.

An alternative view regards a graphical display as an act of communication—like a narrative, or even a poetic text or work of art. This perspective places the greatest emphasis on the desired communication goal, and judges the effectiveness of a graphical display in how well that goal is achieved. Kosslyn (1985, 1989) has articulated this perspective most clearly from a cognitive perspective.

In this view, an effective graphical display, like good writing, requires an understanding of its purpose—what aspects of the data are to be communicated to the viewer. In writing, we communicate most effectively when we know our audience and tailor the message appropriately. So too, we may construct a graph in different ways to use ourselves, to present at a conference or meeting of our colleagues, or to publish in a research report, or a communication to a general audience (Friendly, 1991, chapter 1).

This chapter discusses the effectiveness of both black and white graphics and color graphics. The examples of graphics presented here are shown in black and white only. Examples of color graphics could not be presented here, but these examples can be seen on my web page at http://www.math.yorku.ca/ SCS/Papers/casm/. If the discussion provided in this chapter persuades you that the color graphics might be useful for some of your own analyses, you are invited to look at the examples on my web page.

Figure 20.1 shows one organization of visualization methods in terms of the

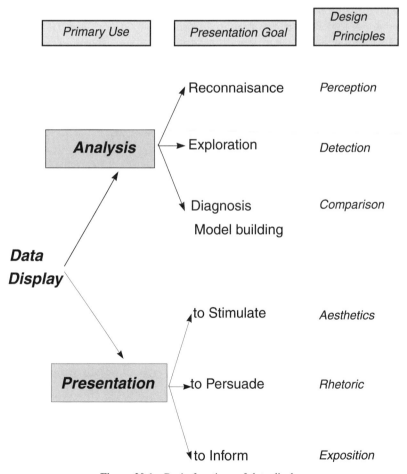

Figure 20.1   Basic functions of data display.

primary use or intended communication goal, the functional presentation goal, and suggested corresponding design principles. The first distinction identifies *Analysis* or *Presentation* as the primary communication goal of a data graphic (with the understanding that a given graph may serve both purposes, or neither). Among graphical methods designed to help study or understand a body of data, I distinguish those designed for:

- *reconnaissance*—a preliminary examination, or an overview of a possibly complex terrain;
- *exploration*—graphs designed to help detect patterns or unusual circumstances, or to suggest hypotheses, analyses or models;
- *diagnosis-graphs* designed to summarize or critique a numerical statistical summary.

Presentation graphics have different presentation goals as well. We may wish to stimulate, or to persuade, or simply to inform. As in writing, it is usually a good idea to know what it is you want to say with a graph, and tailor its message to that goal. (In what follows, the presentation goal is primarily didactic.) As is in writing, as well as research, it is good to investigate the effectiveness of methods, as has been done for the graphics presented here.

20.2 SOME GRAPHICAL METHODS FOR CATEGORICAL DATA

One-way frequency tables may be conveniently displayed in a variety of ways: typically as bar charts (although the bars should often be ordered by frequency, rather than by bar-label), dot charts (Cleveland, 1993b) or pie charts (when percent of total is important). For two- (and higher-) way tables, however, the design principles of perception, detection, and comparison imply that we should try to show the observed frequencies in the cells in relation to what we would expect those frequencies to be under a reasonable null model—for example, the hypothesis that the row and column variables are unassociated.

To this end, several schemes for representing contingency tables graphically are based on the fact that when the row and column variables are independent, the estimated expected frequencies, m_{ij}, are products of the row and column totals (divided by the grand total): $m_{ij} = n_{i+}n_{+j}/n_{++}$. Then, each cell can be represented by a rectangle whose area shows the cell frequency, n_{ij}, or the deviation from independence.

20.2.1 Sieve Diagrams

Table 20.1 shows data on the relation between hair color and eye color among 592 subjects (students in a statistics course) collected by Snee (1974). The Pearson χ^2 for these data is 138.3 with nine degrees of freedom, indicating

Table 20.1 Hair- and Eye-Color Data

Eye Color	Hair Color				Total
	Black	Brown	Red	Blond	
Green	5	29	14	16	64
Hazel	15	54	14	10	93
Blue	20	84	17	94	215
Brown	68	119	26	7	220
Total	108	286	71	127	592

substantial departure from independence. Assume the goal is to understand the *nature* of the association between hair and eye color. For any two-way table, the expected frequencies under independence can be represented by rectangles whose widths are proportional to the total frequency in each column, n_{+j}, and whose heights are proportional to the total frequency in each row, n_{i+}, the area of each rectangle is then proportional to m_{ij}. Figure 20.2 shows the expected frequencies for the hair- and eye-color data. Each rectangle is ruled proportionally to the expected frequency, and we note that the visual densities are equal in all cells.

Riedwyl and Schüpbach (1983, 1994) proposed a *sieve diagram* (later called a *parquet diagram*) based on this principle. In this display the area of each rectangle is proportional to expected frequency, as in Figure 20.2, but observed frequency is shown by the number of squares in each rectangle. Hence, the difference between observed and expected frequency appears as the density of shading, using color to indicate whether the deviation from independence is positive or negative. (In monochrome versions, positive residuals are shown by solid lines, negative by broken lines.) The sieve diagram for hair color and eye color is shown in Figure 20.3.

20.2.2 Mosaic Displays for *n*-way Tables

The mosaic display, proposed by Hartigan and Kleiner (1981, 1984), and extended by Friendly (1992, 1994b) represents the counts in a contingency table directly by tiles whose area is proportional to the *observed* cell frequency. One important design goal is that this display should apply extend naturally to three-way and higher-way tables. Another design feature is to serve both exploratory goals (by showing the pattern of observed frequencies in the full table), and model building goals (by displaying the residuals from a given log-linear model).

One form of this plot, called the *condensed mosaic display*, is similar to a divided bar chart. The width of each column of tiles in Figure 20.4 is proportional to the marginal frequency of hair colors; the height of each tile is determined by the conditional probabilities of eye color in each column. Again, the area of each box is proportional to the cell frequency, and independence is shown when the tiles in each row all have the same height.

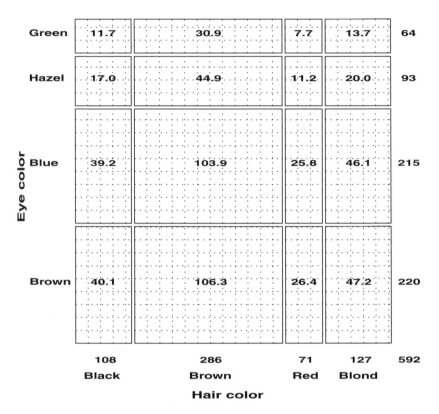

Figure 20.2 Expected frequencies under independence. Each box has area equal to its expected frequency, and is cross-ruled proportionally to the expected frequency.

Enhanced Mosaics The enhanced mosaic display (Friendly, 1992, 1994b) achieves greater visual impact by using color and shading to reflect the size of the residual from independence and by reordering rows and columns to make the pattern of association more coherent. The resulting display shows both the observed frequencies and the pattern of deviations from a specified model.

Figure 20.5 gives the extended mosaic plot, showing the standardized (Pearson) residual from independence, $d_{ij} = (n_{ij} - m_{ij})/\sqrt{m_{ij}}$, by the color and shading of each rectangle: cells with positive residuals are outlined with solid lines and filled with slanted lines; negative residuals are outlined with broken lines and filled with gray scale. The absolute value of the residual is portrayed by shading density: cells with absolute values less than 2 are empty; cells with $|d_{ij}| \geq 2$ are filled; those with $|d_{ij}| \geq 4$ are filled with a darker pattern.[1] Under the assumption of independence, these values roughly correspond to two-tailed

[1]Color versions use blue and red at varying lightness to portray both sign and magnitude of residuals.

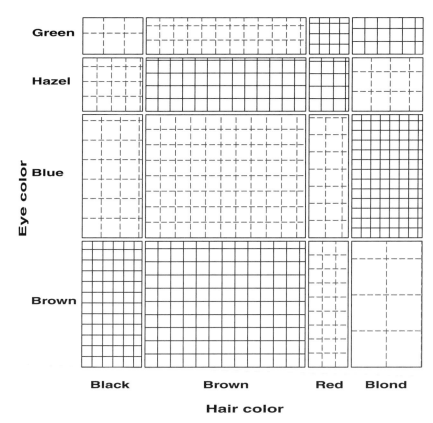

Figure 20.3 Sieve diagram for hair-color, eye-color data. Observed frequencies are equal to the number squares in each cell, so departure from independence appears as variations in shading intensity.

probabilities $p < 0.05$ and $p < 0.0001$ that a given value of $|d_{ij}|$ exceeds 2 or 4. For exploratory purposes, we do not usually make adjustments (e.g., Bonferroni) for multiple tests because the goal is to display the pattern of residuals in the table as a whole. However, the number and values of these cutoffs can be easily set by the user.

When the row or column variables are unordered, we are also free to rearrange the corresponding categories in the plot to help show the nature of association. For example, in Figure 20.5, the eye-color categories have been permuted so that the residuals from independence have an opposite-corner pattern, with positive values running from bottom-left to top-right corners, negative values along the opposite diagonal. Coupled with size and shading of the tiles, the excess in the black–brown and blond–blue cells, together with the underrepresentation of brown-haired blonds and people with black hair and blue eyes is now quite apparent. Though the table was reordered based on the d_{ij} values, both dimensions in Figure 20.5 are ordered from dark to light, suggesting an explanation for the association. [In this example the eye-color categories could

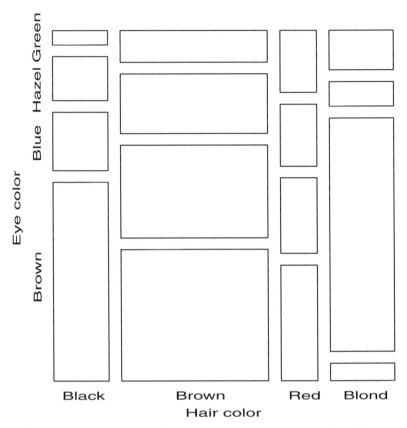

Figure 20.4 Condensed mosaic for hair-color, eye-color data. Each column is divided according to the conditional frequency of eye color given hair color. The area of each rectangle is proportional to observed frequency in that cell.

be reordered by inspection. A general method (Friendly, 1994b) uses category scores on the largest correspondence analysis dimension.]

Multi-Way Tables   Like the scatterplot matrix for quantitative data, the mosaic plot generalizes readily to the display of multi-dimensional contingency tables. Imagine that each cell of the two-way table for hair and eye color is further classified by one or more additional variables—sex and level of education, for example. Then each rectangle in Figure 20.5 can be divided vertically to show the proportion of males and females in that cell, and each of those portions can be subdivided again to show the proportions of people at each educational level in the hair–eye–sex group.

Fitting Models   When three or more variables are represented in the mosaic, we can fit several different models of independence and display the residuals from that model. We treat these models as null or baseline models, which may

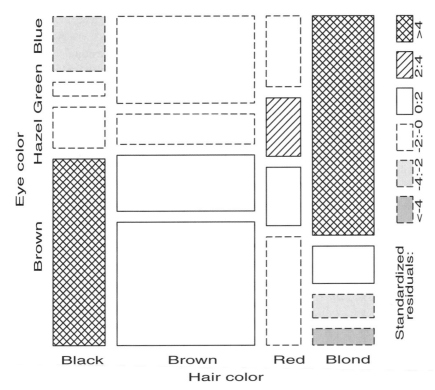

Figure 20.5 Condensed mosaic, reordered, and shaded. Deviations from independence are shown by color and shading. The two levels of shading density correspond to standardized deviations greater than 2 and 4 in absolute value. This form of the display generalizes readily to multiway tables.

not fit the data particularly well. The deviations of observed frequencies from expected, displayed by shading, will often suggest terms to be added to an explanatory model that achieves a better fit. Two examples are:

- *Complete independence:* The model of complete independence asserts that all joint probabilities are products of the one-way marginal probabilities:

$$\pi_{ijk} = \pi_{i++}\pi_{+j+}\pi_{++k} \qquad (1)$$

for all i, j, k in a three-way table. This corresponds to the log-linear model $[A]$ $[B]$ $[C]$. Fitting this model puts all higher terms, and hence, all association among the variables, into the residuals, displayed in the mosaic.

- *Joint independence:* Another possibility is to fit the model in which variable C is jointly independent of variables A and B,

$$\pi_{ijk} = \pi_{ij+}\pi_{++k} \tag{2}$$

This corresponds to the log-linear model $[AB]$ $[C]$. Residuals from this model show the extent to which variable C is related to the combinations of variables A and B, but they do not show any association between A and B.

For example, with the data from Table 20.1 broken down by sex, fitting the joint independence model [HairEye][Sex] allows us to see the extent to which the joint distribution of hair color and eye color is associated with sex. For this model, the likelihood ratio G^2 is 19.86 on 15 df ($p = 0.178$), indicating an acceptable overall fit. The three-way mosaic, shown in Figure 20.6, highlights two cells: among blue-eyed blonds, there are more females (and fewer males) than would be if hair color and eye color were jointly independent of

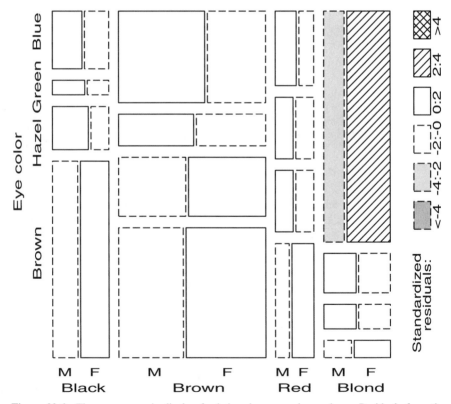

Figure 20.6 Three-way mosaic display for hair color, eye color, and sex. Residuals from the model of joint independence, $[HE]$ $[S]$ are shown by shading. $G = 19.86$ on 15 df. The only lack of fit is an overabundance of females among blue-eyed blondes.

sex. Except for these cells, hair color and eye color would not be understood as associated with sex.

For higher-way tables, there are many more possible models that can be fit. However, they have the characteristic that, for any given table (or marginal subtable), the size of tiles in the mosaic always shows the same observed frequencies, while the shading (showing the sign and magnitude of the residuals) varies from one model to another. Because a good-fitting model will have all or mostly small residuals, and these are shown unfilled, a search for an adequate explanatory model can be thought of as "cleaning the mosaic."

Color Scales for Residuals   The design of the mosaic display shares some features with the use of statistical maps to display quantitative information, and it is appropriate (and perhaps instructive) to consider my use of color and shading in an area-based display. When we display *signed* magnitudes, such as residuals from a fitted model or differences between time points on a mosaic or a map, the goal is usually to convey both magnitude (how big a difference?) and direction (is it more or less?). For this use, a double-ended color scale, using opposing colors such as red and blue, with darker, more saturated colors at the extremes, is usually most effective. In the mosaic display we use the color scheme Cleveland (1993b) refers to as "two-hues, varying lightness," and which Carr (1994) uses quite effectively in the display of residuals on a map.

The choice of opposing colors and assignment to positive and negative, requires care and consideration of the intended audience, however. I use red for negative "deficit" values and blue for positive "excesses"; Carr (1994), in displays of rates of disease, assigns red to positive "hot spots" and cooler blue to negative values. Various government agencies have other conventions about the assignment of color. Unfortunately, black and white reproduction of red/blue displays folds red and blue to approximately equal gray levels. In the mosaic displays, we usually prepare different versions–using gray level and pattern fills for figures to be shown in black and white. Color versions of the figures shown here are available on my web site at http://www.math.yorku.ca/ SCS/ Papers/casm/

20.2.3 Fourfold Display

A third graphical method based on area as the visual mapping of cell frequency is the "fourfold display" (Friendly 1994a,c; Fienberg, 1975) designed for the display of 2×2 (or $2 \times 2 \times k$) tables. In this display the frequency n_{ij} in each cell of a fourfold table is shown by a quarter circle, whose radius is proportional to $\sqrt{n_{ij}}$, so the area is proportional to the cell count.

For a single 2×2 table the fourfold display described here also shows the frequencies by area, but scaled in a way that depicts the sample odds ratio, $\hat{\theta} = (n_{11}/n_{12}) \div (n_{21}/n_{22})$. An association between the variables $\theta \neq 1$ is shown by the tendency of diagonally opposite cells in one direction to differ in size from those in the opposite direction, and the display uses color or shading to show this direction. Confidence rings for the observed theta allow a visual test of the hypothesis H_0: $\theta = 1$. They have the property that the rings for adjacent quadrants overlap *iff* the observed counts are consistent with the null hypothesis.

As an example, Figure 20.7 shows aggregate data on applicants to graduate school at Berkeley for the six largest departments in 1973 classified by admission and sex. At issue is whether the data show evidence of sex bias in admission practices (Bickel, Harnmel, and O'Connell, 1975). The figure shows the observed cell frequencies numerically in the corners of the display. Thus, there were 2691 male applicants, of whom 1198 (44.4%) were admitted, compared with 1855 female applicants of whom 557 (30.0%) were admitted. Hence, the sample odds ratio, Odds (*Admit*|*Male*)/(*Admit*|*Female*) is 1.84 indicating that males were almost twice as likely to be admitted.

The frequencies displayed graphically by shaded quadrants in Figure 20.7 are not the raw frequencies. Instead, the frequencies have been standardized (by iterative proportional fitting) so that all table margins are equal, while preserving the odds ratio. Each quarter circle is then drawn to have an area proportional to this standardized cell frequency. This makes it easier to see the association

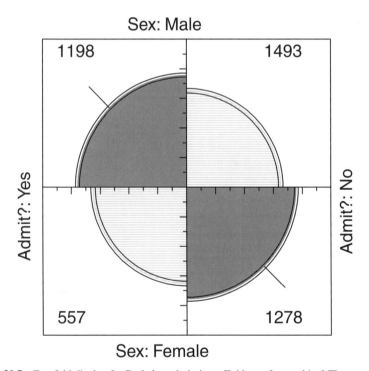

Figure 20.7   Fourfold display for Berkeley admissions: Evidence for sex bias? The area of each shaded quadrant shows the frequency, standardized to equate the margins for sex and admission. Circular arcs show the limits of 99% confidence interval for the odds ratio.

between admission and sex without being influenced by the overall admission rate or the differential tendency of males and females to apply. With this standardization the four quadrants will align when the odds ratio is 1, regardless of the marginal frequencies.

The shaded quadrants in Figure 20.7 do not align and the 99% confidence rings around each quadrant do not overlap, indicating that the odds ratio differs significantly from 1—putative evidence of gender bias. The width of the confidence rings gives a visual indication of the precision of the data—if we stopped here, we might feel quite confident of this conclusion.

Multiple Strata In the case of a $2 \times 2 \times k$ table, the last dimension typically corresponds to "strata" or populations, and it is typically of interest to see if the association between the first two variables is homogeneous across strata. The fourfold display is designed to allow easy visual comparison of the pattern of association between two dichotomous variables across two or more populations.

For example, the admissions data shown in Figure 20.7 were obtained from a sample of six departments; Figure 20.8 displays the data for each department. The departments are labeled so that the overall acceptance rate is highest for Department A and decreases steadily to Department F. Again each panel is standardized to equate the marginals for sex and admission. This standardization also equates for the differential total applicants across departments, facilitating visual comparison.

Surprisingly, Figure 20.8 shows that, for five of the six departments, the odds of admission is essentially identical for men and women applicants. Department A appears to differ from the others, with women approximately 2.86 $(= (313/19)/(512/89))$ times *more* likely to gain admission. This appearance is confirmed visually by the confidence rings, which in Figure 20.8 are *joint* 99% intervals for θ_c, $c = 1, \ldots, 6$. This result, which contradicts the display for the aggregate data in Figure 20.7, is a nice example of Simpson's paradox. The resolution of this contradiction can be found in the large differences in admission rates among departments. Men and women apply to different departments differentially, and in these data women apply in larger numbers to departments that have a low acceptance rate. The aggregate results are misleading because they falsely assume men and women are equally likely to apply in each field.[2]

Visualization Principles An important principle in the display of large, complex datasets is *controlled comparison*—we want to make comparisons against a clear standard, with other things held constant. The fourfold display differs from a pie chart in that it holds the angles of the segments constant and varies the radius, whereas the pie chart varies the angles and holds the radius constant. An important consequence is that we can quite easily compare a series

---

[2]This explanation ignores the possibility of structural bias against women, e.g., lack of resources allocated to departments that attract women applicants.

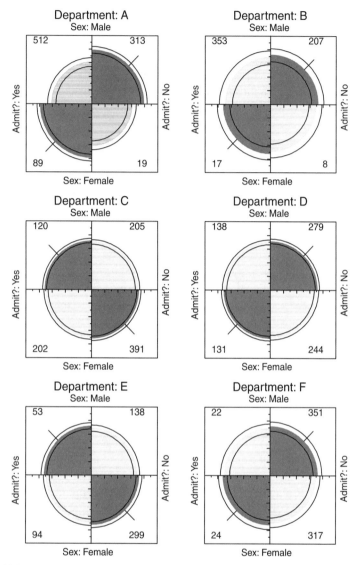

Figure 20.8 Fourfold display for Berkeley admissions, by department. In each panel the confidence rings for adjacent quadrants overlap if the odds ratio for admission and sex do not differ significantly from 1. The data in each panel have been standardized as in Figure 20.7.

of fourfold displays for different strata, since corresponding cells of the table are always in the same position. As a result, an array of fourfold displays serve the goals of comparison and detection better than an array of pie charts. Moreover, it allows the observed frequencies to be standardized by equating either the row or column totals, while preserving the odds ratio. In Figure 20.8, for example, the proportion of men and women, and the proportion of accepted applicants were equated visually in each department. This provides a clear standard which also greatly facilitates controlled comparison.

Another principle is *visual impact*—we want the important features of the display to be easily distinguished from the less important (Tukey, 1993). Figure 20.8 distinguishes the one department for which the odds ratio differs significantly from 1 by shading intensity, even though the same information can be found by inspection of the confidence rings.

20.3   EXAMPLE: NAEP 1992 GRADE 12 MATHEMATICS

The previous section used simple examples to illustrate these somewhat novel techniques themselves. This section illustrates the use of these graphical methods in attempting to understand a large, complex dataset of the sort often found in public policy research, and which often defies standard visualization methods.

Table 20.2 is a four-way classification of over 3.5 million examinees from the 1992 National Assessment of Educational Proficiency (Mullis, Dossey, Owen, and Phillips, 1993; Wainer, 1997), classified by their high school program (Academic, Nonacademic), their economic status (Not Poor vs. Poor), and ethnic group (White, Asian, Hispanic, Black). Students in the background groupings are then classified by their performance on the NAEP Mathematics test, recorded here as Advanced, Proficient, Basic, or Below Basic. One glance at Table 20.2 shows two things: First, very few students achieve Advanced level, none in Nonacademic Programs (except for a few Not Poor white students); as a result, the analysis below combines Advanced and Proficient, labeled "Proficient+." Second, the usefulness of this table as a data display is nil, except as a record of the results.

Mosaic displays are constructed sequentially, with the variables arranged in a particular order, and it usually makes sense to order the variables in a quasi-causal, or predictor–response fashion. Here we consider Ethnicity and Poverty as (partial) determinants of academic Program, and all three of these as potential predictors of achievement level.

20.3.1   Analysis of [Ethnicity, Poverty, Achievement Level]

For simplicity, we begin with an analysis of the marginal table of Ethnicity, Poverty and Level, collapsing over Program. Figure 20.9 shows the mosaic for (the marginal table of) Ethnicity and Poverty, fitting the independence model. If Poverty were unrelated to Ethnic group, the tiles would all be equally wide in each column. There is, of course, a pronounced association between Poverty and Ethnic Group ($G^2(3) = 193.95$), as shown by the shading pattern of the residuals: Asians and Whites are more frequently Not Poor and Hispanics and Blacks are more frequently Poor, than would be the case if these variables were independent.[3]

---

[3]Blacks and Hispanics are interchanged from the original table, in accord with association ordering described below.

Table 20.2 National Assessment of Educational Proficiency, 1992 Mathematics. Counts of Achievement Levels by Program, Poverty, and Ethnicity

	Not Poor				Poor			
	White	Asian	Black	Hispanic	White	Asian	Black	Hispanic
Academic program								
Advanced	23403	1711	0	1050	6814	542	0	633
Proficient	224723	15145	4591	5716	96442	5654	3308	7500
Basic	679319	30590	32937	26709	439109	27276	64451	42726
Below basic	102849	3645	23983	20596	135919	7974	132421	52327
Nonacademic program								
Advanced	1866	0	0	0	0	0	0	0
Proficient	24317	2518	1039	570	716	557	0	482
Basic	208710	13142	14271	16338	151369	5478	9229	23551
Below basic	244240	7149	60868	35233	254734	3315	134874	81356

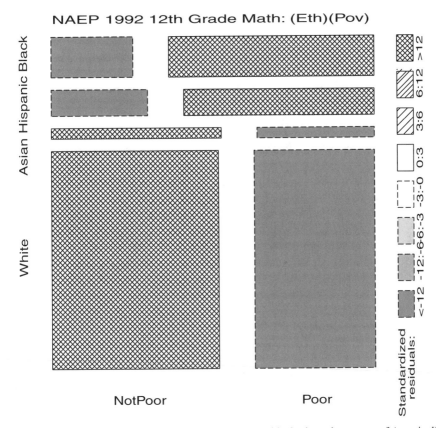

Figure 20.9   Mosaic display of ethnicity and poverty. Residuals show the pattern of (marginal) association between these two variables.

Figure 20.10 shows the relation between Achievement Level and Ethnicity and Poverty, fitting a model [(EthPov)(Level)] which says that Level is independent of the combinations of Ethnicity and Poverty jointly. The shading pattern shows how violently this model is contradicted by the data ($G^2(14) = 553,568$):

- Among Not Poor Whites, an overabundance are in Basic level or higher; for Poor Whites, however, frequencies significantly greater than expected under this model of joint independence occur only in the Basic level.
- Among Asians, there are greater than expected frequencies in all but the Below Basic category, independent of Poverty.
- Among both Hispanics and Blacks, there are greater than expected frequencies in the Below Basic level.

It is depressing how few 12th-grade children are classified in the Advanced or Proficient categories.

336   VISUALIZING CATEGORICAL DATA

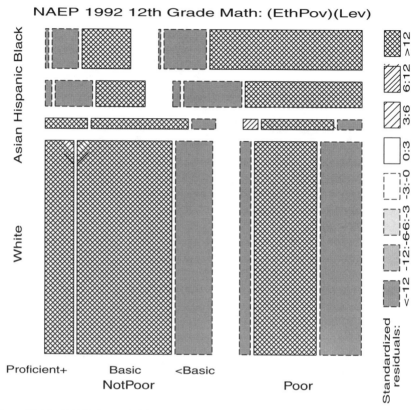

Figure 20.10   NAEP data, ethnicity and poverty vs. achievement level. Residuals display the dependence of level on both ethnicity and poverty.

20.3.2   Analysis of the Full Table

With some understanding among the relations among Poverty, Ethnicity, and Achievement Level, we proceed to fitting sequential models of joint independence to the full table. The analysis goal is not to provide an adequate model, but simply to remove the associations we have already seen, thereby revealing the associations which remain.[4]

Figure 20.11 shows the relation between Academic Program and the combinations of Ethnicity and Poverty; residuals show how Program is associated with the categories of the other two variables:

- Not Poor Whites are more likely to be in Academic programs, while the reverse is true for Poor Whites.

---

[4]All of these models fit very badly, partly due to the enormous sample size, but they are to be regarded only as baseline models. When a model of joint independence, say (Ethnicity, Poverty) (Program), is fit, the association between Ethnicity and Poverty is fit *exactly*, and so does not appear in the residuals.

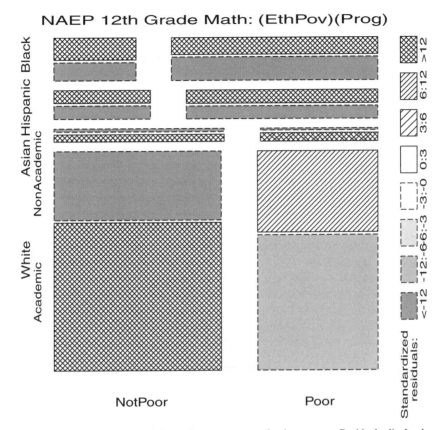

Figure 20.11 NAEP data, ethnicity and poverty vs. academic program. Residuals display how academic program depends on both ethnicity and poverty.

- Asians are more likely to be in Academic programs. This result holds regardless of Poverty.
- Hispanics and Blacks are more likely to be in Nonacademic programs, also independent of Poverty.

Figure 20.12 shows the relation between NAEP Achievement Level and the combinations of Ethnicity, Poverty, and academic Program jointly. The associations among the three background variables just observed (Figure 20.11) have been eliminated and the residuals now show how Achievement Level is associated with the combinations of the other three variables. This is admittedly a complex display, but a few moments study reveals the following:

- Not Poor Whites in Academic programs are more likely to achieve Basic level or above; Not Poor Whites in Nonacademic programs are more likely to be classified Below Basic level.

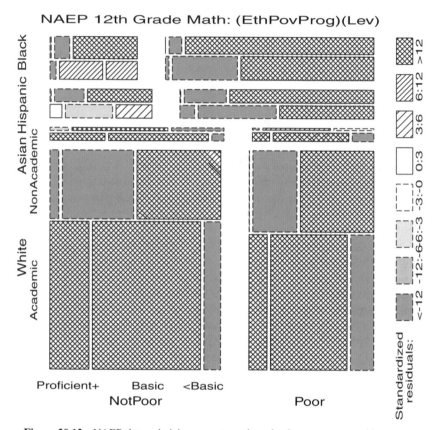

Figure 20.12   NAEP data, ethnicity, poverty, and academic program vs. achievement.

- Approximately the same pattern obtains for Poor Whites. (More detailed analysis shows them less likely than Not Poor Whites to achieve Advanced level.)

- Asians in Academic programs are more likely to achieve Basic level or above; Asians in Nonacademic programs are most likely to achieve just Basic level. Both statements apply independently of Poverty.

The patterns for Blacks and Hispanics diverge here for the first time:

- Blacks are most likely to achieve Below Basic level, independent of Program and Poverty, except that Not Poor Blacks in Academic programs are more likely to achieve in Basic level.

- Hispanics are similar, except that Not Poor Hispanics in Academic programs are more likely to achieve Advanced standing than the model of joint independence predicts.

20.3.3   Other Displays

The NAEP data is a $2 \times 2 \times 4 \times 4$ table, which we can regard as a collection of 2×2 tables, showing the relation between Poverty and Academic Program for each of the 16 Ethnicity by Achievement Level groups. Figure 20.13 shows one fourfold display of these data. In each panel, the lower-left quadrant represents the most advantaged students, and the upper-right represents the least advantaged. Thus, a positive association between the Poverty and Nonacademic Program is shown by tick marks in the $45°$ direction, and we see that nearly all associations are positive.

In order to adjust for the great imbalance in the numbers of examinees in the different Poverty–Ethnic group combinations, the data in Table 20.2 were standardized to equate the column totals in that table. These standardized frequencies are displayed directly in Figure 20.13, scaled so that the largest such frequency has unit radius. As a result, the distributions across the columns and rows may be readily compared. It is immediately apparent that the greatest proportion of all students performed at the lowest achievement levels, and that,

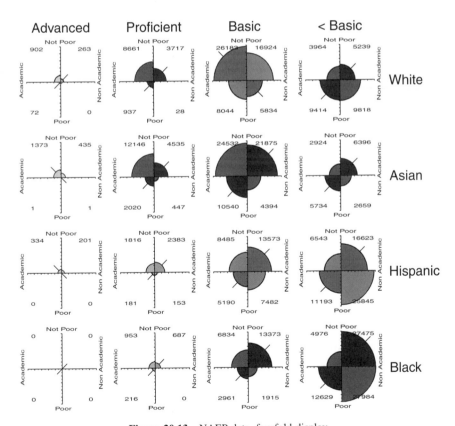

Figure 20.13   NAEP data, fourfold display.

at the very lowest level the largest proportion are in Nonacademic programs, particularly among Hispanics and Blacks. Second, those who achieve Proficient level are much more likely Not Poor, and taking Academic programs. Finally (and sadly) the dearth of students achieving Advanced level stands out clearly.

In closing this section, we will comment on more traditional statistical analyses and graphical displays of such data. One statistical analysis would be to fit a more meaningful log-linear model than the baseline models used earlier. When this is done, we find that the only (barely) tenable model (short of the saturated model, which must fit perfectly) is the all-three-way-interaction model, symbolized as

$$(\text{EthPovProg}) \ (\text{EthPovLev}) \ (\text{EthProgLev}) \ (\text{PovProgLev})$$

If Ethnicity, Poverty, and Program are all regarded as predictors of achievement Level, this model says that achievement level depends on the combinations of Ethnicity and Poverty, the combinations of Ethnicity and academic Program, and the combinations of Poverty and Program. Unfortunately, this analysis is not helpful in understanding *how* achievement Level depends on those factors, although tables of parameter estimates from this model could be interpreted to reveal some of the effects shown in the mosaic displays.

Alternatively, a logistic regression might attempt to model achievement level in relation to the predictors. As an example, consider a model for the probability that a student achieves at least Basic level. The logistic model of equivalent complexity to the log-linear model can be expressed as

$$Pr \ (Basic \ or \ better) \sim \text{Poverty} + \text{Ethnicity} + \text{Program} +$$
$$\text{Poverty} \times \text{Ethnicity} + \text{Poverty} \times \text{Program} + \text{Ethnicity} \times \text{Program} \qquad (3)$$

where $\sim$ means "is modeled by." These results provide a sensible and relatively simple way to display these data, which may serve a presentation goal better than the novel displays shown previously. The idea is to plot the predicted probabilities or predicted log-odds from the model. Figure 20.14 shows these results in one possible format, plotting the log-odds directly, but with a probability scale on the right for those who are more comfortable with probabilities than logits.

This figure shows most of the relations between Achievement Level and the predictors discussed earlier, but shows none of the relations among the predictors. The interaction terms in model (3) stem from the fact that the profiles in Figure 20.14 are not parallel. For students in Nonacademic programs, the effect of Poverty increases as achievement level decreases and is far greater for Hispanics than for Asians. For students in Academic programs, the effect of Poverty is larger for both Asians and Hispanics than for Whites and Blacks. This graph displays these more subtle aspects of the fitted logistic model, but

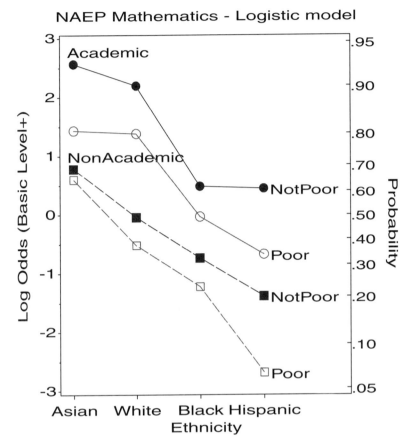

Figure 20.14   NAEP data, logistic regression.

the dominant message is clear: An Academic program greatly increases the likelihood of at least Basic achievement; being Poor greatly decreases it.

20.4   EFFECT ORDERING FOR DATA DISPLAYS

One reason why graphs of quantitative data are effective is that graphing values on quantitative axes automatically orders those values so that "less" and "more" are given visually ordered positions in the display. However, when data values are classified by "factors," the ordering of the levels of the factor variables has considerable impact on graphical display. Ordered factors (such as age, level of education, etc.) are usually (though not always) most sensibly arranged in their natural order for all presentation goals.

Unordered factors (disease classification, geographic region) deserve more careful thought. For a geographic classification (states, provinces) is it common to arrange the units alphabetically or (as is common in Canada) from east to west. When the goal of presentation is detection or comparison (as opposed to table lookup), this is almost always a bad idea.

Instead, we suggest a general rule for arranging the levels of unordered factors in visual displays—tables as well as graphs: *sort the data by the effects to be observed.* Sorting has both global and local effects: globally, a more coherent pattern appears, making it easier to spot exceptions; locally, effect-ordering brings similar items together, making them easier to compare. See de Falguerolles, Fredrich, and Sawitzki (1997) for related ideas.

The use of this principle is illustrated by the following:

- *Main-effects ordering:* For quantitative data where the goal is to see "typical" values, sort the units in boxplots, dotplots and tables by means, medians, or by row and column effects. (If the goal is to see differences in variability, sort by standard deviation or interquartile range.) Wainer (1993) and Carr and Olsen (1996) have presented similar arguments for the effectiveness of such orderings on detection.

- *Discriminant ordering:* For multivariate data, where the goal is to compare different groups in their means on a (possibly large) number of variables, arrange the *variables* in tables or visual displays according to the weights of those variables on the dimensions which best discriminate among the groups (canonical discriminant dimensions).

- *Correlation ordering:* A related idea is that in the display of multivariate data by glyph plots, star plots, parallel coordinate plots, and so forth, the variables should be ordered according to the largest principal component or biplot (Gabriel, 1980, 1981) dimension(s). This arrangement brings similar variables together, where similarity is defined in terms of patterns of correlation.

- *Association ordering:* For categorical data, where the goal is to understand the pattern of association among variables, order the levels of factors according to their position on the largest correspondence analysis dimension (Friendly, 1994b).

20.5   MOSAIC MATRICES AND COPLOTS FOR CATEGORICAL DATA

Different graphics foster different cognitive models of data. Thus, a second reason for the wide usefulness of graphs of quantitative data has been the recognition that combining multiple views of data into a single display allows detection of patterns which could not readily be discerned from a series of separate graphs. The scatterplot matrix shows all pairwise (marginal) views of a set of variables in a coherent display, whose design goal is to show the interdependence among the collection of variables as a whole. The conditioning plot, or *coplot* (Cleveland, 1993b) shows a collection of (conditional) views of several variables, conditioned by the values of one or more other variables. The design goal is to visualize how a relationship depends on or changes with one or more additional factors. These ideas can be readily extended to categorical data.

One analog of the scatterplot matrix for categorical data is a matrix of mosaic displays showing some aspect of the bivariate relation between all pairs of variables. The simplest case shows, for each pair of variables, the marginal relation, summed over all other variables. For example, Figure 20.15 shows the pairwise marginal relations among the variables Admit, Gender, and Department in the Berkeley data shown earlier in fourfold displays (Figures 20.7 and 20.8). The panel in row 2, column 1 shows that Admission and Gender are strongly asso-

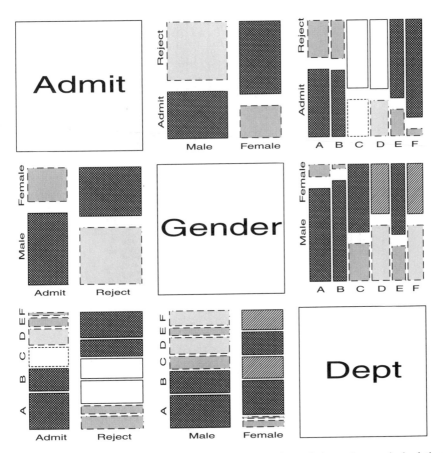

Figure 20.15 Mosaic matrix of Berkeley admissions. Each panel shows the marginal relation, fitting an independence model between the row and column variable, collapsed over other variable(s).

ciated marginally, as we saw in Figure 20.7, and overall, males are more often admitted. The diagonally-opposite panel (row 1, column 2) shows the same relation, splitting first by gender. The panels in the third column (and third row) illuminate the explanation for the paradoxical result (Figure 20.8) that, within all but department A, the likelihood of admission is equal for men and women. The (1, 3) panel shows the marginal relation between Admission and Department; departments A and B have the greatest overall admission rate, departments E and F the least. The (2, 3) panel shows that men apply in much greater numbers to departments A and B, while women apply in greater numbers to the departments with the lowest overall rate of admission.

Several further extensions are now possible. First, we need not show the marginal relation between each pair of variables in the mosaic matrix. For example, Figure 20.16 shows the pairwise *conditional* relations among these

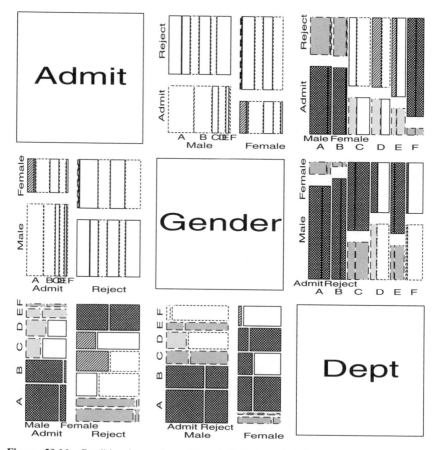

Figure 20.16 Conditional mosaic matrix of Berkeley admissions. Each panel shows the conditional relation, fitting a model of conditional independence between the row and column variable, controlling for other variable(s).

variables, in each case fitting a model of conditional independence with the remaining variable controlled. Thus, the (1, 2) and (2, 1) panels show the fit of the model [Admit, Dept] [Gender, Dept], which asserts that Admission and Gender are independent, given (controlling for) department. Except for Department A, this model fits quite well, again indicating lack of gender bias.

Second, the framework of the scatterplot matrix can now be used as a general method for displaying marginal or conditional relations among a mixture of quantitative and categorical variables. For marginal plots, pairs of quantitative variables are shown as a scatterplot, while pairs of categorical variables are shown as a mosaic display. Pairs consisting of one quantitative and one categorical variable can be shown as a set of boxplots for each level of the categorical variable. For conditional plots, we fit a model predicting the row variable from the column variable, partialing out (or conditioning on) all other variables from each. For quantitative variables, this is just the partial regression plot.

Finally, an analog of the coplot for categorical data is an array of plots of the dependence among two or more variables, conditioned (or stratified) by the values of one or more given variables. Each panel then shows the *partial* associations among the foreground variables; the collection of plots shows how these as the given variables vary. Figures 20.8 and 20.13 are two examples of this idea, using the fourfold display to represent the association in 2 × 2 tables.

Figures 20.17 and 20.18 show two further examples, using the mosaic display to show the partial relations [Admit] [Dept] given Gender, and [Admit] [Gender] given Dept, respectively. Figure 20.18 shows the same results displayed in Figure 20.8: no association between Admission and Gender, except in Dept A, where females are relatively more likely to gain admission.

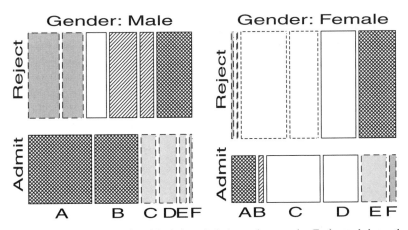

Figure 20.17   Mosaic coplot of Berkeley admissions, given gender. Each panel shows the partial relation, fitting a model of independence model between admission and department.

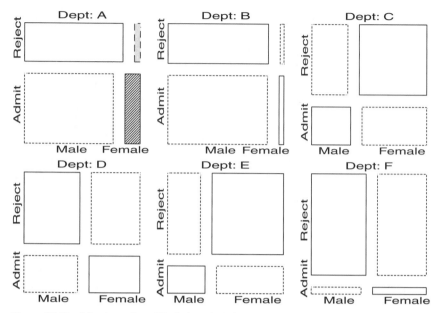

Figure 20.18   Mosaic coplot of Berkeley admissions, given department. Each panel shows the partial relation, fitting a model of independence between admission and gender.

Figure 20.17 shows that there is a very strong association between Admission and Department—different rates of admission, but also shows two things not seen in other displays: First, the *pattern* of association is qualitatively similar for both men and women; second the association is quantitatively stronger for men than women–larger differences in admission rates across departments.

In this chapter, we have shown how graphical methods can lead to a more accurate understanding of data than numeric methods. We have also shown how careful use of graphic methods can bring out effects that might be missed altogether by graphic or statistical analyses. Survey data bear on important societal issues. It is critical, therefore, that presentations of data enable statisticians, survey professionals, and the public to develop a proper cognitive model of phenomena that the data address. Graphical methods (such as sieve diagrams, condensed mosaic patterns, fourfold displays, star plots, and more) should be part of the procedures used by survey professional to elucidate the cognitive aspects of survey design.

ACKNOWLEDGMENTS

Preparation of this chapter was aided by the National Sciences and Engineering Reesarch Council of Canada, Grant OGP0138748.

REFERENCES

Bickel, P. J., Hammel, J. W., and O'Connell, J. W. (1975). Sex bias in graduate admissions: Data from Berkeley. *Science, 187,* 398–403.

Carr, D. (1994). Color perception, the importance of gray and residuals, on a chloropleth map. *Statistical Computing and Statistical Graphics Newsletter, 5(1),* 16–21.

Carr, D., and Olsen, A. (1996). Simplifying visual appearance by sorting: An example using 159 AVHRR classes. *Statistical Computing and Statistical Graphics Newsletter, 7(1),* 10–16.

Cleveland, W. S. (1993a). A model for studying display methods of statistical graphics. *Journal of Computational and Statistical Graphics, 2,* 323–343.

Cleveland, W. S. (1993b). *Visualizing Data.* Summit, NJ: Hobart Press.

Cleveland, W. S., and McGill, R. (1984). Graphical perception: Theory, experimentation and application to the development of graphical methods. *Journal of the American Statistical Association, 79,* 531–554.

Cleveland, W. S., and McGill, R. (1985). Graphical perception and graphical methods for analyzing scientific data. *Science, 229,* 828–833.

Falguerolles, A. D., Fredrich, F., and Sawitzki, G. (1997). A tribute to J. Bertin's graphical data analysis. In W. Bandilla and F. Faulbaurn (Eds.), *Softstat '97, Advances in Statistical Software,* pp. 11–20. Stuttgart: Lucius and Lucius.

Fienberg, S. E. (1975). Perspective Canada as a social report. *Social Indicators Research, 2,* 153–174.

Friendly, M. (1991). *SAS System for Statistical Graphics,* 1st ed. Cary, NC: SAS Institute Inc.

Friendly, M. (1992). Mosaic displays for loglinear models. *Proceedings on the Section on Statistical Graphics, American Statistical Association,* pp. 61–68.

Friendly, M. (1994a). *A fourfold display for 2 by 2 by k tables* (Tech. Rep. No. 217). Department of Psychology, York University.

Friendly, M. (1994b). Mosaic displays for multi-way contingency tables. *Journal of the American Statistical Association, 89,* 190–200.

Friendly, M. (1994c). SASAML graphics for fourfold displays. *Observations, 3(4),* 47–56.

Gabriel, K. R. (1980). Biplot. In N. L. Johnson and S. Kotz (Eds.), *Encyclopedia of Statistical Sciences,* Vol. 1, pp. 263–271. New York: Wiley.

Gabriel, K. R. (1981). Biplot display of multivariate matrices for inspection of data and diagnosis. In V. Barnett (Ed.), *Interpreting Multivariate Data,* pp. 147–173. London: Wiley.

Hartigan, J. A., and Kleiner, B. (1981). Mosaics for contingency tables. In W. F. Eddy (Ed.), *Computer Science and Statistics: Proceedings of the 13th Symposium on the Interface,* pp. 286–273. New York: Springer-Verlag.

Hartigan, J. A., and Kleiner, B. (1984). A mosaic of television ratings. *The American Statistician, 38,* 32–35.

Jenkins, C. R., and Dillman, D. A. (1997). Towards a theory of self-administered questionnaire design. In L. Lyberg, P. Biemer, A. Collins, E. de Leeuw, C. Dippo, N. Schwarz, and D. Trewin (Eds.), *Survey Measurement and Process Quality*, pp. 165–196. New York: Wiley.

Kosslyn, S. M. (1985). Graphics and human information processing: A review of five books. *Journal of the American Statistical Association, 80*, 499–512.

Kosslyn, S. M. (1989). Understanding charts and graphs. *Applied Cognitive Psychology, 3*, 185–225.

Lewandowsky, S., and Spence, I. (1989). The perception of statistical graphs. *Sociological Methods & Research, 18*, 200–242.

Mullis, I. V. S., Dossey, J. A., Owen, E. H., and Phillips, G. W. (1993). *NAEP 1992: Mathematics report card for the nation and the states* (Tech. Rep. No. 23-ST02). Washington, DC: National Center for Education Statistics.

Riedwyl, H., and Schüpbach, M. (1983). *Siebdiagramme: Graphische darstellung von kontingenztafeln* (Tech. Rep. No. 12). Institute for Mathematical Statistics, University of Bern, Bern, Switzerland.

Riedwyl, H., and Schüpbach, M. (1994). Parquet diagram to plot contingency tables. In F. Faulbaum (Ed.), *Softstat '93: Advances in Statistical Software*, pp. 293–299. New York: Gustav Fischer.

Snee, R. D. (1974). Graphical display of two-way contingency tables. *The American Statistician, 28*, 9–12.

Spence, I. (1990). Visual psychophysics of simple graphical elements. *Journal of Experimental Psychology: Human Perception and Performance, 16*, 683–692.

Tufte, E. R. (1990). *Envisioning Information*. Cheshire, CT: Graphics Press.

Tufte, E. R. (1993). *The Visual Display of Quantitative Information*. Cheshire, CT: Graphics Press.

Tukey, J. W. (1993). Graphic comparisons of several linked aspects: Alternative and suggested principles. *Journal of Computational and Statistical Graphics, 2(1)*, 1–33.

Wainer, H. (1993). Tabular presentation. *Chance, 6*, 52–56.

Wainer, H. (1997). Some multivariate displays for NAEP results. *Psychological Methods, 2(1)*, 34–63.

CHAPTER 21

Statistical Graphs and Maps: Higher Level Cognitive Processes

Stephan Lewandowsky
University of Western Australia

The visual display and analysis of data has become increasingly important in most research endeavors. In some cases, individual graphs have demonstrably had a profound influence on the progress of an entire discipline (e.g., Spence and Garrison, 1993). The growing importance of graphs has been accompanied by an increasing recognition among experimental psychologists that the visual display of data is a nontrivial activity whose success relies on knowledge of people's processing capabilities and limitations. Accordingly, empirically-based guidelines for the construction of "user-friendly" graphs has been provided by several authors (e.g., Cleveland and McGill, 1984a; Lohse, 1991; for a review, see Lewandowsky and Spence, 1989).

To date, research has primarily focused on low-level perceptual and psychophysical phenomena, such as the degree of perceptual error associated with basic visual tasks (e.g., Spence, 1990). However, few investigations have addressed the more complex cognitive tasks that can arise even with seemingly simple statistical displays. For example, line graphs often represent the results from factorial experiments, in which case graph readers need not only encode individual data values or trends, but need to form a complex abstract representation of the relations among several experimental variables. Similarly, statistical maps are frequently used to detect larger trends, such as clusters of high or low data values, that also require the formation of aggregate cognitive representations. This chapter selectively reviews research into these more complex tasks.

Cognition and Survey Research, Edited by Monroe G. Sirken, Douglas J. Herrmann, Susan Schechter, Norbert Schwarz, Judith M. Tanur, and Roger Tourangeau.
ISBN 0-471-24138-5 © 1999 John Wiley & Sons, Inc.

21.1 STATISTICAL GRAPHS

For present purposes, a complex cognitive task is defined as requiring the translation of visual features into abstract conceptual relations before the data can be interpreted (Shah and Carpenter, 1995, p. 45). Although many such tasks exist, the focus here is on the interpretation of factorial experiments, in particular, of interactions between experimental variables. The research literature abounds with factorial experiments, and the results are often displayed in line graphs with one of the experimental variables assigned to the x-axis and the other (known as the z-variable) serving as parameter.

Figure 21.1 shows the results of an experiment on graphical perception reported by Spence and Lewandowsky (1991): The two panels provide two alternative views of the data, with different variables serving as x- and z-variables in each case, and the top panel replicating the original presentation format. The impact of the form of the graph on interpretation of the data is apparent. The bottom panel seemingly obscures the result emphasized by Spence and Lewandowsky, namely, that the pie chart is superior to the bar chart only for comparisons involving two pairs of data values, whereas it is indistinguishable for the simpler comparisons. Conversely, the bottom panel emphasizes that performance with the table is far more variable across tasks than with the other displays, a result not emphasized by Spence and Lewandowsky.

Surprisingly, a search of the literature reveals only a single paper that has investigated the comprehension of line graphs with more than one experimental variable (Shah and Carpenter, 1995). Shah and Carpenter found that people have difficulty extracting any precise quantitative information from the z-variable. In one experimental task, people did not readily distinguish between linear and exponential spacing of the levels of the z-variable. In another task, when asked to decide whether two graphs depicted the same or different results, people erroneously considered the same data to be different on nearly half the trials involving two different perspectives, akin to the two panels in Figure 21.1. Similarly, when instructed to draw a previously presented graph from memory, people were quite accurate when perspective remained the same, but performance dropped to about 24 percent when the alternative perspective had to be drawn.

On the basis of that single study, it is clear that people's understanding of standard experimental outcomes is significantly affected by the way in which the data are displayed. Moreover, little abstract guidance exists on which perspective of the data is the "correct" or "better" one. For example, there are no a priori reasons to prefer one panel in Figure 21.1 over the other. By implication, situations may arise in which data are interpreted very differently by different researchers, depending on the perspective chosen for presentation.

Can interactions be displayed in ways that do not engender such perspective-bound interpretations? One possibility, of course, involves the use of "3-D" graphs, in which both the x- and z-variables have metric properties, with the z-axis inflected to simulate visual depth. Notwithstanding the popularity of such

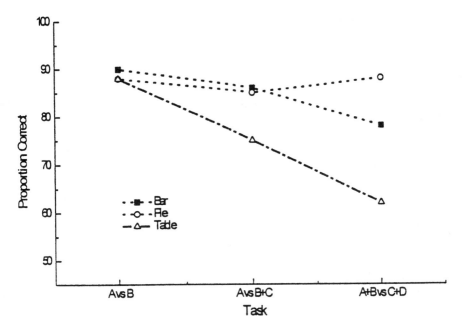

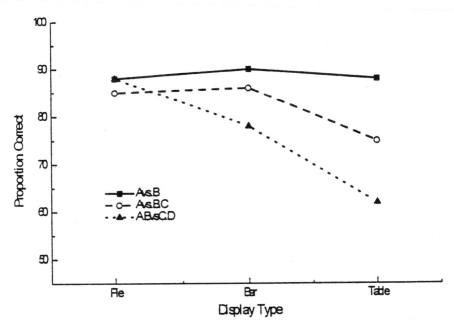

Figure 21.1 Results of Spence and Lewandowsky (1991), showing performance as a function of type of display and task in two different perspectives.

3-D graphs, Shah and Carpenter (1995, p. 59) presented several arguments against their use; a point underscored empirically by Lewandowsky and Myers (1993), who found that thresholds for simple differences between data values were dramatically greater for inflected (z) axes than horizontal (x) axes.

A more promising alternative may be the geometric representations for contingency tables developed by Fienberg (e.g., Fienberg and Gilbert, 1970), which are, at least in principle, perspective-invariant. Specifically, for a 2×2 table, any arbitrary set of four numbers can be shown to map into a single point contained within a tetrahedron whose vertices represent the four (normalized) tables with a nonzero entry in only one of the four cells. Thus, the tables 0–0–0–1, 0–0–1–0, 0–1–0–0, and 1–0–0–0 each correspond to a vertex (tables are normalized such that cell entries are in the range 0–1). Furthermore, any pattern of independence between the two variables, that is, the absence of an interaction, maps into a point located on a saddle-shaped surface within the tetrahedron. Departures from independence translate into points away from that surface. Critically, the display does not involve any representational choices concerning variables and axes, and any given table of means corresponds to a unique graph. However, the utility of Fienberg's technique remains to be established by experimentation, and it must be shown that observers are capable of interpreting tetrahedral displays reliably and in a perspective-invariant manner.

In the absence of validation of such alternative displays, researchers may be well advised to present their data to the audience in several different perspectives. Correspondingly, all readers of graphs may be well advised to hone their skills of recognizing the relationships between different perspectives. A training aid that supports enhanced visual processing of factorial data has been made available on the world wide web by J. T. Behrens of Arizona State University at: http://research.ed.asu.edu/vrml/

21.2   CLUSTERS ON STATISTICAL MAPS

Statistical maps are indispensable tools for the analysis and presentation of measurements whose geographic distribution is of interest. However, maps present some unique challenges to the graph designer that deserve to be clarified. First, unlike the earlier line graphs, maps necessarily use the two dimensions of the plane to represent geographical information, thus eliminating the most easily interpretable graphical codes (horizontal or vertical position) for representation of magnitudes. By implication, perception of any magnitude information in a statistical map is akin to interpreting the z-variable in two-dimensional line graphs. Second, it is not unusual for a map to show many hundreds or indeed thousands of data points, for example, when plotting a variable at the level of U.S. counties. This poses great demands on the map reader, who may be asked to form an integrated perception of the overall pattern. Third, when a variable is represented by shading or coloring of geographic regions (the choropleth technique), the perceived magnitude of each data value is confounded with the, typically unrelated, size of the region. (This review is limited to choropleth maps because they represent the vast majority of published statistical maps; a discussion of other coding schemes can be found in Herrmann and Pickle, 1996 or Lewandowsky and Behrens, 1996). Finally, the use of color to represent magnitude requires awareness of the intricate interactions between brightness, saturation, and hue.

Notwithstanding these challenges, statistical maps, especially those representing mortality data, have had a unique influence on public policy. The earliest known case involved mapping of cholera deaths in London in 1854, which allowed the water-borne illness to be traced to a contaminated water pump (discussed in Gilbert, 1958; Wainer, 1992). More recently, publication of the U.S. Cancer Atlas (Pickle, Mason, Howard, Hoover, and Fraumeni, 1987) identified a particularly high incidence of cervical cancer in West Virginia, which prompted state legislators to allocate extra funds towards early detection and treatment of this often curable disease, with the result that mortality subsequently declined (Maher, 1995). In this case, as in the majority of epidemiological applications, emphasis was on the detection of "clusters," contiguous areas of particularly high (or low) mortalities that represent unusual situations. Once clusters are detected, corrective efforts can be applied immediately (as in the case of West Virginia) or analysis can focus on discovering epidemiologically relevant correlates (such as a contaminated water pump).

Visual analysis is often the preferred way to detect clusters because mathematical approaches cannot model all relevant features of geographical space (Marshall, 1991). The critical role of visual inspection stands in sharp contrast to the scarcity of relevant psychological research. The absence of such research may be partly responsible for the chaotic variability among some 50 mortality atlases sampled by Walter and Birnie (1991). Most atlases used hue to code magnitude, with four colors (orange, brown, blue, and green) variously used to represent each possible numeric category from highest to lowest.

21.2.1 Choropleth Maps: A Priori Considerations

Most choropleth maps aggregate data values into few (5–8) class intervals that subdivide the overall range of magnitudes. Monochrome choropleth maps use increasingly dense shadings of a single color to represent increasing numerical magnitude. Double-ended choropleth maps, also known as bipolar or two-opposing colors maps, use increasingly dense shadings of one color to represent high (above-midrange) magnitudes and increasingly dense shadings of another color to represent low (below-midrange) magnitudes. Categorical color or multi-hue maps, finally, use a different fully-saturated hue for each magnitude.

Even without considering psychological results, the use of color-coded maps carries several risks. First, any color information will be eliminated or rendered misleading in black-and-white photocopies of the original. Similarly, roughly 7 percent of males are color blind and thus unable to differentiate between certain sets of hues. Lest one think this is a minor point, the reader is invited to photocopy the cancer maps in Boyle, Muir, and Grundmann (1989), most of which lose all intelligibility in black and white. As a further example, consider the maps in Figure 21.2 which represent mortality rates for the same disease (age adjusted death rates for white males from asthma during 1986–1988) using monochrome reproductions of color originals. For the top map, hues were assigned to magnitudes in order of increasing brightness, thus ensuring that the data are conveyed accurately even in monochrome reproductions. Comparison with the remaining two maps, which use a spectral categorical arrangement of hues and a red–blue double-ended scale, respectively, reveals the potential difficulties of relying exclusively on color to communicate magnitude.

On these grounds alone, caution is advised when using color coding. Carswell, Kinslow, Pickle, and Herrmann (1995) explored one solution to the reproduction problem by providing double-ended scales that continued to be interpretable on the basis of brightness alone. A related issue concerns the need for legends that arises from the use of color coding: whereas ordinal understanding of a monochrome map does not require a legend, a categorical or double-ended scale is meaningless without accompanying explanation.

The requirement for a legend may not be an obstacle when the map is intended for detailed inspection of a single data set, but legends may be a nuisance when numerous maps are to be inspected during exploratory analysis of a number of data sets.

21.2.2   Choropleth Mortality Maps: Empirical Findings

Double-Ended vs. Monochrome Scales   Turning first to the double-ended scale, the case for color has been stated very eloquently by Carswell et al. (1995). Given that humans typically cannot differentiate more than 3–5 levels of gray or brightness (Sanders and McCormick, 1993), the double-ended scale provides an opportunity to virtually double the number of available categories through concatenation of two hues, each with its own set of shadings. In addition, because low and high magnitudes are typically represented by identical levels of saturation, the double-ended scale affords the opportunity to correct the known human bias of attending primarily to the confirming presence of critical information (i.e., high mortalities), even though its absence (i.e., low mortalities) may be equally relevant to understanding the etiology of a disease.

On the other hand, valid psychological reasons can also be cited against the double-ended scale, in particular, if one assumes that preattentive processes engaged in visual search also underlie the cognitively more elaborate domain of map inspection. There is widespread agreement among cognitive psychologists that visual detection of a target is qualitatively more difficult and time-consuming if conjunctions of features must be formed (e.g., Treisman, 1991). For example, if the features "red" and "triangle" must be detected and combined to identify a target because red squares and green triangles are also present, a qualitatively different perceptual process appears to be engaged than when

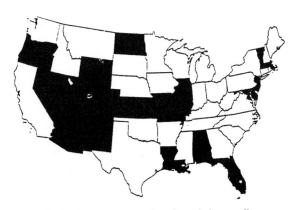

Figure 21.2   Some representative choropleth mortality maps.

the task is to pick out a red object among green distractors or a triangle among squares. Applied to the context of mortality maps, it follows that a monochrome scale may be more readily perceived than the double-ended scale, which relies on formation of the conjunction of saturation and hue.

The available empirical evidence mirrors the theoretical ambivalence surrounding the double-ended scales. Consider first the more negative results. In a study by Lewandowsky et al. (1993), participants had to simultaneously identify and mark high and low mortality clusters. The double-ended scale was found to lead to somewhat greater perceptual variability (i.e., less agreement among participants about cluster locations, suggesting less consistent data communication) than two monochrome scales involving shades of blue and black. At least in part, that result was due to the nonnegligible number of scale reversals for the double-ended scale, exactly as one might expect from the visual search literature. The observed scale reversals are illustrated in Figure 21.3: Solid squares indicate the perceived location of high-mortality clusters, whereas open diamonds show the perceived locations of low-mortality clusters. Even though the map omits the underlying mortality data, it is clear that four (out of 31) participants perceived a high-mortality cluster in regions that were most often identified as having the lowest incidence of the disease. These observations are best interpreted as scale reversals.

Perceptual difficulties associated with the double-ended scale are not confined to a single study: Mersey (1990) also reported occasional scale reversals among participants, and Cuff (1973) found that participants often chose the lightest class interval; that is, the midpoint of the double-ended scale when asked to identify the region with the lowest data value. Although each of these studies in isolation is subject to limitations arising from the choice of participants or details of the design (Carswell et al., 1995), there can be little doubt overall that double-ended scale reversals occur in a nonnegligible number of cases.

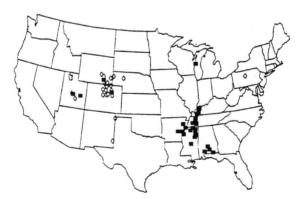

Figure 21.3   Scale reversals with the double-ended scale observed by Lewandowsky et al. (1993).

Turning to the more supportive evidence, Carswell et al. (1995) conducted a particularly thorough investigation of several different double-ended schemes in comparison to a gray scale. Using a variety of measures, including a perceptual variability index akin to Lewandowsky et al. (1993), the results showed that performance with a red–yellow opposing scale was consistently better than with a monochrome gray scale. On the other hand, a blue–yellow opposing scale was found to be inferior to shades of gray, with a third double-ended scale, red–blue, taking up the middle-ground.

Fortunately, Carswell et al. (1995) were able to provide principled explanations for the observed differences between color pairings, which otherwise would have formed a confusing array of seemingly arbitrary outcomes. Carswell et al. (1995) suggested that the success of the red–yellow scale was in large part due to the decoupling of perceived brightness and saturation. Specifically, for that scale, brightness increased monotonically from the lowest (saturated red) to the highest category (saturated yellow), thus providing an additional perceptual cue for estimation of magnitudes and also eliminating the problems otherwise associated with black-and-white reproduction. The failure of the blue–yellow scale at first appears puzzling because it, too, decoupled brightness and saturation. Carswell et al. (1995) suggested that the blue–yellow scale failed because of its conflicting use of two opposing graphical conventions; "darker-for-more" (shades of blue) and "warmer-for-more" (shades of yellow).

On balance, the available experimental results are best summarized as follows:

1. Double-ended scales that decouple brightness and saturation are suitable for cluster detection. Those scales also circumvent the problems associated with black-and-white reproduction and color blindness (Carswell et al., 1995).
2. Other double-ended scales are at best comparable to monochrome scales (Carswell et al., 1995; Lewandowsky et al., 1993).
3. The occurrence of scale reversals with a double-ended scale cannot be ruled out (Cuff, 1973; Lewandowsky et al., 1993; Mersey, 1990).

To conclude on a theoretical note, the double-ended scale provides a graphic analog to the residual approach to data analysis in general, and statistical maps in particular. In this approach, a tentative model is applied to the data and focus of the analysis is on the residuals. Accordingly, the user can focus on the most salient aspects of the data without having to perform the mental computation of determining both the overall effect, and the distance of important deviations from it. The double-ended scale supports such residual analysis, by subtracting the mean of the data and emphasizing deviations from that "model."

Categorical Color Scales Unlike monochrome scales, sets of different colors are not immediately perceived as an ordinal sequence of magnitudes: Green

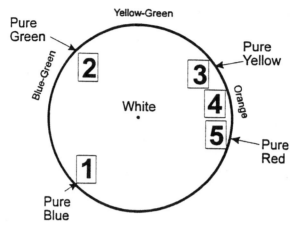

Figure 21.4   The color circle. Color for use in a categorical color map should be chosen so as to maximize separation along the circle.

does not necessarily represent "more" (or "less") than blue or red. In consequence, for most categorical color scales, hues are assigned to magnitudes either arbitrarily or in spectral sequence. The evidence shows that, for a given set of colors, assignment to magnitudes has no effect on performance in a variety of tasks (Hastie, Hammerle, Kerwin, Croner, and Herrmann, 1996).

However, assignment to magnitudes aside, the choice of the total set of colors is important. Cleveland and McGill (1983) showed that the number and size of red areas was consistently overestimated in comparison to areas filled in green. Similarly, using a cluster detection task, Lewandowsky et al. (1993) found that performance with a categorical color scale differed depending on which set of colors observers had to attend to. Good performance was associated with areas that were predominantly blue and green, whereas less satisfactory performance arose with orange and red areas. The likely reason for this performance discrepancy can be illustrated using the perceptual color circle in Figure 21.4, which represents a convenient way to summarize human color perception (e.g,. McBurney and Collings, 1977, p. 44). In general, discriminability between colors corresponds to the distance that separates them along the circle. The bold numbers in the figure represent the ordinal arrangement, by magnitude, of the colors used by Lewandowsky et al. (1993). Considering the distance along the circle between categories, it is readily apparent why judgments involving categories 1 and 2 would be more accurate than those involving 4 and 5.

21.3   CONCLUSIONS

This chapter has focused on two aspects of graphical cognition: The conceptual interpretation of interactions presented in simple graphs, and the detection of clusters in mortality maps. This limited focus of the chapter must not be taken to imply that other aspects of graphical cognition have been neglected by researchers. On the contrary, there has been a considerable amount of work on the perception of simple univariate displays (e.g., Cleveland and McGill, 1987; Spence and Lewandowsky, 1991; for a review see Spence and Lewandowsky, 1990), scatterplots (e.g., Wainer and Thissen, 1979; Cleveland and McGill, 1984b, Lewandowsky and Spence, 1989), and complex multivariate displays (Wang, 1978).

21.3.1 The Role of Observer Expertise

Most of the studies reviewed here involved undergraduate participants, and one might therefore be legitimately concerned about the generalizability of the results to the more expert audiences that are likely to be exposed to published graphs and mortality maps. Interestingly, the few comparisons between experts and novices have not reported any large and pervasive performance differences. Shah and Carpenter (1995) tested advanced graduate students in addition to novices, plus a sample of spatially trained architecture students, and found the same performance deficits in all groups of subjects. In the context of mortality maps, Maher (1995) compared the performance of statisticians, epidemiologists, and Congressional staffers on seven measures of map reading performance, including cluster detection. The only reliable difference to emerge was that epidemiologists and statisticians mentioned more relevant detail when describing map titles and legends than the Congressional staffers. Results along the same lines were reported by Carswell et al. (1995).

21.3.2 A Prospectus for Future Research

A concluding note of caution arises from one principle of graphical perception, illustrated by Spence and Lewandowsky (1991), which holds that questions about what might be "the best statistical graph" are meaningful only within the context of a particular cognitive task. Thus, while one graph may be superior to another when observers need to, say, identify global trends, the reverse may be true when observers must report single data values. In the present context, this implies that the conclusions concerning interpretation of interactions and detection of clusters are bound to the particular displays (line graphs and choropleth maps) that were used to gather the data.

This empirical state of affairs is unsatisfactory in the long run; the inventory of possible cognitive tasks is nearly unlimited, and so is the number of potential outcomes that the practitioner must bear in mind to guide graph construction. What is needed instead is a robust and universal theory of graph cognition that can predict and explain human performance for any arbitrary combination of task and display. This ambitious undertaking is by no means complete, although promising initial work in this direction has been reported by Gillan and Lewis (1994), Lohse (1991), and Simkin and Hastie (1987), to cite but a few existing models of graph perception.

All of these models attempt to explain overall performance on the basis of visual or cognitive "primitives," simple perceptual operations that detect and discriminate graphical elements for subsequent cognitive processing. Lohse's (1991) model has been particularly successful: It predicts the precise sequence and durations of gaze fixations while observers inspect a graph. The model has been found to account for a substantial proportion of the variance among participants' fixation times. Similarly, Gillan and Lewis's (1994) componential model of graph processing accounts for more than 70 percent of the variance among response times when observers were asked to identify, compare, and add values presented in three different simple graphs.

However, I know of no theory that explicitly relates those perceptual processes, which have been investigated and modeled with some success, to the conceptual processes that are so essential to understanding data such as interactions and clusters, but which have so far largely escaped theoretical attention. Future research should seek an integrated examination of perceptual and conceptual processes to provide a complete theoretical framework of graph cognition.

ACKNOWLEDGMENTS

Preparation of this chapter was facilitated by Large Research Grant No. A79600016 from the Australian Research Council and continued previous support from the National Center for Health Statistics (NCHS). Segments of this chapter have been derived from a paper in the 1995 Proceedings of the American Statistical Association, with John Behrens as coauthor.

REFERENCES

Boyle, P., Muir, C. S., and Grundmann, E. (1989) (Eds.). *Cancer Mapping*. Berlin: Springer-Verlag.

Carswell, C. M., Kinslow, H. S., Pickle, L. W., and Herrmann, D. J. (1995). *Using color to represent magnitude in statistical maps: The case for double-ended scales*. Unpublished manuscript.

Cleveland, W. S., and McGill, R. (1983). A color-caused optical illusion on a statistical graph. *The American Statistician, 37,* 101–105.

Cleveland, W. S., and McGill, R. (1984a). Graphical perception: Theory, experimentation, and application to the development of graphical methods. *Journal of the American Statistical Association, 79,* 531–554.

Cleveland, W. S., and McGill, R. (1984b). The many faces of a scatterplot. *Journal of the American Statistical Association, 79,* 807–822.

Cleveland, W. S., and McGill, R. (1987). Graphical perception: The visual decoding of quantitative information on graphical displays of data. *Journal of the Royal Statistical Society, 150A,* 192–229.

Cuff, D. J. (1973). Colour on temperature maps. *The Cartographic Journal, 10*, 17–21.

Fienberg, S. E., and Gilbert, J. P. (1970). The geometry of a two-by-two contingency table. *Journal of the American Statistical Association, 65*, 694–701.

Gilbert, E. W. (1958). Pioneer maps of health and disease in England. *Geographical Journal, 124*, 172–173.

Gillan, D. J., and Lewis, R. (1994). A componential model of human interaction with graphs: 1. Linear regression modeling. *Human Factors, 36*, 419–440.

Hastie, R., Hammerle, O., Kerwin, J., Croner, C. M., and Herrmann, D. J. (1996). Human performance reading statistical maps. *Journal of Experimental Psychology: Applied, 2*, 3–16.

Herrmann, D. J., and Pickle, L. W. (1996). A cognitive subtask model of statistical map reading. *Visual Cognition, 3*, 165–190.

Lewandowsky, S., and Behrens, J. T. (1996). Visual detection of clusters in statistical maps. *Proceedings of the Section on Statistical Graphics, American Statistical Association*, 8–17.

Lewandowsky, S., Herrmann, D. J., Behrens, J. T., Li, S.-C., Pickle, L., and Jobe, J. B. (1993). Perception of clusters in statistical maps. *Applied Cognitive Psychology, 7*, 533–551.

Lewandowsky, S., and Myers, W. E. (1993). Magnitude judgments in 3-D bar charts. In R. Steyer, K. F. Wender, and K. F. Widaman (Eds.), *Psychometric methodology. Proceedings of the 7th European Meeting of the Psychometric Society in Trier*, pp. 266–271. Stuttgart: Gustav Fischer Verlag.

Lewandowsky, S., and Spence, I. (1989). Discriminating strata in scatterplots. *Journal of the American Statistical Association, 84*, 682–688.

Lewandowsky, S., & Spence, I. (1989). The perception of statistical graphs. *Sociological Methods and Research, 18*, 200–242.

Lohse, G. L. (1991). *A cognitive model for understanding graphical perception.* Technical Report No. 39, Cognitive Science and Machine Intelligence Laboratory, University of Michigan.

Maher, R. J. (1995). The interpretation of statistical maps as a function of the map reader's profession. In L. W. Pickle and D. J. Herrmann (Eds.), *Cognitive aspects of statistical mapping*, pp. 249–274. National Center for Health Statistics, Working Paper Series, No. 18.

Marshall, R. J. (1991). A review of methods for the statistical analysis of spatial patterns of disease. *Journal of the Royal Statistical Society A, 154*, 421–441.

McBurney, C., and Collings, V. (1977). *Introduction to Sensation/Perception.* Englewood Cliffs, NJ: Prentice-Hall.

Mersey, J. E. (1990). Colour and thematic map design: The role of colour scheme and map complexity in choropleth map communication. *Cartographica, 27(3), Monograph No. 41*, 1–182.

Pickle, L. W., Mason, T. J., Howard, N., Hoover, R., and Fraumeni, J. F., Jr. (1987). *Atlas of US cancer mortality among whites: 1950–1980.* Washington, DC: National Institutes of Health.

Sanders, M. S., and McCormick, E. J. (1993). *Human Factors in Engineering and Design.* New York: McGraw-Hill.

Shah, P., and Carpenter, P. A. (1995). Conceptual limitations in comprehending line graphs. *Journal of Experimental Psychology: General, 124,* 43–61.

Simkin, D. K., and Hastie, R. (1987). An information-processing analysis of graph perception. *Journal of the American Statistical Association, 82,* 454–465.

Spence, I. (1990). Visual psychophysics of simple graphical elements. *Journal of Experimental Psychology: Human Perception and Performance, 16,* 683–692.

Spence, I., and Garrison, R. F. (1993). A remarkable scatterplot. *The American Statistician, 47,* 12–19.

Spence, I., and Lewandowsky, S. (1990). Graphical perception. In J. Fox and S. Long (Eds.), *Modern Methods of Data Analysis,* pp. 13–57. Newbury Park, CA: Sage.

Spence, I., and Lewandowsky, S. (1991). Displaying proportions and percentages. *Applied Cognitive Psychology, 5,* 61–77.

Treisman, A. M. (1991). Search, similarity, and integration of features between and within dimensions. *Journal of Experimental Psychology: Human Perception and Performance, 17,* 652–676.

Wainer, H. (1992). Understanding graphs and tables. *Educational Researcher, 21,* 14–23.

Wainer, H., and Thissen, D. (1979). On the robustness of a class of naïve estimators. *Applied Psychological Measurement, 3,* 543–551.

Walter, S. D., and Birnie, S. E. (1991). Mapping mortality and morbidity patterns: An international comparison. *International Journal of Epidemiology, 20,* 678–689.

Wang, P. C. C. (1978) (Ed.). *Graphical Representation of Multivariate Data.* New York: Academic Press.

CHAPTER 22

Toward a Research Agenda: Future Development and Applications of Cognitive Sciences to Surveys

Elizabeth Martin
Bureau of the Census

Clyde Tucker
Bureau of Labor Statistics

Throughout the Conference on Cognitive Aspects of Survey Methodology, working groups met to consider new areas of interdisciplinary work which may enrich scientific knowledge about surveys, and to recommend expanded applications of cognitive methods to survey measurement problems. All conference participants were members of one of eight working groups, each of which was asked to develop 2–4 proposals for research related to its topic area. On the last day of the conference, a chair of each working group gave an informal report of the group's deliberations and recommendations, which were discussed by conference participants as a whole. Subsequently, brief summary reports were prepared by the working group chairs, and they are published in a separate conference proceedings.[1] This chapter summarizes the deliberations of the eight

---

[1] See Sirken et al. (in press) for the summary report of each working group as well as the listing of working group members. The eight working groups, and their cochairs, are: *CASM in a Changing Survey Environment* (Kent Marquis and Dan Kasprzyk); *Exploring the Interview Process* (Nora Cate Schaeffer and Patricia Royston); *Different Disciplinary Perspectives on Cognition in the Question and Answer Process* (Stanley Presser and Tracy Wellens); *Applying Cognitive Methods to New Areas of the Survey Process* (Catherine Dippo and Cynthia Clark); *Income Measurement* (Martin David and Jeffrey Moore); *Integrating Cognitive Research into Household Survey Design* (Judith Lessler and Jennifer Rothgeb); *The Measurement of Disability* (Nancy Mathiowetz and Ronald Wilson); *Adapting Cognitive Techniques to Establishment Surveys* (David Cantor and Polly Phipps).

Cognition and Survey Research, Edited by Monroe G. Sirken, Douglas J. Herrmann, Susan Schechter, Norbert Schwarz, Judith M. Tanur, and Roger Tourangeau.
ISBN 0-471-24138-5 © 1999 John Wiley & Sons, Inc.

working groups, and is based on the summary reports, as well as the record of the presentations and discussions at the conference. We also have drawn on our own knowledge and opinions in commenting on the status of research on cognitive aspects of survey methodology.

In the first section, we summarize proposals for new research to apply current cognitive methods to address survey measurement problems. We begin by focusing on household surveys; in particular, we address research proposals to support improved measures of income and disability and also address how cognitive research can better be integrated into the survey design process. We then set forth an agenda for expanding cognitive research to increase knowledge of the response process in establishment surveys.

The second section is concerned with unresolved issues and basic methodological questions about the interview process itself. Standardization is a recurring issue, particularly the tradeoff between interviewer flexibility versus standardized interviewing. The third section looks ahead to broader changes in the survey world—especially those due to the tremendous changes in the organization and technology of survey data collection and distribution. Research is proposed to apply cognitive methods to aspects of the survey process which have not been explored using these methods, and to consider new methods which are called for by technological changes in survey operations.

22.1 PRACTICING THE STATE OF THE ART

As any scientific enterprise matures, a set of rules or procedures are adopted as standard practice and accepted as the "state of the art," because they have proved their worth over time under various circumstances. They are not necessarily considered exhaustive or perfect, but at least have been deemed useful in carrying out the enterprise. As new techniques are adopted as the state of the art and experience is gained with them, they also will be challenged and, eventually, modified or supplanted.

In the survey field, the new cognitively-based methods for evaluating questionnaires and investigating cognitive aspects of the response process have increasingly defined the state of the art. This section first explores the application of state of the art techniques to new survey measurement topics, including income and disability, but then proposes a systematic evaluation of the utility of these techniques. In a final section, it is suggested that the application of these techniques be extended to establishment surveys.

22.1.1 Applying Current Methods to New Areas of Measurement

Most conference participants probably could agree that cognitively-based methods have been applied more intensively and extensively to questionnaire design problems than to any other area of survey research. Several prominent examples in which these techniques were applied as part of the survey design (or redesign) process were the Current Population Survey (CPS), the Consumer Expenditure Survey Program (CE), the Survey of Income and Program Participation (SIPP), the National Crime Victimization Survey (NCVS), recent projects investigating different ways of measuring race and ethnicity, and, of course, the long and short forms used in the decennial Census. Among the methods used by cognitive scientists to evaluate questionnaire design are intensive one-on-one interviews using think-aloud and probing, standardized respondent debriefings and vignettes, focus groups, expert reviews, behavior coding of respondent–interviewer interactions, and sorting and rating tasks. While research needs to begin in other areas of the survey process and from other disciplinary perspectives, much is left to do in the realm of questionnaire design. Two critical areas involve the monitoring of socioeconomic and health trends in the population. The measurement of income requires serious attention, and for health researchers, the measurement of disability is particularly problematic. During the conference, a work group was devoted to each of these topics.

The work group addressing the income measurement problem began by recognizing that the data collected on income in federal surveys are critical to the formulation and evaluation of a variety of economic policies, including taxation and income transfer programs. The validity and usefulness of these data, however, are affected by a number of sources of error and bias, including recall problems, the sensitivity of the questions, concerns about confidentiality, the respondent's lack of knowledge, burden, and other causes yet to be identified. Errors are especially high for certain types of income, such as means-tested transfer payments, income from self-employment, dividends, interest, and interfamily transfers.

Cognitive research into the measurement of income could improve current knowledge about the magnitude and sources of errors in income data, and could suggest better measures which yield improved income estimates. Among the problems that can be directly examined are the respondent's understanding of terms, recall problems, and task complexity in terms of the amount of information needed and the extent of calculation. Some of this work would involve small-scale laboratory experiments, but other studies would have to be conducted in the field.

The work group proposed studies ranging from the purely qualitative to sophisticated quantitative analyses. Ethnographic research is proposed to *gather information about how different subpopulations view and carry out the data reporting task.* This research would determine the cognitive frameworks which respondents use to understand and organize the task, their understanding of terms, and access to and use of records, as well as motivational issues that might lead them to conceal or misreport income. Ethnographic research is proposed with subgroups in the general population who are presented with the task of reporting income for the first time, and with respondents to the CPS Income Supplement or SIPP. In the first instance, cognitive schemas and organization methods might be the focus. In the latter case, respondents would be asked to reflect on the process they have just completed with an eye to uncovering problems with the current procedures.

A second proposal was to *conduct linguistic analysis to identify language problems in the current questions on income and assets.* Citing Graesser's work (see Chapter 13, this book), the group proposed that a computerized linguistic analysis be undertaken. This analysis would require no field work, and it could identify questions which were awkwardly constructed or ambiguous. Questions that demand too much from respondents at one time might be revised or split into two or more questions.

The work group was particularly concerned about the high level of missing data in income reports, a problem which may be exacerbated by survey requirements for exact amounts. The group recommended research to *develop alternative questioning strategies to obtain respondents' estimates of values* when exact amounts are unknown. To this end, studies examining the tradeoffs between missing data and estimation were proposed. Accompanying these studies should be an investigation (similar to Sirken, Willis, and Nathan, 1991) of the respondent's perceived burden and commitment to producing exact answers relative to estimates. To validate estimation techniques, large-scale field tests must follow development.

The foregoing studies point to tailoring questions and adapting procedures to the different reporting situations and cognitive frameworks encountered in the general population. The group proposed an *experimental evaluation of the effects of tailoring*, which would include an evaluation of the screening methods used to assign respondents to alternative tailored question sequences.

Any attempt to reduce measurement error ultimately must be held to the "gold standard" of moving closer to truth. Of course, if researchers could easily find the truth, respondents and surveys would not be needed. Because truth is so elusive and record-check studies are expensive and do not generalize very well, the work group recommended research to *develop improved validity measures and criteria for evaluating improvements in data quality*, to replace the current rule of thumb that "more is better," in terms of income reporting. Data quality indicators, such as those developed by Silberstein (1991) and Tucker (1992), might be embedded in survey instruments on an ongoing basis.

The income work group approached their measurement problem generally, not considering any specific income questions. In contrast, the work group concerned with the measurement of disability focused on two particular measures, the ability to perform normal daily activities and the extent of limitations of on-the-job performance. The difficulties of adequately measuring disability are especially acute, due to the variety of definitions of disability as well as the complexity of its conceptualization. The group's review of studies examining the measurement of limitations on job performance and daily activities uncovered problems of unreliability (see, e.g., Mathiowetz and Lair, 1994; Ofstedal et al., 1995). Measurements show wide variability due to a mixture of true change or variation, and artifactual variation due to survey mode, variable question wording and context, and changes in respondent. Changes and inconsistencies in disability constructs and measures contribute to uncertainties about the interpretation of disability estimates.

To improve measurements of disability, the group proposed three areas for future interdisciplinary research, involving both survey methodologists and cognitive scientists. The first and perhaps most pressing need is *research to determine the sources of variability in the answers to the work limitation and functional limitation questions.* Research should begin with a meta analysis of the literature on this subject and additional analyses of existing data. Using reinterview programs that are already in place as well as new laboratory and field experiments, systematic tests of the effects of various factors on estimates must also be undertaken.

Also needed is research to *investigate respondent difficulties arising from comprehension problems, and from limitations on working memory.* Both problems have been exacerbated by a growing tendency to gather more information in less time in an attempt to reduce costs and limit conditioning effects. This has resulted in the compression of items to the point that data on all family members and multiple limitations are gathered simultaneously. At the same time, disability concepts, never well-defined, have become more complicated and politicized, leading to even greater confusion. Whether the problem has to do with comprehension or memory, cognitive methods such as vignettes and think-aloud interviews could be used. Studies would evaluate the respondent's understanding of specific terms or whole questions as well as the complexity of processing at multiple levels (across members of the family and across different functional levels).

22.1.2 Evaluating the State of the Art

The development and adoption of cognitively-based methods have had a profound effect on questionnaire development for household surveys, and survey sponsors increasingly accept and demand cognitive testing of questionnaires. Although the purpose for applying cognitive methods in the development of household questionnaires is to improve the reliability and validity of survey estimates, there have been few attempts to validate the results of cognitive testing. Do questions identified as problematic in laboratory tests actually yield data of poorer quality or cause more difficulties for respondents and interviewers in the field? (See Willis, DeMaio, and Harris-Kotejn, Chapter 9, this book, and Tucker, 1997, for discussions of this issue.) One work group proposed a *multiphase experiment to validate questionnaire evaluation methods.* In order to evaluate whether applying cognitive methods results in better questions which elicit more accurate data, some assessment of error is needed. The group identified a set of indicators of the quality of survey responses, including the reasonableness of the results, themselves, as well as item nonresponse rates, data inconsistencies, rates of unusable responses, break-off rates, response variance calculated from reinterviews, information from respondent debriefings, and comparisons to external estimates. These measures can be applied to evaluate a variety of cognitive methods, including: (a) intensive cognitive interviews, (b) focus groups, (c) interaction or behavior coding, (d) expert appraisals, (e) vignettes, (f) sorting and rating tasks, and (g) small-scale laboratory experiments.

Questions would be evaluated using the different methodologies, and the results would be used to predict data quality, as measured by the indicators above. A field test would then be conducted to determine which predictions were correct using the measures of data quality. So that the investigations would be unbiased, separate research teams would be randomly assigned to questions and cognitive methods. In addition to performance, the different methods would be evaluated in terms of cost and level of effort.

Additionally, the group proposed to *incorporate findings from cognitive science and other sources of information into models to more accurately estimate population characteristics.* Besides the wording of questions, the probability of error depends on the subject, the context, and a respondent's characteristics (including cognitive abilities and motivation). Based on laboratory studies of how these factors affect data quality, model-based estimates would be produced to adjust for inaccuracies in aggregate statistics or individual reports. Thus, the results of cognitive research would be incorporated into statistical models which adjust for known errors and biases to improve the quality of the data. The group suggested embedding surveys with measures of cognitive functioning to examine how they are related to response quality, possibly incorporating them into error models.

22.1.3 Extending the Application of Cognitive Methods to Establishment Surveys

Although most of the discussion at the conference had to do with household surveys, there was considerable interest in expanding the use of cognitive methods to establishment surveys. For this purpose, a work group was formed which included survey methodologists with experience conducting establishment surveys. Most of the group's members were government employees, because the federal statistical agencies are the primary collectors of establishment data. Much of this work is done using forms with only labeled data fields and no questions. These forms are often completed by the respondents themselves and usually returned by mail. When interviewers are used, they are subject matter specialists, and these specialists may be free to formulate their own questions and collect information in any order that meets their needs or the needs of the establishment.

Thus, establishment surveys are very different from household ones. Contacts may be made not only with data providers but also those authorized to release the information. In many cases, the respondent will have to access record systems to obtain the information. That information is often technical in nature, and involves terminology which may vary across organizations and hence must be clearly defined. Unlike households, certain establishments will be selected with certainty, so that they will always be in the sample, and are likely to be reporting in other surveys. Furthermore, the establishments actually may be users of the data collected in the survey.

To carry out a research agenda applicable to establishment surveys, an inter-disciplinary perspective is needed. For example, cognitive psychology is relevant to the way the respondent performs his or her task. However, theories from organizational psychology will be useful for understanding the way an organization goes about approving the request for information and designating a respondent. Both social psychology and organizational psychology, aided by ethnographic methods, could provide insight into how communication flows and lines of authority operate in different types of establishments. Thus, establishment surveys, largely ignored by the CASM movement, could benefit greatly from research using the methods common to this movement.

One area to be explored is *how organizations decide to cooperate with a survey request, and the best strategies for gaining cooperation and selecting a respondent.* Is this decision controlled by management or the person with the information? Moreover, who knows what information is available and how it is stored (in one location or several)? Which organizational characteristics determine the answers to these questions? How should the organization be approached, and how can the respondent's commitment be obtained? How can these commitments be maintained in longitudinal surveys, and how can measurement quality be assured?

To answer these and other questions, a collaboration between organization theorists and survey methodologists is needed. The group called for studies of organizational attributes and their relationship to the distribution of authority for allowing access as well as the hierarchical nature of data systems (Edwards and Cantor, 1991). This information could be used, in conjunction with firm size, to inform the data collection design. Experiments could then be used to evaluate the effectiveness of the alternative approaches to establishments in obtaining cooperation.

The group also recommended research to explore *the implications of using expert interviewers in establishment surveys.* In contrast to household surveys, interviewers in establishment surveys often have subject matter and technical expertise, and the interview process is more collaborative, less standardized, and different in other ways from the typical household interview. How do these experts assist respondents in interpreting technical terms and reporting data in a form that can be used by the survey organization? How do the interviewers establish rapport, come to a common understanding of technical issues, and negotiate with the respondent to obtain the best possible data? How do the interviewers determine reporting errors or mismatches between the available data and the data needed?

Much of the research needed to answer these questions is qualitative in nature. The group recommended studying the conversational patterns in the exchanges between the expert interviewers and respondents. This work should be followed up with an assessment of the quality of the data provided by the respondents through intensive interviews or debriefings. Based on this first phase of research, small-scale field experiments should be conducted which systematically vary the content of the interviewer's communications with respondents. The final step in this research program would be the codification of rules for conducting expert interviewing, including the technical information needed in particular situations and the ways it should be used. The interviewing protocols and training materials that would result could then be tested in the field.

A third topic proposed for research is the *study of the records-retrieval process*. What is in the records, and how well does the content match survey definitions? How do respondents access the records, and how do they use them to answer survey questions? Answers to these questions would be obtained during visits with establishment respondents to investigate how respondents formulate answers, especially when the information requested does not match what is available in records.

In its final proposal, the work group suggested studying *different techniques for improving response rates and their effects on data quality*. Initial interviews with previous survey respondents and nonrespondents would explore the questions of how the organization typically handles survey requests and decides whether or not to participate. A second stage of experimental studies would examine which design features most affect the organization's actions. Experiments which vary design features, such as the use of prenotification letters and incentives, and the mode of interview, would be used to evaluate their effects on response rates and the quality of the data provided. The data quality measures would be obtained either from reinterviews or respondent debriefings.

22.2   ADDRESSING UNRESOLVED ISSUES

It is ironic, in light of the first CASM Conference's focus on the survey interview and the critique of standardization that was inspired by the conference (Suchman and Jordan, 1990), that the issue of whether, and how much and what kinds of, flexibility ought to be permitted to interviewers in household surveys remains largely unresearched. The work group charged with exploring the interview process took as its starting point the common criticism that practices of standardization may interfere with respondents' ability and motivation to understand survey questions and provide accurate data. Based on their observations of the barriers to communication created by standardization, Suchman and Jordan (1990) proposed that survey researchers adopt a more collaborative interviewing method, while Mishler (1986) advocated replacing standardized interviews with the collection of narratives. As the group notes, however, no published research has yet compared the results obtained using procedures like those suggested by Suchman and Jordan to the results of standardized interviews. Such research would need to examine a number of potential outcomes of more flexible interviewing in order to balance costs, in the form of more time consuming interviews, the need for more highly skilled interviewers with more training, and possible increases in variability due to interviewers, against benefits, such as improved data quality, reduced respondent burden, and improved respondent motivation. Indeed, there is a lack of basic information on just how "standardized interviewing" is actually implemented in practice by different survey organizations. Fowler and Mangione (1990) provide a concise statement of one method of standardization that is widely invoked as a standard. However, organizations involved in survey data collection—including academic, government, and commercial organizations—vary widely in how they actually implement or interpret standardization, and standards of practice likely vary for surveys of opinions and surveys of factual matters. (For example, organizations vary in their practices of allowing interviewers to verify answers or encouraging them to probe "don't know" or undecided answers.)

Thus, a starting point for research in this area is *to describe the current practice of standardization*. This includes describing how interviewers are trained, and the rules of standardization they are trained to follow, as well as describing actual interviewing practices and how they accord with principles of standardization. The goal of this research is to provide a realistic benchmark against which to test the effects of introducing various degrees of flexibility into standardized interviewing procedures. Additionally, when there are clear differences in practice among different organizations, it may be possible to use existing data to examine the effects of more flexible practices on data quality in surveys otherwise comparable in content. The group proposes to collect written policies, training materials, and manuals, to conduct observations of interviewer training, and to document procedures for monitoring and quality control to establish the official norms of standardized interviewing in different organizations. Additionally, focus groups would be conducted with interviewers to provide their perspectives on their organizations' rules and solicit their opinions and recommendations about what works and what does not. Systematic data on actual interviewing practice might be based on observations or transcriptions of interviews as well as surveys of interviewers' practices and beliefs about standardization.

The group notes that transcripts of interviews might be used to analyze interactions in standardized interviews, to identify interviewer and respondent behaviors that are likely to be associated with data quality. Such analysis might identify "flexible" behaviors which appear to be promising tools for improving respondent–interviewer communication and which might be systematically evaluated in a second, experimental phase of research. Analysis of transcripts might also shed light on when and how respondents seek help in understanding questions, and help identify indicators of comprehension problems (e.g., delays).

A second phase of the research would *measure the costs and benefits of varying degrees of interviewer flexibility within the standardized format*. The hypothesis being tested is that, although standardization decreases interviewer variability, it also decreases the validity of at least some responses, because (by some organizations' rules) interviewers are not permitted to repair respondents' misunderstandings of question meaning and intent. Thus the proposal is to evaluate the effects of giving the interviewer more flexibility in detecting and responding to comprehension problems in particular. Experimentation would be conducted to measure the effects on data quality of varying degrees of flexibility in response to requests for clarification (such as, neutral probing, providing official definitions to respondents who request help, etc.). This research builds on work by Schober and Conrad (1997) suggesting that when interviewers were allowed to provide survey definitions, response accuracy was improved. Other types of flexibility—for example, allowing respondents to control the order in which they answer questions—may also merit further research.

The issues involved are complex. Different interviewing styles and rules may be appropriate for opinion and for factual surveys: it may be impossible to permit interviewers to "explain" attitude or opinion items without destroying comparability. Additionally, research shows that not only respondents but also interviewers themselves may misinterpret key survey concepts (Campanelli, Martin, and Rothgeb, 1991). Permitting interviewers to explain survey concepts assumes they can explain them; as the work group points out, flexible interviewing requires more and better training, and a more highly skilled and highly paid work force, than strictly standardized interviewing. The group raises the question, "Can all interviewers be trained to do flexible interviewing well?" to which the answer is almost certainly, "No." A second issue concerns the extent to which comprehension problems are, or can be, known to interviewers or even to respondents themselves. Invisible problems of comprehension presumably cannot be remedied by flexible interviewing tools. A respondent debriefing study conducted in conjunction with the Current Population Survey revealed that only 8 percent of respondents interpreted "work" as intended by the survey, yet most respondents who answered the survey question were apparently unaware that they were interpreting the construct too broadly or narrowly (Campanelli, et al., 1991). When surveys use common words to refer to technical concepts, there may be no signal to respondents or interviewers that a misunderstanding has occurred.

Another work group was charged with examining different disciplinary perspectives on the question and answer process. The group took as its starting point the observation that the tremendous growth in laboratory testing of survey questionnaires undoubtedly has contributed to the improvement of individual questionnaires, but has contributed little to a general understanding of the question and answer process. In other words, the findings of cognitive research on questionnaires have not cumulated nor have general principles of questionnaire design and construction emerged as a result of cognitive testing. In part, this is true because the results are not published, and hence, not easily available to survey researchers. The group advocates that *the reports of these tests be collected, coded, and archived in a data base made available through the world wide web.*[2] The work group notes that such a generally accessible data

---

[2]Many results of cognitive testing conducted by the Census Bureau's Center for Survey Methods Research are available as *Working Papers in Survey Methodology* at http://www.census.gov/srd/www/byyear.html

base would offer two advantages. First, the results of previous testing could be used by researchers who intend to use similar items in a new survey. This would allow survey researchers to build on the results of prior testing, avoid the use of items which have been demonstrated to be flawed, and incorporate improvements into new surveys. A second advantage is that the archive would stimulate systematic research into questions such as: What characteristics of questions are identified by cognitive interviewing as creating particular kinds of problems? What testing features are associated with discovering different problem types? What sorts of solutions are adopted in response to various classes of problems?

A second proposal is to *promote cognitive research on survey anomalies* by writing up survey findings and making them known to cognitive scientists, perhaps through papers presented at professional meetings, in articles of newsletters, or small-scale conferences devoted to interdisciplinary discussions of relevant measurement issues, such as those sponsored by the Social Science Research Council after the first CASM conference.

A third proposal is to *conduct an ethnography of the survey interview.* The group notes that survey researchers know little about how responses are affected by respondents' understandings of the larger survey context. Research into social representations of the question and answer process is needed to address issues such as: Why do people think they are being interviewed, and what do they think will be done with the answers? Do people have scripts or schemas about the survey interview? Do people assume a survey interview is like a normal conversation, or that it is more like a test? Similarly, they point out, we have only limited information about these questions from interviewers' perspectives. For example, how do they construe the purposes and aims of survey interviews, and how do they understand the meaning of confidentiality assurances they give respondents? Available evidence suggests that interview situations, and participants' understandings of them, may depart considerably and in surprising ways from survey designers' intentions.

In a classic study, Cannell and his colleagues reinterviewed respondents in a health survey and found that most did not know what was expected of them, nor did they understand why the information was collected or even recall who had conducted the survey (Cannell, Fowler, and Marquis, 1968). Biderman (1975) suggested that the survey interview has become increasingly institutionalized over time, although what understandings respondents have of it, based on direct experience or exposure to media presentations of polls and survey results, is anybody's guess (see also Turner and Martin, 1984). It is certain, however, that many respondents, as well as interviewers, express skepticism about the confidentiality of individual information, at least in government surveys and the census. In a 1976 survey, only 5 percent of respondents were sure that census records were fully confidential (that is, that they were not open to the public or available to other government agencies "even if they really tried" to gain access; National Academy of Sciences, 1979). In a 1988 survey, a majority (53 percent) of Census Bureau interviewers believed that other government agencies could *not* gain access to Census Bureau information about individual respondents "if they really tried," but a sizable minority (21 percent) thought they could, and 26 percent were not sure (Lavin, 1989). Clearly, there is a need for additional research to address the extent to which respondents' and interviewers' understandings of the interview situation match survey designers' intentions, and if possible to devise improved methods and training to attempt to create more appropriate expectations.

22.3   LOOKING TO THE FUTURE

Even as the application of methods and ideas derived from cognitive science continues to grow, the survey world is changing rapidly. This section considers two major, related developments. The first is radically changing technology for survey data collection, which has confronted survey researchers with new problems to solve in the area of questionnaire design and survey administration. The growth of automation requires novel applications of current techniques and invites the development of new methods. The second is the recent application of cognitive techniques to other parts of the survey process, driven in part by the development of new technologies for data distribution.

22.3.1   Accommodating the New Technology

A working group charged with examining cognitive aspects of surveys in a changing environment identified three critical areas in which continuing research is desirable. First, *research is needed to develop a fuller understanding of respondent tasks in self-administered, computer-assisted questionnaires*, and how they can best be communicated and structured. Ideally, the design of the interfaces in self-administered questionnaires should be based on models of human cognition and performance, and evaluated and revised through usability testing. Novices and experts may interact with electronic questionnaires differently, which implies that the design requirements may differ for automated instruments which are self-administered rather than interviewer administered. Navigation is a particularly difficult problem in automated instruments with long question sequences or complex branching patterns, so the user may easily get lost, especially if an erroneous response is entered which sends him or her down a path of inappropriate questions. Usability testing, which combines cognitive psychology, human factors engineering, and (in some cases) computer science, provides tools and protocols for testing and evaluating problems users have navigating and interacting with computerized questionnaires. There is a need for interface design principles that predict respondent performance and can be the basis for designing improved interfaces. Another issue is how to motivate respondents to cooperate with self-administered electronic questionnaires. To date, such surveys have not achieved very high response rates.

Second, *research is needed to understand the changed role of the interviewer in automated data collection*, and how automation impacts the interview process. Both procedural and technical improvements are needed. Future automated instrument designs may permit the traditional stricture that a rigidly standardized question sequence must be followed to be relaxed, allowing interviewers and respondents more control over the sequencing of questions and tasks. Technology may potentially make the interviewer's job easier, by handling many tasks automatically. On the other hand, poorly designed automated instruments may make the interviewer's job much more difficult, and, as Couper graphically demonstrated (Couper, 1997), may completely disrupt the interview as respondent and interviewer both focus their attention on trying to make a balky computer cooperate. Research is needed to improve designs to reduce the cognitive load involved in coordinating multiple tasks and technologies, improve technologies for inputting data (e.g., pen-based, voice recognition), develop principles of error recovery, develop interface guidelines and style sheets, and improve training in use of automated instruments.

Third, *research and development are needed to develop better technologies for creating and reviewing questionnaires*, and other tools to improve respondent and interviewer performance. The group finds promise in the application of computer models to evaluate questions to identify cumbersome syntax, rare words, ambiguous terms, misleading presuppositions, memory overload conditions, and out-of-context items. Other areas of research include general software to analyze skip pattern consistency and branching in complex instruments, implementation of graphical user interface (GUI) tools in government CAPI interviewing applications, and alternative structures for organizing research, design, and implementation of electronic data collections.

22.3.2 Moving Beyond Questionnaire Design

The survey process is more than questionnaire development. Other parts of the process include concept development, the creation of coding and classification rules, and the dissemination of the survey results. One work group examined these areas, as well as the assessment of the statistical literacy of the community of data users.

With respect to concept development, the problem is usually one of definition—concepts may be defined too vaguely to permit valid or reliable measures to be developed. Subject matter experts often do not realize the inherent complexity of their concepts. In some cases, they do not even have a notion of how the concept might be measured. Currently, survey methodologists approach these problems in an ad hoc manner. The work group proposed *developing a general model for helping subject matter experts refine concepts and their measures*.

One example of concept development examined by the group was classification systems, such as the industrial and occupational classification systems. These systems can be designed for respondents, interviewers, or coders. In many instances they are based on intuition or common-sense judgments with no reference to empirical research and sometimes without the benefit of theoretical support. Thus, the product is often incomplete, inconsistent, and, at its worst, arbitrary or idiosyncratic. In the past, cognitive psychologists have used multidimensional scaling to define concepts and the criteria for their measurement (Rips, Shoben, and Smith, 1973; Conrad and Tonn, 1993). In addition to multidimensional scaling, other methods for modeling mental taxonomies were identified by the group (see Rips and Conrad, 1989). The results from this work could be evaluated using measures of intercoder reliability, accuracy with respect to expert judgments, and subjective ratings of satisfaction by the users.

The work group also believed cognitive methods would be useful for designing data dissemination and visualization products, particularly for the Internet. Work in this area would include identifying the right display tool in a specific situation. Cognitive research can be applied to develop the new tools needed to convey difficult concepts to audiences with varying levels of statistical sophistication. The group proposed concentrating efforts on the new website developed and maintained by the Federal Interagency Council on Statistical Policy (http://www.fedstats.gov). Research techniques such as laboratory usability studies, web surveys, and the analysis of user logs could be used to answer questions about the efficiency of various search tools, ways to provide flexible methods for the retrieval and display of data, the design of convenient links to metadata, and easy-to-use online analysis techniques.

To ensure that these products are actually used in the most appropriate ways, the group considered the issue of statistical literacy. As a first step, the needs of users should be determined. Next, the levels of understanding among users with respect to the tools required to meet their needs must be assessed. For example, how well do users distinguish between particular instances or anecdotes versus generalizable aggregate statistics? Both online surveys and focus groups could be used to answer such questions. Finally, educational tools should be developed for users based on principles from cognitive psychology and educational psychology.

22.4   IMPLICATIONS AND CONCLUSIONS

Although the working groups covered a great diversity of topics, a number of common themes emerged from their discussions and recommendations. One dominant theme is the need for continued development and evaluation of indicators of data quality. The cognitively based methods which have been developed over approximately 20 years permit more fine-grained examination of different aspects of the response process. However, there has been very little work to correlate assessments based on cognitive methods with traditional measures of data quality, such as reinterview reliability, consistency with external records or other data sources, etc. (See Hess and Singer (1996) and Presser and Blair (1994) for two such evaluations.) Field experiments which administer multiple methods to the same individuals are needed to examine consistency across methods, and to evaluate how well cognitive testing methods predict data quality and performance in the field.

Improved measures of data quality are also needed to evaluate improvements achieved by different questioning strategies. For example, in order to decide which of two measurement methods results in better income data, researchers need a better criterion for improvement than the common rule of thumb that "more is better." One promising proposal is to incorporate measures of data quality into the statistical estimation process itself, in order to adjust estimates for known biases and errors. Another idea is to build in cognitive assessments of, e.g., short term memory, in order to measure and test the effects of cognitive factors that may influence respondents' performance on survey items.

A second theme is a critical need for research on interviewers and the interview process. There is insufficient knowledge of how interviewers and respondents understand the interview and their roles in it, how they negotiate (or fail to negotiate) shared understandings of survey questions, and how their understandings influence data quality.

Interestingly, the need for attention to and adaptation of the interview process was voiced both for household surveys and for establishment surveys, but from very different points of view. In household surveys, current thinking is that standardization undermines the communication process and makes it more difficult for interviewers and respondents to reach a shared understanding of questions, thus reducing the quality of the data. In establishment surveys, interviewers are subject matter experts and the interview process is more collaborative and less standardized, involving negotiation about the meaning of questions and how they should be answered. This description seems similar to what has been proposed as an improved, more flexible interviewing style in surveys of households. Yet, the concern in establishment surveys is that the interview process is not codified and not under control. The pressure to codify and standardize interviewing practice in establishment surveys arises in part from cost concerns—more standardized and structured interviews would make it possible to hire less skilled, and cheaper, interviewing staff. This transition in establishment surveys appears to follow a path taken in household surveys some 50 years ago (see, e.g., Converse, 1987). In both areas, research is proposed to examine the conversational dynamics in actual interviews, and to document current standards and practice as a starting point for evaluating possible reforms. It would be wise to coordinate such efforts in these two areas, since it may be possible to learn about the pitfalls and the advantages of flexible or conversational interviewing from how interviews are currently conducted in establishment surveys. Conversely, it may serve well to study the history of the transition to more structured and standardized interviewing in household surveys in order to guide this process intelligently in establishment surveys.

Perhaps it would be fruitful to bring together different perspectives on the interview process in a conference that examined norms and practices of interviewing in both establishment surveys and household surveys, and that included research on interviewing in other, nonsurvey contexts as well—for example, clinical interviews (in which disclosure has been researched), or police interviews (in which accuracy of eye-witness accounts has been investigated), and other types of interview situations.

Concern about the interview process is not recent: it was a focus of the first CASM Conference, as well as the deliberations and recommendations of the National Academy of Sciences Panel on Surveying Subjective Phenomena (see, for example, Turner and Martin, 1984, p. 61 and Jabine et al., 1984, p. 9, for similar recommendations). It is curious to speculate why research to investigate interviewers' and respondents' understandings of the survey interview process is so often recommended and so seldom undertaken. CASM I raised questions about whether the strictures of standardized interviewing should be relaxed to allow for more flexibility, but very little research on the topic ensued. Why not? There was some interesting discussion on this point at the conference. The bottom line was that we do not pay interviewers enough to attract large numbers of the highly trained and skilled staff needed to carry out this type of interviewing in household surveys. Perhaps another reason is that research on interviewers is an expensive proposition in organizations in which field staff is hard put to keep up with production in a time of growing demand. However, we continue to neglect this critical area at our peril, since the interviewers are at the heart of any survey's ability to get response and quality data.

A third theme points to another area of relative neglect. Several groups called, in different ways, for improvements in the process of documenting and cumulating knowledge about survey methods and practice, and about sources of variation in survey measurements. The proposal mentioned earlier in the chapter to document norms and actual practice concerning how standardization is implemented by interviewers in household surveys is laudable, but in a way it is astonishing that it is necessary. We in the survey business do not know exactly how interviewers actually administer questionnaires, and in fact, we are not really very sure of what we mean by standardization in the first place. Standardization is itself not standardized in meaning. A second area in which a recommendation points to a lack of documentation is the call for an archive of results of pretesting questions. This surely makes good sense, and would provide a basis for building more general knowledge about the question and answer process. A side benefit would be to help survey researchers avoid asking and testing the same bad questions again and again.

A fourth theme concerns technology, and its impact on all aspects of the survey process. There is ambivalence, and perhaps even a degree of schizophrenia, concerning the enormous technological changes affecting the survey world. On one hand, the promise of automated solutions continues to beguile survey researchers with the promise of nearly effortless improvements in how we conduct our work. An example is the new development of automated questionnaire evaluation and testing software, which promises to yield better questionnaires without time-consuming review and testing. On the other hand, technological developments once implemented often bedevil the survey process, adding layers of complexity and creating unanticipated organizational and communication problems. Most technological changes are not fully or experimentally tested before implementation, so their effects on the data are never fully understood or documented. However, it is promising that changes in technology are giving rise to new areas of research and new evaluation methods. There is broad applicability for a new kind of research focused on people's interactions with computers—usability testing, as it is called. This is a relatively new area combining aspects of psychology and operations research, and it is needed to explore human-computer interfaces in a number of different areas—respondents interacting with computerized self-administered questionnaires, interviewers administering questionnaires, individuals using and obtaining data via the Internet. Usability testing may facilitate a more systematic process of designing, evaluating, and proving in new technological developments that may reduce the difficulties survey organizations currently experience in making such conversions.

A final common theme concerns statistical literacy and the importance of research to investigate how the public understands the purposes of surveys and the uses made of the data. Interestingly, this was viewed as critical at both ends of the survey process: we need to know more about how respondents interpret their role and the uses made of the data they provide because we believe it may influence their willingness to respond and the way they answer questions. At the other end it also affects public use and interpretation—and the credibility—of survey results. The decennial Census is perhaps an excellent example of just how critical public understanding and acceptance of survey methods are in affecting an organization's ability to carry out a survey. We know that respondents who understand the uses made of census data and the reasons for the census are more likely to mail back their forms (Fay, Bates, and Moore, 1991). Even more critical at the current time is the fact that the public at large, including many members of Congress, do not fully accept sampling, and are skeptical that it can be carried out in an objective and unbiased fashion. The current problems the Census Bureau is having in making its case for sampling might be lessened if statistical sampling were better understood.

REFERENCES

Biderman, A. (1975). The survey method as an institution and the survey institution as a method. In H. W. Sinaiko and L. A. Broedling (Eds.), *Perspectives on Attitude Assessment: Surveys and their Alternatives*, pp. 48–62. Washington, DC: Smithsonian Institution.

Campanelli, P., Martin, E. A., and Rothgeb, J. M. (1991). The use of respondent and interviewer debriefing studies as a way to study response error in survey data. *The Statistician, 40,* 253–264.

Cannell, C., Fowler, J., and Marquis, K. (1968). The influence of interviewer and respondent psychological and behavioral variables on the reporting in household interviews. *Vital and Health Statistics*, Series 2, No. 26. Washington, DC: U.S. Government Printing Office.

Conrad, F., and Tonn, B. (1993). Intuitive notions of occupation. *Proceedings of International Conference on Occupational Classification*, pp. 169–178. Washington, DC: Bureau of Labor Statistics.

Converse, J. M. (1987). *Survey Research in the United States*. Berkeley: University of California Press.

Couper, M. (June, 1997). *The application of cognitive science to computer-assisted interviewing*. Presented at the Second Advanced Seminar on the Cognitive Aspects of Survey Methodology, Charlottesville, VA.

Edwards, W. S., and Cantor, D. (1991). Toward a response model in establishment surveys. In P. Biemer, R. Groves, L. Lyberg, N. Mathiowetz, and S. Sudman (Eds.), *Measurement Error in Surveys*, pp. 211–236. New York: Wiley.

Fay, R. E., Bates, N. A., and Moore, J. (1991). Lower mail response in the 1990 census: A preliminary interpretation. *Proceedings of the 1991 Annual Research Conference*, pp. 3–32. Washington, DC: U.S. Bureau of the Census.

Fowler, F. J., and Mangione, T. W. (1990). *Standardized Survey Interviewing: Minimizing Interviewer-Related Error*. Newbury Park, CA: Sage.

Hess, J., and Singer, E. (1996). Predicting test-retest reliability from behavior coding. *Proceedings of the Section on Survey Research Methods, American Statistical Association*, 1004–1009.

Jabine, T., Straf, M., Tanur, J., and Tourangeau, R. (Eds.) (1984). *Cognitive Aspects of Survey Methodology: Building a Bridge Between Disciplines*. Washington, DC: National Academy Press.

Lavin, S. (1989). *Results of the nonsampling error measurement survey of field interviewers*. Unpublished U.S. Bureau of the Census report, March 1989.

Mathiowetz, N. A., and Lair, T. J. (1994). Getting better? Changes or errors in the measurement of functional limitations. *Journal of Economic and Social Measurement*, 20, 237–262.

Mishler, E. (1986). *Research Interviewing*. Cambridge, MA: Harvard.

National Academy of Sciences. (1979). *Privacy and Confidentiality as Factors in Survey Response*. Washington, DC: National Academy of Sciences.

Ofstedal, M. B., Feldman, J. J., Lentzner, H. R., and Walke, T. (1998). *A comparison of national estimates of functional limitation among the elderly*. Unpublished research report, National Center for Health Statistics.

Presser, S., and Blair, J. (1994). Survey pretesting: Do different methods produce different results? In P. V. Marsden (Ed.), *Sociological Methodology: Vol. 24*, pp. 73–104. Washington, DC: American Sociological Association.

Rips, L. J., and Conrad, F. G. (1989). Folk psychology of mental activities. *Psychological Review*, 96, 187–207.

Rips, L. J., Shoben, E. J., and Smith, E. E. (1973). Semantic distance and the verification of semantic relations. *Journal of Verbal Learning and Verbal Behavior*, 12, 1–20.

Schober, M. F., and Conrad, F. G. (1997). Does conversational interviewing reduce survey measurement error? *Public Opinion Quarterly*, 60, 576–602.

Silberstein, A. R. (1991). Response performance in the Consumer Expenditure Diary Survey. *Proceedings of the Section on Survey Research Methods, American Statistical Association*, 338–343.

Sirken, M., Jabine, T., Willis, G., Martin, E., and Tucker, C. (in press). A new agenda for interdisciplinary survey research methods: *Proceedings of the CASM II Seminar*. National Center for Health Statistics.

Sirken, M. G., Willis, G. B., and Nathan, G. (1991). Cognitive aspects of answering sensitive survey questions: *Proceedings of the International Association of Survey Statisticians*, International Statistical Institute, 130–131.

Suchman, L., and Jordan, B. (1990). Interactional troubles in face-to-face survey interviews. *Journal of the American Statistical Association, 85(409)*, 232–241.

Tucker, C. (1992). The estimation of instrument effects on data quality in the Consumer Expenditure Diary Survey. *Journal of Official Statistics, 8*, 41–61.

Tucker, C. (1997). Measurement issues surrounding the use of cognitive methods in survey research. *Bulletin de Methodologie Sociologique, 55*, 67–92.

Turner, C. F., and E. Martin (Eds.) (1984). *Surveying Subjective Phenomena*. New York: Russell Sage Foundation.

Index

Aborn, Murray, 16–17, 49
Accessible material, context effects and, 116, 125–126
Acculturating agents, 22
Adam, S., 282
Addressee, in conversations, 79
Advanced Research Seminar, 15
AFDC (Aid to Families with Dependent Children), 158–159, 165
Aitchison, J., 186
Akmajian, A., 186, 305
Allan, K., 186
Altmann, A., 289
Ambiguity
  in interviewer-respondent interaction, 85
  in questions, 192–193
American Sociological Association, 34
American Statistical Association, 25, 30
Andersen, B., 167
Anderson, J. R., 252, 308
Anderson, N., 117
Andrews, F., 246
Applied CASM research, 2–4
Archival materials, 70
Armed Forces Qualification Test, 24
Aronoff, M., 186
Artificial intelligence (AI), 267
Asch, S., 111
Assets, income reporting, 165–166, 169–170
Assimilation, context effects
  attitude and, 121
  defined, 50, 113
  judgment, 117–118
  question comprehension, 113
  retrieval-based, 115–116
Association
  networks of, 230–231
  ordering, 342
Astute observations, 24

Asymmetrical grounding, 84
Attitude
  belief-sampling model, context effects, 126
  context effects and, 120–121
  measurement, 67–68
  questions, 111–112, 127
Audience design, questions and
  benefits of, 78, 90
  spontaneous conversation, 78–80
  surveys, 82–83, 85–86
Autobiographical memory
  characteristics of, 67, 69
  components of, generally, 96
  event representations, 97–98
  retrieval
    event representations, 97–98
    generally, 261
    uniqueness, specificity, and faithfulness, 98–101
  significance of, 5, 95
  year's events
    calendar effect, 102
    recall peaks, 102–106

Bagnara, S., 289
Bar charts, 322
Barsalou, L. W., 101
Basic CASM research, 2–4
Basic research, 70
Basic Instructions for Interviewers, 52
Bassili, J. N., 147
Baumgartner, R., 161
Bayesian hierarchical model, 7
Beatty, Paul, 6, 82, 138, 144, 178, 273
Behavioral-frequency questions, 134
Behavioral reports, 67
Behavioral theory, 237–239
Behavior coding, 71, 144, 146
Behaviorism, 267–268

383

392

WILEY SERIES IN PROBABILITY AND STATISTICS
ESTABLISHED BY WALTER A. SHEWHART AND SAMUEL S. WILKS

Editors
Vic Barnett, Noel A. C. Cressie, Nicholas I. Fisher,
Iain M. Johnstone, J. B. Kadane, David G. Kendall, David W. Scott,
Bernard W. Silverman, Adrian F. M. Smith, Jozef L. Teugels;
Ralph A. Bradley, Emeritus, J. Stuart Hunter, Emeritus

Probability and Statistics Section

*ANDERSON · The Statistical Analysis of Time Series
ARNOLD, BALAKRISHNAN, and NAGARAJA · A First Course in Order Statistics
ARNOLD, BALAKRISHNAN, and NAGARAJA · Records
BACCELLI, COHEN, OLSDER, and QUADRAT · Synchronization and Linearity:
    An Algebra for Discrete Event Systems
BASILEVSKY · Statistical Factor Analysis and Related Methods: Theory and
    Applications
BERNARDO and SMITH · Bayesian Statistical Concepts and Theory
BILLINGSLEY · Convergence of Probability Measures
BOROVKOV · Asymptotic Methods in Queuing Theory
BOROVKOV · Ergodicity and Stability of Stochastic Processes
BRANDT, FRANKEN, and LISEK · Stationary Stochastic Models
CAINES · Linear Stochastic Systems
CAIROLI and DALANG · Sequential Stochastic Optimization
CONSTANTINE · Combinatorial Theory and Statistical Design
COOK · Regression Graphics
COVER and THOMAS · Elements of Information Theory
CSÖRGŐ and HORVÁTH · Weighted Approximations in Probability Statistics
CSÖRGŐ and HORVÁTH · Limit Theorems in Change Point Analysis
DETTE and STUDDEN · The Theory of Canonical Moments with Applications in
    Statistics, Probability, and Analysis
*DOOB · Stochastic Processes
DRYDEN and MARDIA · Statistical Analysis of Shape
DUPUIS and ELLIS · A Weak Convergence Approach to the Theory of Large Deviations
ETHIER and KURTZ · Markov Processes: Characterization and Convergence
FELLER · An Introduction to Probability Theory and Its Applications, Volume 1,
    Third Edition, Revised; Volume II, *Second Edition*
FULLER · Introduction to Statistical Time Series, *Second Edition*
FULLER · Measurement Error Models
GHOSH, MUKHOPADHYAY, and SEN · Sequential Estimation
GIFI · Nonlinear Multivariate Analysis
GUTTORP · Statistical Inference for Branching Processes
HALL · Introduction to the Theory of Coverage Processes
HAMPEL · Robust Statistics: The Approach Based on Influence Functions
HANNAN and DEISTLER · The Statistical Theory of Linear Systems
HUBER · Robust Statistics
IMAN and CONOVER · A Modern Approach to Statistics
JUREK and MASON · Operator-Limit Distributions in Probability Theory
KASS and VOS · Geometrical Foundations of Asymptotic Inference
KAUFMAN and ROUSSEEUW · Finding Groups in Data: An Introduction to Cluster
    Analysis

*Now available in a lower priced paperback edition in the Wiley Classics Library.

*Now available in a lower priced paperback edition in the Wiley Classics Library.

*Now available in a lower priced paperback edition in the Wiley Classics Library.

*Now available in a lower priced paperback edition in the Wiley Classics Library.

*Now available in a lower priced paperback edition in the Wiley Classics Library.

Applied Probability and Statistics (Continued)
WOOLSON · Statistical Methods for the Analysis of Biomedical Data
*ZELLNER · An Introduction to Bayesian Inference in Econometrics

Texts and References Section

AGRESTI · An Introduction to Categorical Data Analysis
ANDERSON · An Introduction to Multivariate Statistical Analysis, *Second Edition*
ANDERSON and LOYNES · The Teaching of Practical Statistics
ARMITAGE and COLTON · Encyclopedia of Biostatistics: Volumes 1 to 6 with Index
BARTOSZYNSKI and NIEWIADOMSKA-BUGAJ · Probability and Statistical Inference
BERRY, CHALONER, and GEWEKE · Bayesian Analysis in Statistics and
  Econometrics: Essays in Honor of Arnold Zellner
BHATTACHARYA and JOHNSON · Statistical Concepts and Methods
BILLINGSLEY · Probability and Measure, *Second Edition*
BOX · R. A. Fisher, the Life of a Scientist
BOX, HUNTER, and HUNTER · Statistics for Experimenters: An Introduction to
  Design, Data Analysis, and Model Building
BOX and LUCEÑO · Statistical Control by Monitoring and Feedback Adjustment
BROWN and HOLLANDER · Statistics: A Biomedical Introduction
CHATTERJEE and PRICE · Regression Analysis by Example, *Second Edition*
COOK and WEISBERG · An Introduction to Regression Graphics
COX · A Handbook of Introductory Statistical Methods
DILLON and GOLDSTEIN · Multivariate Analysis: Methods and Applications
DODGE and ROMIG · Sampling Inspection Tables, *Second Edition*
DRAPER and SMITH · Applied Regression Analysis, *Third Edition*
DUDEWICZ and MISHRA · Modern Mathematical Statistics
DUNN · Basic Statistics: A Primer for the Biomedical Sciences, *Second Edition*
FISHER and VAN BELLE · Biostatistics: A Methodology for the Health Sciences
FREEMAN and SMITH · Aspects of Uncertainty: A Tribute to D. V. Lindley
GROSS and HARRIS · Fundamentals of Queueing Theory, *Third Edition*
HALD · A History of Probability and Statistics and their Applications Before 1750
HALD · A History of Mathematical Statistics from 1750 to 1930
HELLER · MACSYMA for Statisticians
HOEL · Introduction to Mathematical Statistics, *Fifth Edition*
HOLLANDER and WOLFE · Nonparametric Statistical Methods, *Second Edition*
HOSMER and LEMESHOW · Applied Survival Analysis: Regression Modeling of
  Time to Event Data
JOHNSON and BALAKRISHNAN · Advances in the Theory and Practice of Statistics: A
  Volume in Honor of Samuel Kotz
JOHNSON and KOTZ (editors) · Leading Personalities in Statistical Sciences: From the
  Seventeenth Century to the Present
JUDGE, GRIFFITHS, HILL, LÜTKEPOHL, and LEE · The Theory and Practice of
  Econometrics, *Second Edition*
KHURI · Advanced Calculus with Applications in Statistics
KOTZ and JOHNSON (editors) · Encyclopedia of Statistical Sciences: Volumes 1 to 9
  wtih Index
KOTZ and JOHNSON (editors) · Encyclopedia of Statistical Sciences: Supplement
  Volume
KOTZ, REED, and BANKS (editors) · Encyclopedia of Statistical Sciences: Update
  Volume 1
KOTZ, REED, and BANKS (editors) · Encyclopedia of Statistical Sciences: Update
  Volume 2
LAMPERTI · Probability: A Survey of the Mathematical Theory, *Second Edition*

*Now available in a lower priced paperback edition in the Wiley Classics Library.

Texts and References (Continued)
LARSON · Introduction to Probability Theory and Statistical Inference, *Third Edition*
LE · Applied Categorical Data Analysis
LE · Applied Survival Analysis
MALLOWS · Design, Data, and Analysis by Some Friends of Cuthbert Daniel
MARDIA · The Art of Statistical Science: A Tribute to G. S. Watson
MASON, GUNST, and HESS · Statistical Design and Analysis of Experiments with
    Applications to Engineering and Science
MURRAY · X-STAT 2.0 Statistical Experimentation, Design Data Analysis, and
    Nonlinear Optimization
PURI, VILAPLANA, and WERTZ · New Perspectives in Theoretical and Applied
    Statistics
RENCHER · Methods of Multivariate Analysis
RENCHER · Multivariate Statistical Inference with Applications
ROSS · Introduction to Probability and Statistics for Engineers and Scientists
ROHATGI · An Introduction to Probability Theory and Mathematical Statistics
RYAN · Modern Regression Methods
SCHOTT · Matrix Analysis for Statistics
SEARLE · Matrix Algebra Useful for Statistics
STYAN · The Collected Papers of T. W. Anderson: 1943–1985
TIERNEY · LISP-STAT: An Object-Oriented Environment for Statistical Computing
    and Dynamic Graphics
WONNACOTT and WONNACOTT · Econometrics, *Second Edition*

WILEY SERIES IN PROBABILITY AND STATISTICS
ESTABLISHED BY WALTER A. SHEWHART AND SAMUEL S. WILKS

Editors
Robert M. Groves, Graham Kalton, J. N. K. Rao, Norbert Schwarz,
Christopher Skinner

Survey Methodology Section

BIEMER, GROVES, LYBERG, MATHIOWETZ, and SUDMAN · Measurement
    Errors in Surveys
COCHRAN · Sampling Techniques, *Third Edition*
COUPER, BAKER, BETHLEHEM, CLARK, MARTIN, NICHOLLS, and O'REILLY
    (editors) · Computer Assisted Survey Information Collection
COX, BINDER, CHINNAPPA, CHRISTIANSON, COLLEDGE, and KOTT (editors) ·
    Business Survey Methods
*DEMING · Sample Design in Business Research
DILLMAN · Mail and Telephone Surveys: The Total Design Method
GROVES and COUPER · Nonresponse in Household Interview Surveys
GROVES · Survey Errors and Survey Costs
GROVES, BIEMER, LYBERG, MASSEY, NICHOLLS, and WAKSBERG ·
    Telephone Survey Methodology
*HANSEN, HURWITZ, and MADOW · Sample Survey Methods and Theory,
    Volume 1: Methods and Applications
*HANSEN, HURWITZ, and MADOW · Sample Survey Methods and Theory,
    Volume II: Theory

*Now available in a lower priced paperback edition in the Wiley Classics Library.